Richard Wormsbecker

Understanding Non-equilibrium Thermodynamics

G. Lebon • D. Jou • J. Casas-Vázquez

Understanding Non-equilibrium Thermodynamics

Foundations, Applications, Frontiers

Prof. Dr. David Jou
Universitat Autònoma de Barcelona
Dept. Fisica - Edifici Cc
Grup Fisica Estadistica
Bellaterra 08193
Catalonia, Spain
David.Jou@uab.es

Prof. Dr. José Casas-Vázquez
Universitat Autònoma de Barcelona
Dept. Fisica - Edifici Cc
Grup Fisica Estadistica
Bellaterra 08193
Catalonia, Spain
Jose.Casas@uab.es

Prof. Dr. Georgy Lebon
Université de Liège
Dept. d'Astrophysique
Géophysique et Océanographie
B5 Sart Tilman
Liege 4000
Belgium
g.lebon@ulg.ac.be

Cover image: Thermal radiation leaving the Earth, seen by the EOS-Terra satellite (NASA). Image from www.visibleearth.nasa.gov. Owner: NASA.

ISBN: 978-3-642-09359-3 e-ISBN: 978-3-540-74252-4

Cover design: WMX Design GmbH, Heidelberg

Printed on acid-free paper

9 8 7 6 5 4 3 2 1

springer.com

Preface

Our time is characterized by an explosion of information and by an accéleration of knowledge. A book cannot compete with the huge amount of data available on the Web. However, to assimilate all this information, it is necessary to structure our knowledge in a useful conceptual framework. The purpose of the present work is to provide such a structure for students and researchers interested by the current state of the art of non-equilibrium thermodynamics. The main features of the book are a concise and critical presentation of the basic ideas, illustrated by a series of examples, selected not only for their pedagogical value but also for the perspectives offered by recent technological advances. This book is aimed at students and researchers in physics, chemistry, engineering, material sciences, and biology.

We have been guided by two apparently antagonistic objectives: generality and simplicity. To make the book accessible to a large audience of non-specialists, we have decided about a simplified but rigorous presentation. Emphasis is put on the underlying physical background without sacrificing mathematical rigour, the several formalisms being illustrated by a list of examples and problems. All over this work, we have been guided by the formula: "Get the more from the less", with the purpose to make a maximum of people aware of a maximum of knowledge from a minimum of basic tools.

Besides being an introductory text, our objective is to present an overview, as general as possible, of the more recent developments in non-equilibrium thermodynamics, especially beyond the local equilibrium description. This is partially a *terra incognita*, an unknown land, because basic concepts as temperature, entropy, and the validity of the second law become problematic beyond the local equilibrium hypothesis. The answers provided up to now must be considered as partial and provisional, but are nevertheless worth to be examined.

Chapters 1 and 2 are introductory chapters in which the main concepts underlying equilibrium thermodynamics and classical non-equilibrium thermodynamics are stated. The basic notions are discussed with special emphasis on these needed later in this book.

Several applications of classical non-equilibrium thermodynamics are presented in Chaps. 3 and 4. These illustrations have not been chosen arbitrarily, but keeping in mind the perspectives opened by recent technological advancements. For instance, advances in material sciences have led to promising possibilities for thermoelectric devices; localized intense laser heating used to make easier the separation of molecules has contributed to a revival of interest in thermodiffusion; chemical reactions are of special interest in biology, in relation with their coupling with active transport across membranes and recent developments of molecular motors.

The purpose of Chaps. 5 and 6 is to discuss two particular aspects of classical non-equilibrium thermodynamics which have been the subject of active research during the last decades. Chapter 5 is devoted to finite-time thermodynamics whose main concern is the competition between maximum efficiency and maximum power and its impact on economy and ecology. This classical subject is treated here in an updated form, taking into account the last technological possibilities and challenges, as well as some social concerns. Chapter 6 deals with instabilities and pattern formation; organized structures occur in closed and open systems as a consequence of fluctuations growing far from equilibrium under the action of external forces. Patterns are observed in a multitude of our daily life experiences, like in hydrodynamics, biology, chemistry, electricity, material sciences, or geology. After introducing the mathematical theory of stability, several examples of ordered structures are analysed with a special attention to the celebrated Bénard cells.

Chapters 1–6 may provide a self-consistent basis for a graduate introductory course in non-equilibrium thermodynamics.

In the remainder of the book, we go beyond the framework of the classical description and spend some time to address and compare the most recent developments in non-equilibrium thermodynamics. Chapters 7–11 will be of interest for students and researchers, who feel attracted by new scientific projects wherein they may be involved. This second part of the book may provide the basis for an advanced graduate or even postgraduate course on the several trends in contemporary thermodynamics.

The coexistence of several schools in non-equilibrium thermodynamics is a reality; it is not a surprise in view of the complexity of most macroscopic systems and the fact that some basic notions as temperature and entropy are not univocally defined outside equilibrium. To appreciate this form of multiculturalism in a positive sense, it is obviously necessary to know what are the foundations of these theories and to which extent they are related. A superficial inspection reveals that some viewpoints are overlapping but none of them is rigorously equivalent to the other. A detailed and complete understanding of the relationship among the diverse schools turns out to be not an easy task. The first difficulty stems from the fact that each approach is associated with a certain insight, we may even say an intuition or feeling that is sometimes rather difficult to apprehend. Also some unavoidable differences in the terminology and the notation do not facilitate the communication. Another factor that contributes to the difficulty to reaching a mutual comprehension is that the schools are not frozen in time: they evolve as a consequence of internal dynamics and by contact with others. Our goal is to contribute to a better understanding among the different schools by discussing their main concepts, results, advantages, and limitations. Comparison of different viewpoints may be helpful for a deeper comprehension and a possible synthesis of the many faces of the theory. Such a comparative study is not found in other textbooks.

One problem was the selection of the main representative ones among the wealth of thermodynamic formalisms. Here we have focused our attention on five of them: extended thermodynamics (Chap. 7), theories with internal variables (Chap. 8), rational thermodynamics (Chap. 9), Hamiltonian formulation (Chap. 10), and mesoscopic approaches (Chap.11). In each of them, we have tried to save the particular spirit of each theory.

It is clear that our choice is subjective: we have nevertheless been guided not only by the pedagogical aspect and/or the impact and universality of the different formalisms, but also by the fact that we had to restrict ourselves. Moreover, it is our belief that a good comprehension of these different versions allows for a better and more understandable comprehension of theories whose opportunity was not offered to be discussed here. The common points shared by the theories presented in Chaps. 7–11 are not only to get rid of the local equilibrium hypothesis, which is the pillar of the classical theory, but also to propose new phenomenological approaches involving non-linearities, memory and non-local effects, with the purpose to account for the technological requirements of faster processes and more miniaturized devices.

It could be surprising that the book is completely devoted to macroscopic and mesoscopic aspects and that microscopic theories have been widely omitted. The reasons are that many excellent treatises have been written on microscopic theories and that we decided to keep the volume of the book to a reasonable ratio. Although statistical mechanics appears to be more fashionable than thermodynamics in the eyes of some people and the developments of microscopic methods are challenging, we hope to convince the reader that macroscopic approaches, like thermodynamics, deserve a careful attention and are the seeds of the progress of knowledge. Notwithstanding, we remain convinced that, within the perspectives of improvement and unification, it is highly desirable to include as many microscopic results as possible into the macroscopic framework.

Chapters 7–11 are autonomous and self-consistent, they have been structured in such a way that they can be read independently of each other and in arbitrary order. However, it is highly recommended to browse through all the chapters to better apprehend the essence and the complementarity of the diverse theories.

At the end of each chapter is given a list of problems. The aim is not only to allow the reader to check his understanding, but also to stimulate his interest to solve concrete situations. Some of these problems have been inspired by recent papers, which are mentioned, and which may be consulted for further investigation. More technical and advanced parts are confined in boxes and can be omitted during a first reading.

We acknowledge many colleagues, and in particular M. Grmela (Montreal University), P.C Dauby and Th. Desaive (Liège University), for the discussions on these and related topics for more than 30 years. We also appreciate our close collaborators for their help and stimulus in research and teaching. Drs. Vicenç Méndez and Vicente Ortega-Cejas deserve special gratitude for their help in the technical preparation of this book. We also acknowledge the finantial support of the Dirección General de Investigación of the Spanish Ministry of Education under grants BFM2003-06003 and FIS2006-12296-C02-01, and of the Direcció General de Recerca of the Generalitat of Catalonia, under grants 2001 SGR 00186 and 2005 SGR 00087.

Liège-Bellaterra, March 2007
G. Lebon,
D. Jou,
J. Casas-Vázquez

Contents

Chapter 1
Equilibrium Thermodynamics: A Review

Equilibrium States, Reversible Processes, Energy Conversion

Equilibrium or classical thermodynamics deals essentially with the study of macroscopic properties of matter at *equilibrium*. A comprehensive definition of equilibrium will be given later; here it is sufficient to characterize it as a time-independent state, like a column of air at rest in absence of any flux of matter, energy, charge, or momentum. By extension, equilibrium thermodynamics has also been applied to the description of *reversible processes*: they represent a special class of idealized processes considered as a continuum sequence of equilibrium states.

Since time does not appear explicitly in the formalism, it would be more appropriate to call it *thermostatics* and to reserve the name *thermodynamics* to the study of processes taking place in the course of time outside equilibrium. However, for historical reasons, the name "thermodynamics" is widely utilized nowadays, even when referring to equilibrium situations. We shall here follow the attitude dictated by the majority but, to avoid any confusion, we shall speak about *equilibrium thermodynamics* and designate beyond-equilibrium theories under the name of *non-equilibrium thermodynamics*.

The reader is assumed to be already acquainted with equilibrium thermodynamics but, for the sake of completeness, we briefly recall here the essential concepts needed along this book. This chapter will run as follows. After a short historical introduction and a brief recall of basic definitions, we present the fundamental laws underlying equilibrium thermodynamics. We shall put emphasis on Gibbs' equation and its consequences. After having established the criteria of stability of equilibrium, a last section, will be devoted to an introduction to chemical thermodynamics.

1.1 The Early History

Equilibrium thermodynamics is the natural extension of the older science, Mechanics. The latter, which rests on Newton's law, is essentially concerned with the study of motions of idealized systems as mass-particles and rigid solids. Two important notions, heat and temperature, which are absent in mechanics, constitute the pillars of the establishment of equilibrium thermodynamics as a branch of science. The need to develop a science beyond the abstract approach of Newton's law to cope with the reality of engineer's activities was born in the beginning of nineteenth century. The first steps and concepts of thermodynamics were established by Fourier, Carnot, Kelvin, Clausius, and Gibbs among others. Thermodynamics began in 1822 with Fourier's publication of the *Théorie analytique de la chaleur* wherein is derived the partial differential equation for the temperature distribution in a rigid body. Two years later, in 1824, Sadi Carnot (1796–1832) put down further the foundations of thermodynamics with his renowned memoir *Réflexions sur la puissance motrice du feu et sur les machines propres à développer cette puissance.* Carnot perceived that steam power was a motor of industrial revolution that would prompt economical and social life. Although a cornerstone in the formulation of thermodynamics, Carnot's work is based on several misconceptions, as for instance the identification of heat with a hypothetical indestructible weightless substance, the caloric, a notion introduced by Lavoisier. Significant progresses towards a better comprehension of the subject can be attributed to a generation of outstanding scientists as James P. Joule (1818–1889) who identified heat as a form of energy transfer by showing experimentally that heat and work are mutually convertible. This was the birth of the concept of energy and the basis of the formulation of the first law of thermodynamics. At the same period, William Thomson (1824–1907), who later matured into Lord Kelvin, realized that the work of Carnot was not contradicting the ideas of Joule. One of his main contributions remains a particular scale of absolute temperature. In his paper "On the dynamical theory of heat" appeared in 1851, Kelvin developed the point of view that the mechanical action of heat could be interpreted by appealing to two laws, later known as the first and second laws. In this respect, Rudolf Clausius (1822–1888), a contemporary of Joule and Kelvin, accomplished substantial advancements. Clausius was the first to introduce the words "internal energy" and "entropy", one of the most subtle notions of thermodynamics. Clausius got definitively rid of the notion of caloric, reformulated Kelvin's statement of the second law, and tried to explain heat in terms of the behaviour of the individual particles composing matter. It was the merit of Carnot, Joule, Kelvin, and Clausius to thrust thermodynamics towards the level of an undisputed scientific discipline. Another generation of scientists was needed to unify this new formalism and to link it with other currents of science. One of them was Ludwig Boltzmann (1844–1906) who put forward a decisive "mechanistic" interpretation of heat transport; his major contribution was to link the behaviour of the particles at the microscopic level to their consequences on the macroscopic level. Another prominent scientist, Josiah Williard Gibbs (1839–1903), deserves the credit to have converted thermodynamics into a deductive science. In fact he recognized soon that thermodynamics of the nineteenth century is a pure static science wherein time does not play any role. Among his main contributions, let us point out the theory of stability based on the use of the properties of convex (or concave) functions, the potential bearing his name, and the well-known Gibbs' ensembles. Gibbs' paper "On the equilibrium of the heterogeneous substances" ranks among the most decisive impacts in the developments of modern chemical thermodynamics.

Other leading scientists have contributed to the development of equilibrium thermodynamics as a well structured, universal, and undisputed science since the pioneers laid down its first steps. Although the list is far from being exhaustive, let us mention the names of Caratheodory, Cauchy, Clapeyron, Duhem, Einstein, Helmholtz, Maxwell, Nernst, and Planck.

1.2 Scope and Definitions

Equilibrium thermodynamics is a section of macroscopic physics whose original objective is to describe the transformations of energy in all its forms. It is a generalization of mechanics by introducing three new concepts:

1. The concept of *state*, i.e. an ensemble of quantities, called state variables, whose knowledge allows us to identify any property of the system under study. It is desirable that the state variables are independent and easily accessible to experiments. For example, a motionless fluid may be described by its mass m, volume V, and temperature T.
2. The notion of *internal energy*, complementing the notion of kinetic energy, which is of pure mechanical origin. Answering the question "what is internal energy?" is a difficult task. Internal energy is not a directly measurable quantity: there exist no "energymeters". For the moment, let us be rather evasive and say that it is presumed to be some function of the measurable properties of a system like mass, volume, and temperature. Considering a macroscopic system as agglomerate of individual particles, the internal energy can be viewed as the mean value of the sum of the kinetic and interacting energies of the particles. The notion of internal energy is also related to these temperature and heat, which are absent from the vocabulary of mechanics.
3. The notion of *entropy*. Like internal energy, it is a characteristic of the system but we cannot measure it directly, we will merely have a way to measure its changes. From a microscopic point of view, the notion of entropy is related to disorder: the higher the entropy, the larger the disorder inside the system. There are also connections between entropy and information in the sense that entropy can be considered as a measure of our lack of information on the state of the system. The link between entropy and information is widely exploited into the so-called information theory.

Energy and entropy are obeying two major laws: the first law stating that *the energy of the universe is a constant*, and the second law stating that *the entropy of the universe never decreases*.

At this stage, it is useful to recall some definitions. By *system* is understood a portion of matter with a given mass, volume, and surface. An *open system* is able to exchange matter and energy through its boundaries, a *closed system* exchanges energy but not matter with the outside while an *isolated system* does exchange neither energy nor matter with its surroundings. It is admitted that the universe (the union of system and surroundings) acts as an isolated system. In this chapter, we will deal essentially with *homogeneous* systems, whose properties are independent of the position.

As mentioned earlier, the *state* of a system is defined by an ensemble of quantities, called state variables, characterizing the system. Considering a system evolving between two equilibrium states, A and B, it is important to realize that, by definition, the state variables will not depend on the particular way taken to go from A to B. The selection of the state variables is not a trivial task, and both theoretical and experimental observations constitute a suitable guide. It is to a certain extent arbitrary and non-unique, depending on the level of description, either microscopic or macroscopic, and the degree of accuracy that is required. A delicate notion is that of *equilibrium* state which turns out to be a state, which is time independent and generally spatially homogeneous. It is associated with the absence of fluxes of matter and energy. On the contrary, a non-equilibrium state needs for its description time- and space-dependent state variables, because of exchanges of mass and energy between the system and its surroundings. However, the above definition of equilibrium is not complete; as shown in Sect. 1.3.3, equilibrium of an isolated system is characterized by a maximum of entropy. Notice that the concept of equilibrium is to some extent subjective; it is itself an idealization and remains a little bit indefinite because of the presence of fluctuations inherent to each equilibrium state. It depends also widely on the available data and the degree of accuracy of our observations.

One distinguishes *extensive* and *intensive* state variables; extensive variables like mass, volume, and energy have values in a composite system equal to the sum of the values in each subsystem; intensive variables as temperature or chemical potential take the same values everywhere in a system at equilibrium. As a variable like temperature can only be rigorously defined at equilibrium, one may expect difficulties when dealing with situations beyond equilibrium.

Classical thermodynamics is not firmly restricted to equilibrium states but also includes the study of some classes of processes, namely those that may be considered as a sequence of neighbouring equilibrium states. Such processes are called *quasi-static* and are obtained by modifying the state variables very slowly and by a small amount. A quasi-static process is either *reversible* or *irreversible*. A reversible process $1 \to 2 \to 3$ may be viewed as a continuum sequence of equilibrium states and will take place infinitesimally slowly. When undergoing a reverse transformation $3 \to 2 \to 1$, the state variables take the same values as in the direct way and the exchanges of matter and energy with the outside world are of opposite sign; needless to say that reversible processes are pure idealizations. An irreversible process is a non-reversible one. It takes place at finite velocity, may be mimicked by a discrete series of equilibrium states and in a reverse transformation, input of external energy from the outside is required to go back to its initial state. Irreversible processes are generally associated with friction, shocks, explosions, chemical reactions, viscous fluid flows, etc.

1.3 The Fundamental Laws

The first law, also popularly known as the law of conservation of energy, was not formulated first but second after the second law, which was recognized first. Paradoxically, the zeroth law was formulated the latest, by Fowler during the 1930s and quoted for the first time in Fowler and Guggenheim's book published in 1939.

1.3.1 The Zeroth Law

It refers to the introduction of the idea of empirical temperature, which is one of the most fundamental concepts of thermodynamics. When a system 1 is put in contact with a system 2 but no net flow of energy occurs, both systems are said to be in thermal equilibrium. As sketched in Fig. 1.1a, we take two systems 1 and 2, characterized by appropriate parameters, separated by an adiabatic wall, but in contact (a thermal contact) with the system 3 through a diathermal wall, which allows for energy transfer in opposition with an adiabatic wall. If the systems 1 and 2 are put in contact (see Fig. 1.1b), they will change the values of their parameters in such a way that they reach a state of thermal equilibrium, in which there is no net heat transfer between them.

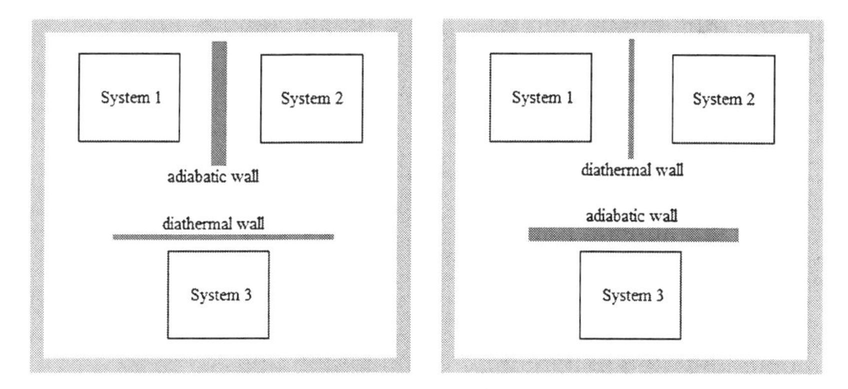

Fig. 1.1 Steps for introducing the empirical temperature concept

The zeroth law of thermodynamics states that *if the systems* 1 *and* 2 *are separately in thermal equilibrium with* 3, *then* 1 *and* 2 *are in thermal equilibrium with one another.* The property of transitivity of thermal equilibrium allows one to classify the space of thermodynamic states in classes of equivalence, each of which constituted by states in mutual thermal equilibrium. Every class may be assigned a label, called *empirical temperature*, and the mathematical relation describing a class in terms of its state variables and the empirical temperature is known as *the thermal equation of state* of the system. For one mole of a simple fluid this equation has the general form $\phi(p, V, \theta) = 0$ where p is the pressure, V the volume, and θ the empirical temperature.

1.3.2 The First Law or Energy Balance

The first law introduces the notion of energy, which emerges as a unifying concept, and the notion of heat, related to the transfer of energy. Here, we examine the formulation of the first law for closed systems.

Consider first a system enclosed by a thermally isolated (adiabatic), impermeable wall, so that the sole interaction with the external world will appear under the form of a mechanical work W, for instance by expansion of its volume or by stirring. Referring to the famous experience of Joule, the work can be measured by the decrease in potential energy of a slowly falling weight and is given by $W = mgh$, where h is the displacement and g the acceleration of gravity. During the evolution of the system between the two given equilibrium states A and B, it is checked experimentally that the work W is determined exclusively by the initial and the final states A and B, independently of the transformation paths. This observation allows us to identify W with the difference $\Delta U = U(B) - U(A)$ of a state variable U which will be given the name of *internal energy*

$$W = \Delta U. \tag{1.1}$$

The above result provides a mean to measure the internal energy of a system, whatever be its nature. Assume now that we remove the adiabatic wall enclosing the system, which again proceeds from state A to state B. When this is accomplished, it is observed that in general $W \neq \Delta U$, and calling Q the difference between these two quantities, one obtains

$$\Delta U - W = Q, \tag{1.2}$$

where Q is referred to as the heat exchanged between the system and its surroundings. Expression (1.2) is the first law of thermodynamics and is usually written under the more familiar form

$$\Delta U = Q + W, \tag{1.3a}$$

or, in terms of differentials,

$$dU = đQ + đW, \tag{1.3b}$$

where the stroke through the symbol "d" means that $đQ$ and $đW$ are inexact differentials, i.e. that they depend on the path and not only on the initial and final states. From now on, we adopt the sign convention that $Q > 0$, $W > 0$ when heat and work are supplied to the system, $Q < 0$, $W < 0$ when heat and work are delivered by the system. Some authors use other conventions resulting in a minus sign in front of $đW$.

It is important to stress that the domain of applicability of the first law is not limited to reversible processes between equilibrium states. The first law remains valid whatever the nature of the process, either reversible or irreversible and the status of the states A and B, either equilibrium or non-equilibrium. Designating by $E = U + K + E_{pot}$ the total energy of the system (i.e. the sum of the internal U, kinetic K, and potential energy E_{pot}), (1.3b) will be cast in the more general form

$$dE = đQ + đW. \tag{1.4}$$

At this point, it should be observed that with respect to the law of energy $\Delta K = W$ as known in mechanics, we have introduced two new notions: internal energy U and heat Q. The internal energy can be modified either by heating the body or by acting mechanically, for instance by expansion or compression, or by coupling both mechanisms. The quantity U consists of a stored energy in the body while Q and W represent two different means to transfer energy through its boundaries. The internal energy U is a state function whose variation is completely determined by the knowledge of the initial and final states of the process; in contrast, Q and W are not state functions as they depend on the particular path followed by the process. It would therefore be incorrect to speak about the heat or the work of a system. The difference between heat and work is that the second is associated with a change of the boundaries of the system or of the field acting on it, like a membrane deformation or a piston displacement. Microscopically, mechanical work is related to coherent correlated motions of the particles while heat represents that part of motion, which is uncorrelated, say incoherent.

In equilibrium thermodynamics, the processes are reversible from which follows that the energy balance equation (1.4) will take the form:

$$dU = đQ_{rev} - p\,dV, \tag{1.5}$$

wherein use is made of the classical result that the reversible work performed by a piston that compress a gas of volume V and pressure p trapped in a cylinder is given by $đW_{rev} = -p\,dV$ (see Problem 1.1). In engineering applications, it is customary to work with the enthalpy H defined by $H = U + pV$. In terms of H, expression (1.5) of the first law reads as

$$dH = đQ_{rev} + V\,dp. \tag{1.6}$$

For an isolated system, one has simply

$$\mathrm{d}U = 0 \tag{1.7}$$

expressing that its energy remains constant.

Note that, when applied to *open systems* with n different constituents, (1.5) will contain an additional contribution due to the exchange of matter with the environment and takes the form (Prigogine 1947)

$$\mathrm{d}U = \mathrm{d}^{\mathrm{i}}Q - p\,\mathrm{d}V + \sum_{k=1}^{n} h_k \mathrm{d}^{\mathrm{e}} m_k; \tag{1.8}$$

note that $\mathrm{d}^{\mathrm{i}}Q$ is not the total amount of heat but only that portion associated to the variations of the thermomechanical properties, T and p, and the last term in (1.8), which is the extra contribution caused by the exchange of matter $\mathrm{d}^{\mathrm{e}} m_k$ with the surroundings, depends on the specific enthalpy $h_k = H/m_k$ of the various constituents.

1.3.3 The Second Law

The first law does not establish any preferred direction for the evolution of the system. For instance, it does not forbid that heat could pass spontaneously from a body of lower temperature to a body of higher temperature, nor the possibility to convert completely heat into work or that the huge energy contained in oceans can be transformed in available work to propel a boat without consuming fuel. More generally, the first law establishes the equivalence between heat and work but is silent about the restrictions on the transformation of one into the other. The role of the second law of thermodynamics is to place such limitations and to reflect the property that natural processes evolve spontaneously in one direction only. The first formulations of the second law were proposed by Clausius (1850, 1851) and Kelvin (1851) and were stated in terms of the impossibility of some processes to be performed. Clausius' statement of the second law is enunciated as follows: *No process is possible whose sole effect is to transfer heat from a cold body to a hot body.* Kelvin's statement considers another facet: *it is impossible to construct an engine which can take heat from a single reservoir, and convert it entirely to work in a cyclic process.* In this book we will examine in detail, the formulations of the second law out of equilibrium. Here, we shall concentrate on some elements that are essential to a good understanding of the forthcoming chapters. We will split the presentation of the second law in two parts. In the first one, we are going to build-up a formal definition of a new quantity, the entropy – so named by Clausius from the Greek words *en* (in) and *trope* (turning) for representing "capacity of change or transformation" – which is as fundamental and universal (for equilibrium systems) as the notion of energy. In the second part, which constitutes truly the essence of the second law, we shall enounce the principle of entropy increase during an irreversible process.

1.3.3.1 The Concept of Entropy

Consider a homogeneous system of constant mass undergoing a reversible transformation between two equilibrium states A and B. The quantity of heat $\int_A^B đQ_{\text{rev}}$ depends on the path followed between states A and B (in mathematical terms, it is an imperfect differential) and therefore cannot be selected as a state variable. However, experimental observations have indicated that by dividing $đQ_{\text{rev}}$ by a uniform and continuous function $T(\theta)$ of an empirical temperature θ, one obtains an integral which is independent of the path and may therefore be identified with a state function, called entropy and denoted S

$$\int_A^B \frac{đQ_{\text{rev}}}{T(\theta)} = \Delta S = S_B - S_A. \tag{1.9}$$

Since in reversible processes, quantities of heat are additive, entropy is also additive and is thus an extensive quantity. A function like $T(\theta)$ which transforms an imperfect differential into a perfect one is called an integrating factor. The empirical temperature is that indicated by a mercury or an alcohol thermometer or a thermocouple and its value depends of course on the nature of the thermometer; the same remark is true for the entropy, as it depends on $T(\theta)$. It was the great merit of Kelvin to propose a temperature scale for T, the absolute temperature, independently of any thermodynamic system (see Box 1.1). In differential terms, (1.9) takes the form

$$\mathrm{d}S = \frac{đQ_{\text{rev}}}{T}. \tag{1.10}$$

This is a very important result as it introduces two new concepts, absolute temperature and entropy. The latter can be viewed as the quantity of heat exchanged by the system during a reversible process taking place at the equilibrium temperature T. Note that only differences in entropy can be measured. Given two equilibrium states A and B, it is always possible to determine their entropy difference regardless of whether the process between A and B is reversible or irreversible. Indeed, it suffices to select or imagine a reversible path joining these initial and final equilibrium states. The question is how to realize a reversible heat transfer. Practically, the driving force for heat transfer is a temperature difference and for reversible transfer, we need only imagine that this temperature difference is infinitesimally small so that $đQ_{\text{rev}} = \lim_{\Delta T \to 0} đQ$. Nevertheless, when the process takes place between non-equilibrium states, the problem of the definition of entropy is open, and actually not yet definitively solved.

Box 1.1 Absolute Temperature

Heat engines take heat from some hot reservoir, deliver heat to some cold reservoir, and perform an amount of work, i.e. they partially transform heat into work. Consider a Carnot's reversible engine (see Fig. 1.2a) operating between a single hot reservoir at the unknown empirical temperature θ_1 and a single cold reservoir at temperature θ_2. The Carnot cycle is accomplished in four steps consisting in two isothermal and two adiabatic transformations (Fig. 1.2b).

During the first isothermal process, the Carnot's engine absorbs an amount of heat Q_1 at temperature θ_1. In the second step, the system undergoes an adiabatic expansion decreasing the temperature from θ_1 to θ_2. Afterwards, the system goes through an isothermal compression at temperature θ_2 (step 3) and finally (step 4), an adiabatic compression which brings the system back to its initial state. After one cycle, the engine has performed a quantity of work W but its total variation of entropy is zero

$$\Delta S_{\text{engine}} = \frac{|Q_1|}{T(\theta_1)} - \frac{|Q_2|}{T(\theta_2)} = 0. \tag{1.1.1}$$

Selecting the reference temperature as $T(\theta_2) = 273.16$, the triple point temperature of water, it follows from (1.1.1)

$$T(\theta_1) = 273.16\frac{|Q_1|}{|Q_2|}. \tag{1.1.2}$$

The ratio $|Q_1| / |Q_2|$ is universal in the sense that it is independent of the working substance. Therefore, Carnot cycles offer the opportunity to reduce temperature measurements to measurements of quantities of heat and to define an absolute scale of positive temperatures, independently of the measurement of temperature on any empirical temperature scale, which depends on thermometric substance.

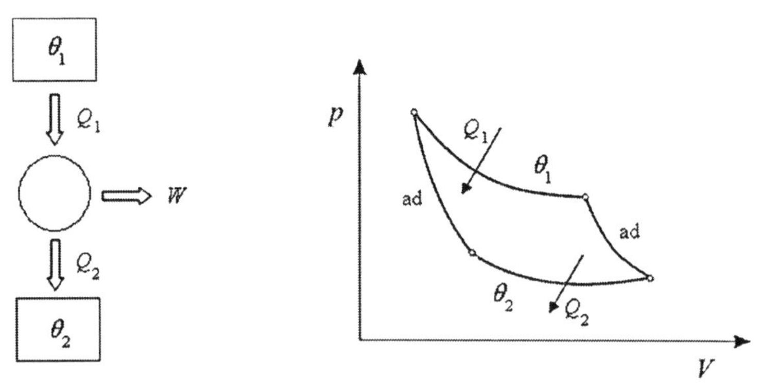

Fig. 1.2 (**a**) Heat engine and (**b**) Carnot diagram

The *efficiency* of a heat engine, in particular that of Carnot, is defined by the ratio of the work produced to the heat supplied

$$\eta = \frac{W}{Q_1} \tag{1.11}$$

for a cycle one has, in virtue of the first law, $W = Q_1 - Q_2$, so that

$$\eta = 1 - \frac{Q_2}{Q_1}. \tag{1.12}$$

Finally, making use of (1.1.2), it is found that the efficiency of a reversible cycle is

$$\eta = 1 - \frac{T_2}{T_1}. \tag{1.13}$$

As it will be seen, this is the maximum value for the efficiency of any heat engine working between the selected heat reservoirs. More considerations about the efficiency of reversible and irreversible cycles are developed in Chap. 5.

1.3.3.2 The Principle of Increase of Entropy

The second law was formulated by Clausius (1865) for isolated systems in terms of the change of the entropy in the form

$$\Delta S \geq 0. \tag{1.14}$$

To illustrate the principle of entropy increase, imagine an arbitrary number of subsystems, for instance three different gases A, B, and C at equilibrium, enclosed in a common *isolated* container and separated each other by adiabatic and rigid walls (Fig. 1.3). Let S_{ini} be the entropy in this initial configuration. Remove then the internal wall separating A and B which are diffusing into each other until a new state of equilibrium characterized by an entropy S_{int}, corresponding to the intermediate configuration, which is larger than S_{ini} is reached. By eliminating finally the last internal constraint between $A \cup B$ and C, and after the final state of equilibrium, corresponding to complete mixing, is reached, it is noted that entropy S_{fin} is still increased: $S_{\text{fin}} > S_{\text{int}} > S_{\text{ini}}$. Figure 1.3 reflects also that disorder is increased by passing from the initial

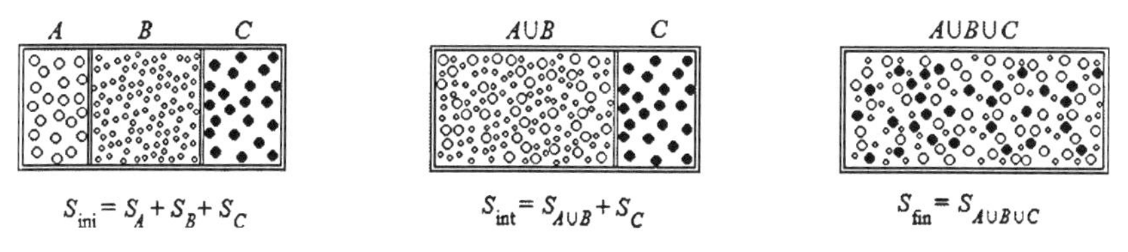

Fig. 1.3 Increase of entropy after removal of internal constraints

to the final configuration, which suggests the use of entropy as a measure of disorder: larger the disorder larger the entropy (Bridgman 1941).

It is therefore concluded that entropy is increased as internal constraints are removed and that entropy reaches a maximum in the final state of equilibrium, i.e. the state of maximum "disorder". In other terms, in isolated systems, one has

$$\Delta S = S_{\text{fin}} - S_{\text{in}} \geq 0 \quad \text{(isolated system)}. \tag{1.15}$$

Thus, entropy is continuously increasing when irreversible processes take place until it reaches a state of maximum value, the equilibrium state, which in mathematical terms is characterized by $\mathrm{d}S = 0, \mathrm{d}^2S < 0$. This statement constitutes the celebrated principle of entropy increase and is often referred to as the *Second Law* of thermodynamics. It follows that a decrease in entropy $\mathrm{d}S < 0$ corresponds to an impossible process. Another consequence is that the entropy of an isolated system remains constant when reversible processes occur in it.

An illustration of the entropy increase principle is found in Box 1.2. When the system is not isolated, as in the case of *closed and open systems*, the entropy change in the system consists in two parts: $\mathrm{d}^{\mathrm{e}}S$ due to exchanges of energy and matter with the outside, which may be positive or negative, and $\mathrm{d}^{\mathrm{i}}S$ due to internal irreversible processes

$$\mathrm{d}S = \mathrm{d}^{\mathrm{e}}S + \mathrm{d}^{\mathrm{i}}S. \tag{1.16}$$

The second law asserts that the entropy production $\mathrm{d}^{\mathrm{i}}S$ can only be greater than or equal to zero

$$\mathrm{d}^{\mathrm{i}}S \geq 0 \quad \text{(closed and open systems)}, \tag{1.17}$$

the equality sign referring to reversible or equilibrium situations. Expression (1.17) is the statement of the second law in its more general form. In the particular case of isolated systems, there is no exchange of energy and matter so that $\mathrm{d}^{\mathrm{e}}S = 0$ and one recovers (1.15) of the second law, namely $\mathrm{d}S = \mathrm{d}^{\mathrm{i}}S \geq 0$. For closed systems, for which $\mathrm{d}^{\mathrm{e}}S = \text{đ}Q/T$, one has

$$\mathrm{d}S \geq \text{đ}Q/T \quad \text{(closed system)}. \tag{1.18}$$

In the particular case of a cyclic process for which $\mathrm{d}S = 0$, one has $\text{đ}Q/T \leq 0$, which is usually identified as the Clausius' inequality.

Box 1.2 Entropy Increase

Consider two different gases A and B at equilibrium, enclosed in a common isolated container and separated each other by an adiabatic and fixed wall (Fig. 1.4). Both gases are characterized by their internal energy U and volume V.

In the initial configuration, entropy $S^{(\mathrm{i})}$ is a function of the initial values of internal energy $U_A^{(\mathrm{i})}$ and volume $V_A^{(\mathrm{i})}$ corresponding to subsystem A, and similarly of $U_B^{(\mathrm{i})}$ and $V_B^{(\mathrm{i})}$ for subsystem B, in such a way that $S^{(\mathrm{i})} = S_A(U_A^{(\mathrm{i})}, V_A^{(\mathrm{i})}) + S_B(U_B^{(\mathrm{i})}, V_B^{(\mathrm{i})})$. If the adiabatic and fixed wall separating both subsystems A and B is replaced by a diathermal and movable wall, a new configuration is attained whose entropy $S^{(\mathrm{f})}$ may be expressed as $S^{(\mathrm{f})} = S_A(U_A^{(\mathrm{f})}, V_A^{(\mathrm{f})}) + S_B(U_B^{(\mathrm{f})}, V_B^{(\mathrm{f})})$; superscript (f) denotes the final values of energy and volume submitted to the closure relations $U_A^{(\mathrm{i})} + U_B^{(\mathrm{i})} = U_A^{(\mathrm{f})} + U_B^{(\mathrm{f})} = U_{\mathrm{total}}$ and $V_A^{(\mathrm{i})} + V_B^{(\mathrm{i})} = V_A^{(\mathrm{f})} + V_B^{(\mathrm{f})} = V_{\mathrm{total}}$ reflecting conservation of these quantities for the composite system $A+B$. The removal of internal constraints that prevent the exchange of internal energy and volume leads to the establishment of a new equilibrium state of entropy $S^{(\mathrm{f})} > S^{(\mathrm{i})}$. The values taken by the (extensive) variables, in the absence of internal constraints, in this case $U_A^{(\mathrm{f})}$, $V_A^{(\mathrm{f})}$ and $U_B^{(\mathrm{f})}$, $V_B^{(\mathrm{f})}$, are those that maximize the entropy over the manifold of equilibrium states (Callen 1985).

In Fig. 1.4 is represented $S^{(\mathrm{f})}/S^{(\mathrm{i})}$ in terms of $x \equiv U_A/U_{\mathrm{total}}$ and $y \equiv V_A/V_{\mathrm{total}}$ using an ideal gas model; the final values of x and y are those corresponding to the maximum of $S^{(\mathrm{f})}/S^{(\mathrm{i})}$. The arbitrary curve drawn on the surface between the initial "i" and final "f" states stands for an idealized process defined as a succession of equilibrium states, quite distinct from a real physical process formed by a temporal succession of equilibrium and non-equilibrium states.

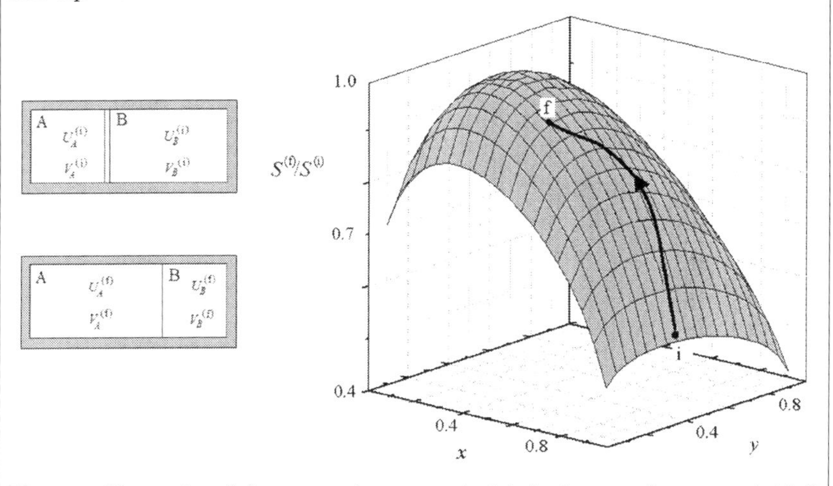

Fig. 1.4 Illustration of the entropy increase principle in the case of two gases initially separated by an adiabatic and fixed wall

1.3.4 The Third Law

The roots of this law appear in the study of thermodynamic quantities as the absolute temperature tends to zero. In 1909, Nernst formulated his heat theorem, later known as the third law of thermodynamics, to better understand the nature of chemical equilibrium. Nernst's formulation was that *the entropy change in any isothermal process approaches zero as the temperature at which the process occurs approaches zero*, i.e.

$$(\Delta S)_{T\to 0} \to 0. \tag{1.19}$$

This statement is sufficient for any thermodynamic development, but sometimes the stronger Planck's statement ($S \to 0$ as $T \to 0$) is preferred. Since the third law is more of quantum statistical essence, it is not of the same nature as the other laws and no further reference will be made to it in this book.

1.4 Gibbs' Equation

Let us now gather the results obtained for the first and second laws. Consider a reversible transformation, taking place in a closed system, for which the first law takes the form

$$\mathrm{d}U = đQ_{\mathrm{rev}} - p\,\mathrm{d}V, \tag{1.20}$$

and combine it with the definition of entropy $đQ_{\mathrm{rev}} = T\,\mathrm{d}S$, resulting in

$$\mathrm{d}U = T\,\mathrm{d}S - p\,\mathrm{d}V. \tag{1.21}$$

Expression (1.21) is known as Gibbs' equation; it is, however, not complete when there are matter exchanges as in open systems, or variations in composition as in chemical reactions. To calculate the reversible work corresponding to a chemical reaction involving n species, it is necessary to devise a reversible process of mixing. This is achieved thanks to van't Hoff's box (Kestin 1968), accordingly the reversible chemical work is given by

$$đW_{\mathrm{rev}}^{\mathrm{ch}} = \sum_{k=1}^{n} \bar{\mu}_k \mathrm{d}m_k, \tag{1.22}$$

where $\bar{\mu}_k$ is defined as the chemical potential of substance k. The properties of the chemical potential will be explicitly examined below. With this additional term, one is led to the generalized Gibbs' equations

$$\mathrm{d}U = T\,\mathrm{d}S - p\,\mathrm{d}V + \sum_{k=1}^{n} \bar{\mu}_k \mathrm{d}m_k, \tag{1.23a}$$

or equivalently,

$$\mathrm{d}S = T^{-1}\mathrm{d}U + pT^{-1}\mathrm{d}V - \sum_{k=1}^{n} T^{-1}\bar{\mu}_k \mathrm{d}m_k. \tag{1.23b}$$

As discussed in the forthcoming sections, the Gibbs' equation plays a fundamental role in equilibrium thermodynamics. We should also mention that Gibbs' equation is one of the pillars of the Classical Theory of Irreversible Processes, as shown in Chap. 2. Let us now examine the main consequences of Gibbs' equation.

1.4.1 Fundamental Relations and State Equations

It follows directly from Gibbs' equation (1.23a) that

$$U = U(S, V, m_1, m_2, \ldots, m_n), \tag{1.24}$$

or, solving with respect to S,

$$S = S(U, V, m_1, m_2, \ldots, m_n). \tag{1.25}$$

Relations like (1.24) or (1.25) expressing that U or S are single-valued functions of extensive state variables are called *fundamental relations* because they contain all thermodynamic information about the system. When U (respectively, S) is expressed as a function of the variables, we are speaking of the "energy representation" (respectively, "entropy representation").

Another consequence of Gibbs' equation (1.23a) is that the intensive variables, represented by temperature, pressure and chemical potentials, can be defined as partial derivatives of U:

$$T = \left(\frac{\partial U}{\partial S}\right)_{V,\{m_k\}} \text{(a)}, \quad p = -\left(\frac{\partial U}{\partial V}\right)_{S,\{m_k\}} \text{(b)}, \quad \bar{\mu}_k = \left(\frac{\partial U}{\partial m_k}\right)_{V,S,\{m_{i\neq k}\}} \text{(c)}, \tag{1.26}$$

where $\{m_k\}$ stands for all m_k constant. Since U is a function of S, V, m_k, the same remains true for T, p, and μ_k so that

$$T = T(S, V, m_1, m_2, \ldots, m_n), \tag{1.27a}$$

$$p = p(S, V, m_1, m_2, \ldots, m_n), \tag{1.27b}$$

$$\bar{\mu}_k = \bar{\mu}_k(S, V, m_1, m_2, \ldots, m_n). \tag{1.27c}$$

Such relationships between intensive and extensive variables are called *state equations*. Elimination of S between (1.27a) and (1.27b) leads to the thermal equation of state $p = p(T, V, m_1, m_2, \ldots, m_n)$; similarly by combining

(1.24) and (1.27a), one obtains the so-called caloric equation $U = U(T, V, m_1, m_2, \ldots, m_n)$. The knowledge of one single state equation is not sufficient to describe the state of a system, which requires the knowledge of *all* the equations of state. For instance in the case of a monatomic perfect gas, $pV = NRT$ does not constitute the complete knowledge of the system but must be complemented by $U = \frac{3}{2}NRT$, R being the gas constant and N the mole number.

1.4.2 Euler's Relation

The extensive property of U implies that, from the mathematical point of view, it is a first-order homogeneous function of the extensive variables:

$$U(\lambda S, \lambda V, \lambda m_1, \ldots, \lambda m_n) = \lambda U(S, V, m_1, \ldots, m_n), \tag{1.28}$$

where λ is an arbitrary scalar. Differentiation of the fundamental relation (1.28) with respect to λ and setting $\lambda = 1$, leads to

$$\left(\frac{\partial U}{\partial S}\right)_{V,\{m_k\}} S + \left(\frac{\partial U}{\partial V}\right)_{S,\{m_k\}} V + \sum_k \left(\frac{\partial U}{\partial m_k}\right)_{V,S,\{m_{j\neq k}\}} m_k = U, \tag{1.29}$$

and, after making use of (1.26), one obtains *Euler's relation*

$$U = TS - pV + \sum_k \bar{\mu}_k m_k. \tag{1.30}$$

1.4.3 Gibbs–Duhem's Relation

A differential equation among the intensive variables can be derived directly from Euler's relation. Indeed, after differentiating (1.30), it is found that

$$\mathrm{d}U = T\,\mathrm{d}S - p\,\mathrm{d}V + \sum_{k=1}^{n} \bar{\mu}_k \mathrm{d}m_k + S\,\mathrm{d}T - V\,\mathrm{d}p + \sum_{k=1}^{n} m_k \mathrm{d}\bar{\mu}_k, \tag{1.31}$$

which, after using Gibbs' equation (1.23a), yields Gibbs–Duhem's relation

$$S\,\mathrm{d}T - V\,\mathrm{d}p + \sum_{k=1}^{n} m_k \mathrm{d}\bar{\mu}_k = 0. \tag{1.32}$$

It follows that the $n + 2$ intensive variables are not independent but related through the Gibbs–Duhem's relation. For a n-component mixture, the number of independent intensive state variables, called *thermodynamic degrees of freedom*, is equal to $n + 1$: for instance, the $n - 1$ chemical potentials plus temperature and pressure. In the case of a one-component fluid, the thermodynamic description of the system requires the knowledge of two independent intensive quantities, generally selected as the temperature T and the pressure p.

1.4.4 Some Definitions

In view of further developments, it is useful to introduce the following definitions of well-known experimental quantities:

- *Coefficient of thermal expansion*:

$$\alpha = \frac{1}{V}\left(\frac{\partial V}{\partial T}\right)_{p,\{m_k\}}. \tag{1.33}$$

- *Isothermal compressibility*:

$$\kappa_T = -\frac{1}{V}\left(\frac{\partial V}{\partial p}\right)_{T,\{m_k\}}. \tag{1.34}$$

- *Heat capacity at constant volume*:

$$c_V = \left(\frac{\partial U}{\partial T}\right)_{V,\{m_k\}} = \left(\frac{đQ_{\mathrm{rev}}}{\mathrm{d}T}\right)_{V,\{m_k\}}. \tag{1.35}$$

- *Heat capacity at constant pressure*:

$$c_p = \left(\frac{\partial H}{\partial T}\right)_{p,\{m_k\}} = \left(\frac{đQ_{\mathrm{rev}}}{\mathrm{d}T}\right)_{p,\{m_k\}}. \tag{1.36}$$

Other partial derivatives may be introduced but generally, they do not have a specific practical usefulness. Relations between these partial derivatives may be derived by equating mixed second-order partial derivatives of U and S. Such expressions have been identified as *Maxwell's relations.*

As a last remark, let us mention that the results established so far in homogeneous systems of total mass m and volume V are still valid when referred per unit mass and unit volume. Analogous to (1.24), the fundamental relation per unit mass is

$$u = u(s, v, \ldots, c_k, \ldots) \tag{1.37}$$

with $u = U/m$, $s = S/m$, $v = V/m$, $c_k = m_k/m$, and $\sum_k c_k = 1$. After differentiation, (1.37) reads as

$$\mathrm{d}u = T\,\mathrm{d}s - p\,\mathrm{d}v + \sum_{k=1}^{n-1}(\bar{\mu}_k - \bar{\mu}_n)\mathrm{d}c_k \tag{1.38}$$

with $T = (\partial u/\partial s)_{v,\{c_k\}}$, $p = -(\partial u/\partial v)_{s,\{c_k\}}$, $\bar{\mu}_k = (\partial u/\partial c_k)_{s,v,\{c_{j\neq k}\}}$. Similarly, the Euler and Gibbs–Duhem's relations (1.30) and (1.32) take the form

$$u = Ts - pv + \sum_{k=1}^{n}\bar{\mu}_k c_k, \quad S\,\mathrm{d}T - v\,\mathrm{d}p + \sum_{k=1}^{n} c_k \mathrm{d}\bar{\mu}_k = 0. \tag{1.39}$$

1.4.5 The Basic Problem of Equilibrium Thermodynamics

To maintain a system in an equilibrium state, one needs the presence of constraints; if some of them are removed, the system will move towards a new equilibrium state. The basic problem is to determine the final equilibrium state when the initial equilibrium state and the nature of the constraints are specified. As illustration, we have considered in Box 1.3 the problem of thermo-diffusion. The system consists of two gases filling two containers separated by a rigid, impermeable and adiabatic wall: the whole system is isolated. If we now replace the original wall by a semi-permeable, diathermal one, there will be heat exchange coupled with a flow of matter between the two subsystems until a new state of equilibrium is reached; the problem is the calculation of the state parameters in the final equilibrium state.

Box 1.3 Thermodiffusion

Let us suppose that an isolated system consists of two separated containers I and II, each of fixed volume, and separated by an impermeable, rigid and adiabatic wall (see Fig. 1.5). Container I is filled with a gas A and container II with a mixture of two non-reacting gases A and B. Substitute now the original wall by a diathermal, non-deformable but semi-permeable membrane, permeable to substance A. The latter will diffuse through the membrane until the system comes to a new equilibrium, of which we want to know the properties. The volumes of each container and the mass of substance B are fixed:

$$V_{\mathrm{I}} = \text{constant}, \quad V_{\mathrm{II}} = \text{constant}, \; m_{\mathrm{II}}^{\mathrm{B}} = \text{constant}, \tag{1.3.1}$$

but the energies in both containers as well as the mass of substance A are free to change, subject to the constraints

$$U_{\mathrm{I}} + U_{\mathrm{II}} = \text{constant}, \quad m_{\mathrm{I}}^{\mathrm{A}} + m_{\mathrm{II}}^{\mathrm{A}} = \text{constant}. \tag{1.3.2}$$

In virtue of the second law, the values of $U_{\mathrm{I}}, U_{\mathrm{II}}, m_{\mathrm{I}}^{\mathrm{A}}, m_{\mathrm{II}}^{\mathrm{A}}$ in the new equilibrium state are such as to maximize the entropy, i.e. $\mathrm{d}S = 0$ and, from the additivity of the entropy in the two subsystems

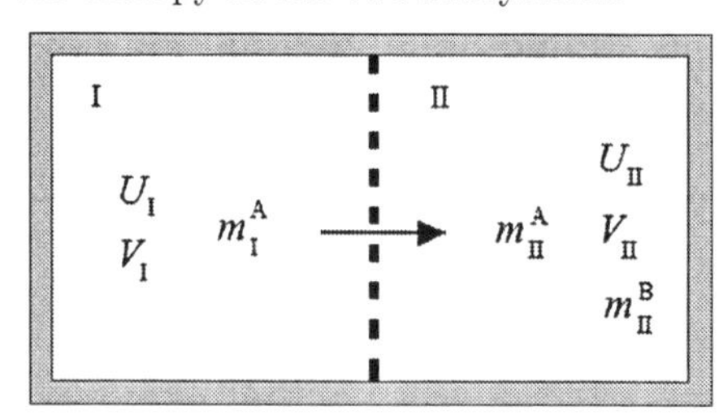

Fig. 1.5 Equilibrium conditions for thermodiffusion

$$dS = dS_{\mathrm{I}} + dS_{\mathrm{II}} = 0. \tag{1.3.3}$$

Making use of the Gibbs' relation (1.23b) and the constraints (1.3.1) and (1.3.2), one may write

$$\begin{aligned} dS &= \frac{\partial S_{\mathrm{I}}}{\partial U_{\mathrm{I}}} dU_{\mathrm{I}} + \frac{\partial S_{\mathrm{I}}}{\partial m_{\mathrm{I}}^{\mathrm{A}}} dm_{\mathrm{I}}^{\mathrm{A}} + \frac{\partial S_{\mathrm{II}}}{\partial U_{\mathrm{II}}} dU_{\mathrm{II}} + \frac{\partial S_{\mathrm{II}}}{\partial m_{\mathrm{II}}^{\mathrm{A}}} dm_{\mathrm{II}}^{\mathrm{A}} \\ &= \left(\frac{1}{T_{\mathrm{I}}} - \frac{1}{T_{\mathrm{II}}} \right) dU_{\mathrm{I}} - \left(\frac{\bar{\mu}_{\mathrm{I}}^{\mathrm{A}}}{T_{\mathrm{I}}} - \frac{\bar{\mu}_{\mathrm{II}}^{\mathrm{A}}}{T_{\mathrm{II}}} \right) dm_{\mathrm{I}}^{\mathrm{A}} = 0. \end{aligned} \tag{1.3.4}$$

Since this relation must be satisfied for arbitrary variations of U_{I} and $m_{\mathrm{I}}^{\mathrm{A}}$, one finds that the equilibrium conditions are that

$$T_{\mathrm{I}} = T_{\mathrm{II}}, \quad \bar{\mu}_{\mathrm{I}}^{\mathrm{A}} = \bar{\mu}_{\mathrm{II}}^{\mathrm{A}}. \tag{1.3.5}$$

The new equilibrium state, which corresponds to absence of flow of substance A, is thus characterized by the equality of temperatures and chemical potentials in the two containers.

In absence of mass transfer, only heat transport will take place. During the irreversible process between the initial and final equilibrium states, the only admissible exchanges are those for which

$$dS = \left(\frac{1}{T_{\mathrm{I}}} - \frac{1}{T_{\mathrm{II}}} \right) đQ_{\mathrm{I}} > 0, \tag{1.3.6}$$

where use has been made of the first law $dU_{\mathrm{I}} = đQ_{\mathrm{I}}$. If $T_{\mathrm{I}} > T_{\mathrm{II}}$, one has $đQ_{\mathrm{I}} < 0$ while for $T_{\mathrm{I}} < T_{\mathrm{II}}, đQ_{\mathrm{I}} > 0$ meaning that heat will spontaneously flow from the hot to the cold container. The formal restatement of this item is the Clausius' formulation of the second law: "no process is possible in which the sole effect is transfer of heat from a cold to a hot body".

Under isothermal conditions ($T_{\mathrm{I}} = T_{\mathrm{II}}$), the second law imposes that

$$dS = \frac{1}{T_{\mathrm{I}}} \left(\bar{\mu}_{\mathrm{II}}^{\mathrm{A}} - \bar{\mu}_{\mathrm{I}}^{\mathrm{A}} \right) dm_{\mathrm{I}}^{\mathrm{A}} > 0 \tag{1.3.7}$$

from which is concluded that matter flows spontaneously from regions of high to low chemical potential.

1.5 Legendre Transformations and Thermodynamic Potentials

Although the fundamental relations (1.24) and (1.25) that are expressed in terms of extensive variables are among the most important, they are not the most useful. Indeed, in practical situations, the intensive variables, like

Box 1.4 Legendre Transformations

The problem to be solved is the following: given a fundamental relation of the extensive variables $A_1, A_2, \ldots, A_n$,

$$Y = Y(A_1, A_2, \ldots, A_k, A_{k+1}, \ldots, A_n) \tag{1.4.1}$$

find a new function for which the derivatives

$$P_i = \frac{\partial Y}{\partial A_i} \quad (i = 1, \ldots, k \leq n) \tag{1.4.2}$$

will be considered as the independent variables instead of $A_1, \ldots, A_k$. The solution is given by

$$Y[P_1, \ldots, P_k] = Y - \sum_{i=1}^{k} P_i A_i. \tag{1.4.3}$$

Indeed, taking the infinitesimal variation of (1.4.3) results in

$$\mathrm{d}Y[P_1, \ldots, P_k] = -\sum_{1}^{k} A_i \mathrm{d}P_i + \sum_{k+1}^{n} P_i \mathrm{d}A_i, \tag{1.4.4}$$

which indicates clearly that $Y[P_1, \ldots, P_k]$ is a function of the independent variables $P_1, \ldots, P_k, A_{k+1}, \ldots, A_n$. With Callen (1985), we have used the notation $Y[P_1, \ldots, P_k]$ to denote the partial Legendre transformation with respect to $A_1, \ldots, A_k$. The function $Y[P_1, \ldots, P_k]$ is referred to as a *Legendre transformation.*

temperature and pressure, are more easily measurable and controllable. In contrast, there is no instrument to measure directly entropy and internal energy. This observation has motivated a reformulation of the theory, in which the central role is played by the intensive rather than the extensive quantities. Mathematically, this is easily achieved thanks to the introduction of *Legendre transformations*, whose mathematical basis is summarized in Box 1.4.

1.5.1 Thermodynamic Potentials

The application of the preceding general considerations to thermodynamics is straightforward: the derivatives $P_1, P_2, \ldots$ will be identified with the intensive variables $T, -p, \mu_k$ and the several Legendre transformations are known as the thermodynamic potentials. Starting from the fundamental relation, $U = U(S, V, m_k)$, replace the entropy S by $\partial U/\partial S \equiv T$ as independent variable, the corresponding Legendre transform is, according to (1.4.3),

$$U[T] \equiv F = U - \left(\frac{\partial U}{\partial S}\right)_{V,\{m_k\}} S = U - TS, \tag{1.40}$$

which is known as *Helmholtz's free energy*. Replacing the volume V by $\partial U/\partial V \equiv -p$, one defines *the enthalpy* H as

$$U[p] \equiv H = U - \left(\frac{\partial U}{\partial V}\right)_{S,\{m_k\}} V = U + pV. \tag{1.41}$$

The Legendre transform which replaces simultaneously S by T and V by $-p$ is the so-called *Gibbs' free energy* G given by

$$U[T,p] \equiv G = U - TS + pV = \sum_{k=1}^{n} m_k \bar{\mu}_k. \tag{1.42}$$

The last equality has been derived by taking account of Euler's relation (1.30). Note that the complete Legendre transform

$$U[T,p,\mu_1,\ldots,\mu_r] = U - TS + pV - \sum_{k=1}^{n} \bar{\mu}_k m_k = 0 \tag{1.43}$$

is identically equal to zero in virtue of Euler's relation and this explains why only three thermodynamic potentials can be defined from U. The fundamental relations of F, H, and G read in differential form:

$$\mathrm{d}F = -S\,\mathrm{d}T - p\,\mathrm{d}V + \sum_{k=1}^{n} \bar{\mu}_k \mathrm{d}m_k, \tag{1.44a}$$

$$\mathrm{d}H = T\,\mathrm{d}S + V\,\mathrm{d}p + \sum_{k=1}^{n} \bar{\mu}_k \mathrm{d}m_k, \tag{1.44b}$$

$$\mathrm{d}G = -S\,\mathrm{d}T + V\,\mathrm{d}p + \sum_{k=1}^{n} \bar{\mu}_k \mathrm{d}m_k. \tag{1.44c}$$

Another set of Legendre transforms can be obtained by operating on the entropy $S = S(U, V, m_1, \ldots, m_n)$, and are called the *Massieu–Planck functions*, particularly useful in statistical mechanics.

1.5.2 Thermodynamic Potentials and Extremum Principles

We have seen that the entropy of an *isolated* system increases until it attains a maximum value: the equilibrium state. Since an isolated system does not exchange heat, work, and matter with the surroundings, it will therefore be characterized by constant values of energy U, volume V, and mass m. In short, for a constant mass, the second law can be written as

$$\mathrm{d}S \geq 0 \text{ at } U \text{ and } V \text{ constant.} \tag{1.45}$$

Because of the invertible roles of entropy and energy, it is equivalent to formulate the second principle in terms of U rather than S.

1.5.2.1 Minimum Energy Principle

Let us show that the second law implies that, in absence of any internal constraint, the energy U evolves to a minimum at S and V fixed:

$$\mathrm{d}U \leq 0 \text{ at } S \text{ and } V \text{ constant.} \tag{1.46}$$

We will prove that if energy is not a minimum, entropy is not a maximum in equilibrium. Suppose that the system is in equilibrium but that its internal energy has not the smallest value possible compatible with a given value of the entropy. We then withdraw energy in the form of work, keeping the entropy constant, and return this energy in the form of heat. Doing so, the system is restored to its original energy but with an increased value of the entropy, which is inconsistent with the principle that the equilibrium state is that of maximum entropy.

Since in most practical situations, systems are not isolated, but closed and then subject to constant temperature or (and) constant pressure, it is appropriate to reformulate the second principle by incorporating these constraints. The evolution towards equilibrium is no longer governed by the entropy or the energy but by the thermodynamic potentials.

1.5.2.2 Minimum Helmholtz's Free Energy Principle

For closed systems maintained at constant temperature and volume, the leading potential is Helmholtz's free energy F. In virtue of the definition of $F(= U - TS)$, one has, at constant temperature,

$$\mathrm{d}F = \mathrm{d}U - T\,\mathrm{d}S, \tag{1.47}$$

and, making use of the first law and the decomposition $\mathrm{d}S = \mathrm{d}^{\mathrm{e}}S + \mathrm{d}^{\mathrm{i}}S$,

$$\mathrm{d}F = \bar{\mathrm{d}}Q - p\,\mathrm{d}V - T\,\mathrm{d}^{\mathrm{e}}S - T\,\mathrm{d}^{\mathrm{i}}S. \tag{1.48}$$

In closed systems $\mathrm{d}^{\mathrm{e}}S = \bar{\mathrm{d}}Q/T$ and, if V is maintained constant, the change of F is

$$\mathrm{d}F = -T\,\mathrm{d}^{\mathrm{i}}S \leq 0. \tag{1.49}$$

It follows that closed systems at fixed values of the temperature and the volume, are driven towards an equilibrium state wherein the Helmholtz's free energy is minimum. Summarizing, at equilibrium, the only admissible processes are those satisfying

$$\mathrm{d}F \leq 0 \text{ at } T \text{ and } V \text{ constant.} \tag{1.50}$$

1.5.2.3 Minimum Enthalpy Principle

Similarly, the enthalpy $H = U + pV$ can also be associated with a minimum principle. At constant pressure, one has

$$\mathrm{d}H = \mathrm{d}U + p\,\mathrm{d}V = đQ, \tag{1.51}$$

but for closed systems, $đQ = T\,\mathrm{d^e}S = T(\mathrm{d}S - \mathrm{d^i}S)$, whence, at fixed values of p and S,

$$\mathrm{d}H = -T\,\mathrm{d^i}S \leq 0, \tag{1.52}$$

as a direct consequence of the second law. Therefore, at fixed entropy and pressure, the system evolves towards an equilibrium state characterized by a minimum enthalpy, i.e.

$$\mathrm{d}H \leq 0 \text{ at } S \text{ and } p \text{ constant.} \tag{1.53}$$

1.5.2.4 Minimum Gibbs' Free Energy Principle

Similar considerations are applicable to closed systems in which both temperature and pressure are maintained constant but now the central quantity is Gibbs' free energy $G = U - TS + pV$. From the definition of G, one has at T and p fixed,

$$\mathrm{d}G = \mathrm{d}U - T\,\mathrm{d}S + p\,\mathrm{d}V = đQ - T(\mathrm{d^e}S + \mathrm{d^i}S) = -T\,\mathrm{d^i}S \leq 0, \tag{1.54}$$

wherein use has been made of $\mathrm{d^e}S = đQ/T$. This result tells us that a closed system, subject to the constraints T and p constant, evolves towards an equilibrium state where Gibbs' free energy is a minimum, i.e.

$$\mathrm{d}G \leq 0 \text{ at } T \text{ and } p \text{ constant.} \tag{1.55}$$

The above criterion plays a dominant role in chemistry because chemical reactions are usually carried out under constant temperature and pressure conditions.

It is left as an exercise (Problem 1.7) to show that the (maximum) work delivered in a reversible process at constant temperature is equal to the decrease in the Helmholtz's free energy:

$$đW_{\mathrm{rev}} = -\mathrm{d}F. \tag{1.56}$$

This is the reason why engineers call frequently F the available work at constant temperature. Similarly, enthalpy and Gibbs' free energy are measures of the maximum available work at constant p, and at constant T and p, respectively.

As a general rule, it is interesting to point out that the Legendre transformations of energy are a minimum for constant values of the transformed intensive variables.

1.6 Stability of Equilibrium States

Even in equilibrium, the state variables do not keep rigorous fixed values because of the presence of unavoidable microscopic fluctuations or external perturbations, like small vibrations of the container. We have also seen that irreversible processes are driving the system towards a unique equilibrium state where the thermodynamic potentials take extremum values. In the particular case of isolated systems, the unique equilibrium state is characterized by a maximum value of the entropy. The fact of reaching or remaining in a state of maximum or minimum potential makes that any *equilibrium state be stable.* When internal fluctuations or external perturbations drive the system away from equilibrium, spontaneous irreversible processes will arise that bring the system back to equilibrium. In the following sections, we will exploit the consequences of equilibrium stability successively in single and multi-component homogeneous systems.

1.6.1 Stability of Single Component Systems

Imagine a one-component system of entropy S, energy U, and volume V in equilibrium and enclosed in an isolated container. Suppose that a hypothetical impermeable internal wall splits the system into two subsystems I and II such that $S = S_{\mathrm{I}} + S_{\mathrm{II}}, U = U_{\mathrm{I}} + U_{\mathrm{II}} = \text{constant}, V = V_{\mathrm{I}} + V_{\mathrm{II}} = \text{constant}$. Under the action of a disturbance, either internal or external, the wall is slightly displaced and in the new state of equilibrium, the energy and volume of the two subsystems will take the values $U_{\mathrm{I}} + \Delta U$, $V_{\mathrm{I}} + \Delta V$, $U_{\mathrm{II}} - \Delta U$, $V_{\mathrm{II}} - \Delta V$, respectively; let ΔS be the corresponding change of entropy. But at equilibrium, S is a maximum so that perturbations can only decrease the entropy

$$\Delta S < 0, \tag{1.57}$$

while, concomitantly, spontaneous irreversible processes will bring the system back to its initial equilibrium configuration. Should $\Delta S > 0$, then the fluctuations would drive the system away from its original equilibrium state with the consequence that the latter would be unstable. Let us now explore the consequences of inequality (1.57) and perform a Taylor-series expansion of $S(U_{\mathrm{I}}, V_{\mathrm{I}}, U_{\mathrm{II}}, V_{\mathrm{II}})$ around the equilibrium state. For small perturbations, we may restrict the developments at the second order and write symbolically

$$\Delta S = S - S_{\mathrm{eq}} = \mathrm{d}S + \mathrm{d}^2 S + \cdots < 0. \tag{1.58}$$

From the property that S is extremum in equilibrium, the first-order terms vanish ($\mathrm{d}S = 0$) and we are left with the calculation of $\mathrm{d}^2 S$: it is found (as detailed in Box 1.5) that

$$\mathrm{d}^2 S = -T^2 C_V (\mathrm{d}T^{-1})^2 - \frac{1}{VT\kappa_T}(\mathrm{d}V)^2 < 0, \tag{1.59}$$

where C_V is the heat capacity at constant volume and κ_T the isothermal compressibility.

Box 1.5 Calculation of $\mathrm{d}^2 S$ for a Single Component System

Since the total energy and volume are constant $\mathrm{d}U_{\mathrm{I}} = -\mathrm{d}U_{\mathrm{II}} = \mathrm{d}U$, $\mathrm{d}V_{\mathrm{I}} = -\mathrm{d}V_{\mathrm{II}} = \mathrm{d}V$, one may write

$$\mathrm{d}^2 S = \frac{1}{2}\left[\left(\frac{\partial^2 S_{\mathrm{I}}}{\partial U_{\mathrm{I}}^2} + \frac{\partial^2 S_{\mathrm{II}}}{\partial U_{\mathrm{II}}^2}\right)_{\mathrm{eq}} (\mathrm{d}U)^2 + 2\left(\frac{\partial^2 S_{\mathrm{I}}}{\partial U_{\mathrm{I}}\partial V_{\mathrm{I}}} + \frac{\partial^2 S_{\mathrm{II}}}{\partial U_{\mathrm{II}}\partial V_{\mathrm{II}}}\right)_{\mathrm{eq}} \mathrm{d}U\mathrm{d}V \right.$$
$$\left. + \left(\frac{\partial^2 S_{\mathrm{I}}}{\partial V_{\mathrm{I}}^2} + \frac{\partial^2 S_{\mathrm{II}}}{\partial V_{\mathrm{II}}^2}\right)_{\mathrm{eq}} (\mathrm{d}V)^2\right] \leq 0. \tag{1.5.1}$$

Recalling that the same substance occupies both compartments, S_{I} and S_{II} and their derivatives will present the same functional dependence with respect to the state variables, in addition, these derivatives are identical in subsystems I and II because they are calculated at equilibrium. If follows that (1.5.1) may be written as

$$\mathrm{d}^2 S = S_{UU}(\mathrm{d}U)^2 + 2S_{UV}\mathrm{d}U\mathrm{d}V + S_{VV}(\mathrm{d}V)^2 \leq 0, \tag{1.5.2}$$

wherein

$$S_{UU} = \left(\frac{\partial^2 S}{\partial U^2}\right)_V = \left(\frac{\partial T^{-1}}{\partial U}\right)_V, \quad S_{VV} = \left(\frac{\partial^2 S}{\partial V^2}\right)_U = \left(\frac{\partial (pT^{-1})}{\partial V}\right)_U,$$
$$S_{UV} = \left(\frac{\partial^2 S}{\partial U \partial V}\right)_V = \left(\frac{\partial T^{-1}}{\partial V}\right)_U = \left(\frac{\partial (pT^{-1})}{\partial U}\right)_V. \tag{1.5.3}$$

To eliminate the cross-term in (1.5.2), we replace the differential $\mathrm{d}U$ by $\mathrm{d}T^{-1}$, i.e.

$$\mathrm{d}T^{-1} = S_{UU}\mathrm{d}U + S_{UV}\mathrm{d}V, \tag{1.5.4}$$

whence

$$\mathrm{d}^2 S = \frac{1}{S_{UU}}(\mathrm{d}T^{-1})^2 + \left(S_{VV} - \frac{S_{UV}^2}{S_{UU}}\right)(\mathrm{d}V)^2 > 0. \tag{1.5.5}$$

Furthermore, since

$$S_{UU} = \left(\frac{\partial T^{-1}}{\partial U}\right)_V = -\frac{1}{T^2}\left(\frac{\partial T}{\partial U}\right)_V = -\frac{1}{T^2 C_V} \tag{1.5.6}$$

and

$$S_{VV} - \frac{S_{UV}^2}{S_{UU}} = \left(\frac{\partial(pT^{-1})}{\partial V}\right)_T = \frac{1}{T}\left(\frac{\partial p}{\partial V}\right)_T = -\frac{1}{VT\kappa_T}, \tag{1.5.7}$$

as can be easily proved (see Problem 1.9), (1.5.5) becomes

$$\mathrm{d}^2 S = -T^2 C_V(\mathrm{d}T^{-1})^2 - \frac{1}{VT\kappa_T}(\mathrm{d}V)^2 \leq 0. \tag{1.5.8}$$

The criterion (1.59) for $\mathrm{d}^2 S < 0$ leads to the following conditions of stability of equilibrium:

$$C_V = (\text{đ}Q_{\text{rev}}/\mathrm{d}T)_V > 0, \quad \kappa_T = -(1/V)(\partial V/\partial p)_T > 0. \tag{1.60}$$

The first criterion is generally referred to as the condition of *thermal stability*; it means merely that, removing reversibly heat, at constant volume, must decrease the temperature. The second condition, referred to as *mechanical stability*, implies that any isothermal increase of pressure results in a diminution of volume, otherwise, the system would explode because of instability. Inequalities (1.60) represent mathematical formulations of *Le Chatelier's principle*, i.e. that any deviation from equilibrium will induce a spontaneous process whose effect is to restore the original situation. Suppose for example that thermal fluctuations produce suddenly an increase of temperature locally in a fluid. From the stability condition that C_V is positive, and heat will spontaneously flow out from this region ($\text{đ}Q < 0$) to lower its temperature ($\mathrm{d}T < 0$). If the stability conditions are not satisfied, the homogeneous system will evolve towards a state consisting of two or more portions, called phases, like liquid water and its vapour. Moreover, when systems are driven far from equilibrium, the state is no longer characterized by an extremum principle and irreversible processes do not always maintain the system stable (see Chap. 6).

1.6.2 Stability Conditions for the Other Thermodynamic Potentials

The formulation of the stability criterion in the energy representation is straightforward. Since equilibrium is characterized by minimum energy, the corresponding stability criterion will be expressed as $\mathrm{d}^2U(S,V) \geq 0$ or, more explicitly,

$$U_{SS} \geq 0, \quad U_{VV} \geq 0, \quad U_{SS}U_{VV} - (U_{SV})^2 \geq 0 \tag{1.61}$$

showing that the energy is jointly a convex function of U and V (and also of N in open systems).

The results are also easily generalized to the Legendre transformations of S and U. As an example, consider the Helmholtz's free energy F. From $\mathrm{d}F = -S\,\mathrm{d}T - p\,\mathrm{d}V$, it is inferred that

$$F_{TT} = -T^{-1}C_V \leq 0, \quad F_{VV} = \frac{1}{V\kappa_T} \geq 0 \tag{1.62}$$

from which it follows that F is a concave function of temperature and a convex function of the volume as reflected by the inequalities (1.62). By concave (convex) function is meant a function that lies everywhere below (above) its family of tangent lines, be aware that some authors use the opposite definition for the terms concave and convex. Similar conclusions are drawn for the enthalpy, which is a convex function of entropy and a concave function of pressure:

$$H_{SS} \geq 0, \quad H_{pp} \leq 0. \tag{1.63a}$$

Finally, the Gibbs' free energy G is jointly a concave function of temperature and pressure

$$G_{TT} \leq 0, \quad G_{pp} \leq 0, \quad G_{TT}G_{pp} - (G_{Tp})^2 \geq 0. \tag{1.63b}$$

1.6.3 Stability Criterion of Multi-Component Mixtures

Starting from the fundamental relation of a mixture of n constituents in the entropy representation, $S = S(U, V, m_1, m_2, \ldots, m_n)$, it is detailed in Box 1.6 that the second-order variation d^2S, which determines the stability, is given by

$$\mathrm{d}^2S = \mathrm{d}T^{-1}\mathrm{d}U + \mathrm{d}(pT^{-1})\mathrm{d}V - \sum_{k=1}^{n} \mathrm{d}(\bar{\mu}_k T^{-1})\mathrm{d}m_k \leq 0. \tag{1.64}$$

At constant temperature and pressure, inequality (1.64) reduces to

$$\sum_{k,l}^{n} \frac{\partial \bar{\mu}_k}{\partial m_l} \mathrm{d}m_k \mathrm{d}m_l \geq 0, \tag{1.65}$$

Box 1.6 Calculation of d^2S for Multi-Component Systems

In an N-component mixture of total energy U, total volume V, and total mass $m = \sum_k m_k$ the second variation of S is

$$
\begin{aligned}
d^2S =& \tfrac{1}{2}\left(S_{UU}\right)_{\rm eq}(dU)^2 + \tfrac{1}{2}\left(S_{VV}\right)_{\rm eq}(dV)^2 + \tfrac{1}{2}\sum_{k,l}\left(S_{m_k m_l}\right)_{\rm eq} dm_k dm_l \\
&+ \left(S_{UV}\right)_{\rm eq} dU dV + \sum_k \left(S_{U m_k}\right)_{\rm eq} dU dm_k + \sum_k \left(S_{V,m_k}\right)_{\rm eq} dV\, dm_k .
\end{aligned}
\tag{1.6.1}
$$

Making use of the general results $\partial S/\partial U = 1/T, \partial S/\partial V = p/T, \partial S/\partial m_k = -\mu_k/T$, the above expression can be written as

$$
\begin{aligned}
d^2S =& \frac{1}{2}\left(\frac{\partial T^{-1}}{\partial U}dU + \frac{\partial T^{-1}}{\partial V}dV + \sum_k \frac{\partial T^{-1}}{\partial m_k}dm_k\right) dU \\
&+\frac{1}{2}\left(\frac{\partial (pT^{-1})}{\partial U}dU + \frac{\partial (pT^{-1})}{\partial V}dV + \sum_k \frac{\partial (pT^{-1})}{\partial m_k}\right) dV \\
&-\frac{1}{2}\sum_k\left(\frac{\partial (\bar{\mu}_k T^{-1})}{\partial U}dU + \frac{\partial (\bar{\mu}_k T^{-1})}{\partial V}dV + \sum_l \frac{\partial (\bar{\mu}_k T^{-1})}{\partial m_l}dm_l\right) dm_k
\end{aligned}
\tag{1.6.2}
$$

from which follows the general stability condition

$$
d^2S = \frac{1}{2}\left[dT^{-1}dU + d(pT^{-1})dV - \sum_{k=1}^{n} d(\bar{\mu}_k T^{-1})dm_k\right] \leq 0. \tag{1.6.3}
$$

to be satisfied whatever the values of dm_k and dm_l from which follows that:

$$
\frac{\partial \bar{\mu}_k}{\partial m_k} \geq 0, \quad \det\left|\frac{\partial \bar{\mu}_k}{\partial m_l}\right| \geq 0. \tag{1.66}
$$

The criteria (1.65) or (1.66) are referred to as the conditions of *stability with respect to diffusion.* The first inequality (1.66) indicates that the stability of equilibrium implies that any increase on mass of a given constituent will increase its chemical potential. This provides another example of the application of Le Chatelier's principle. Indeed, any non-homogeneity which manifests in a part of the system in the form of increase of mass will induce locally an increase of chemical potential. Since the latter is larger than its ambient value, there will be a net flow of matter from high to lower chemical potentials that will tend to eradicate the non-homogeneity.

1.7 Equilibrium Chemical Thermodynamics

In the last part of this chapter, we shall apply the general results of equilibrium thermodynamics to chemical thermodynamics. As an illustration, consider the reaction of synthesis of hydrogen chloride

$$H_2 + Cl_2 \rightleftharpoons 2HCl \tag{1.67}$$

or more generally

$$\sum_{k=1}^{n} \nu_k X_k = 0, \tag{1.68}$$

where the X_ks are the symbols for the n chemical species and ν_k the stoichiometric coefficients; conventionally the latter will be counted positive when they correspond to products and negative for reactants. In the above example, $X_1 = H_2$, $X_2 = Cl_2$, $X_3 = HCl$ and $\nu_1 = -1$, $\nu_2 = -1$, $\nu_3 = 2$, $n = 3$. The reaction may proceed in either direction depending on temperature, pressure, and composition; in equilibrium, the quantity of reactants that disappear is equal to the quantity of products that instantly appear. The change in the mole numbers $\mathrm{d}N_k$ of the various components of (1.68) is governed by

$$\frac{\mathrm{d}N_{H_2}}{-1} = \frac{\mathrm{d}N_{Cl_2}}{-1} = \frac{\mathrm{d}N_{HCl}}{2} \equiv \mathrm{d}\xi, \tag{1.69}$$

where ξ is called the degree of advancement or extent of reaction. At the beginning of the reaction $\xi = 0$; its time derivative $\mathrm{d}\xi/\mathrm{d}t$ is related to the velocity of reaction which vanishes at chemical equilibrium (see Chap. 4). The advantage of the introduction of ξ is that all the changes in the mole numbers are expressed by one single parameter, indeed from (1.69),

$$\mathrm{d}N_k = \nu_k \mathrm{d}\xi, \tag{1.70}$$

and, after integration,

$$N_k = N_k^0 + \nu_k \xi, \tag{1.71}$$

wherein superscript 0 denotes the initial state; observe that the knowledge of ξ completely specifies the composition of the system. When expressed in terms of the mass of the constituents, (1.70) reads as $\mathrm{d}m_k = \nu_k M_k \mathrm{d}\xi$ with M_k the molar mass of k; after summation on k, one obtains the mass conservation law

$$\sum_{k=1}^{n} \nu_k M_k = 0. \tag{1.72}$$

The above results are directly generalized when r chemical reactions are taking place among the n constituents. In this case, (1.70) and (1.72) will, respectively, be of the form

$$\mathrm{d}N_k = \sum_{j=1}^{r} \nu_{jk} \mathrm{d}\xi_j (k = 1, 2, \ldots, n), \quad \sum_{k=1}^{n} \nu_{jk} M_k = 0 \quad (j = 1, 2, \ldots, r). \tag{1.73}$$

In Sects. 1.7.1–1.7.3, we discuss further the conditions for chemical equilibrium and the consequences of stability of equilibrium.

1.7.1 General Equilibrium Conditions

Since chemical reactions take generally place at constant temperature and pressure, it is convenient to analyse them in terms of Gibbs' free energy $G = G(T, p, N_1, N_2, \ldots, N_n)$. At constant temperature and pressure, the change in G associated with the variations $\mathrm{d}N_k$ in the mole numbers is

$$\mathrm{d}G = \sum_{k=1}^{n} \mu_k \mathrm{d}N_k, \tag{1.74}$$

where μ_k is the chemical potential per mole of species k. This μ_k is closely related to that appearing in (1.23), because the mass of the species k is $m_k = M_k N_k$, with M_k the corresponding molar mass. Then, it is immediate to see that $\mu_k = \bar{\mu}_k / M_k$. Since μ_k and $\bar{\mu}_k$ are often found in the literature, it is useful to be acquainted with both of them. After substitution of $\mathrm{d}N_k$ by its value (1.70), one has

$$\mathrm{d}G = \left(\sum_{k=1}^{n} \nu_k \mu_k \right) \mathrm{d}\xi = -\mathcal{A}\, \mathrm{d}\xi, \tag{1.75}$$

wherein, with De Donder, we have introduced the "affinity" of the reaction as defined by

$$\mathcal{A} = -\sum_{k=1}^{n} \nu_k \mu_k. \tag{1.76}$$

Since G is a minimum at equilibrium ($\mathrm{d}G/\mathrm{d}\xi = 0$), the condition of chemical equilibrium is that the affinity is zero:

$$\mathcal{A}_{\mathrm{eq}} = 0. \tag{1.77}$$

In presence of r reactions, equilibrium implies that the affinity of each individual reaction vanishes: $(\mathcal{A}_j)_{\mathrm{eq}} = 0 (j = 1, 2, \ldots, r)$. To better apprehend the physical meaning of the result (1.77), let us express $\mathcal{A}$ in terms of physical quantities. In the case of ideal gases or diluted solutions, the chemical potential of constituent k can be written as

$$\mu_k = \mu_k^0(p, T) + RT \ln x_k, \tag{1.78}$$

where $x_k = N_k/N$ is the mole fraction of substance k, N the total number of moles and $\mu_k^0(p,T)$ is the part of chemical potential depending only on p and T. For non-ideal systems, the above form is preserved at the condition to replace $\ln x_k$ by $\ln(x_k \gamma_k)$ where γ_k is called the activity coefficient and $a_k = \gamma_k x_k$ the activity. By substituting (1.78) in expression (1.76) of the affinity, we can express the equilibrium condition in terms of the mole fractions, which are measurable quantities, and one has

$$\mathcal{A}_{\text{eq}} \equiv \sum_k (-\nu_k \mu_k^0) - RT \sum_k \nu_k \ln x_k = 0. \tag{1.79}$$

Defining the *equilibrium constant* $K(T,p)$ by means of $\ln K(T,p) = -(\sum_k \nu_k \mu_k^0)/RT$, the previous relation can be cast in the simple form

$$\mathcal{A}_{\text{eq}} = RT \ln \frac{K(T,p)}{x_1^{\nu_1} x_2^{\nu_2} \cdots x_n^{\nu_n}} = 0, \tag{1.80}$$

whence

$$x_1^{\nu_1}(\xi_{\text{eq}}) x_2^{\nu_2}(\xi_{\text{eq}}) \cdots x_n^{\nu_n}(\xi_{\text{eq}}) = K(T,p), \tag{1.81}$$

which is called the *mass action law* or *Guldberg and Waage law*. This is the key relation in equilibrium chemistry: it is one algebraic equation involving a single unknown, namely, the value of ξ_{eq} of degree of advancement which gives the corresponding number of moles in equilibrium through (1.71). If $K(T,p)$ is known as a function of T and p for a particular reaction, all the equilibrium mole fractions can be computed by the mass action law. Going back to our illustrative example (1.67) and assuming that each component is well described by the ideal gas model, the law of mass action is

$$\frac{x_{\text{HCl}}^2}{x_{\text{H}_2} x_{\text{Cl}_2}} = K(T,p). \tag{1.82}$$

When a number r of reactions are implied, we will have r algebraic equations of the form (1.81) to be solved for the unknowns $(\xi_1)_{\text{eq}}, (\xi_2)_{\text{eq}}, \ldots, (\xi_r)_{\text{eq}}$.

1.7.2 Heat of Reaction and van't Hoff Relation

Most of the chemical reactions supply or absorb heat, thus the heat of reaction is an important notion in chemistry. To introduce it, let us start from the first law written as

$$đQ = \mathrm{d}H - V\,\mathrm{d}p, \tag{1.83}$$

with the enthalpy H given by the equation of state $H = H(T,p,\xi)$, whose differential form is

$$\mathrm{d}H = \left(\frac{\partial H}{\partial T}\right)_{p,\xi} \mathrm{d}T + \left(\frac{\partial H}{\partial p}\right)_{T,\xi} \mathrm{d}p + \left(\frac{\partial H}{\partial \xi}\right)_{T,p} \mathrm{d}\xi. \tag{1.84}$$

Substitution of (1.84) in (1.83) yields

$$đQ = C_{p,\xi}\mathrm{d}T + h_{T,\xi}\mathrm{d}p - r_{T,p}\mathrm{d}\xi, \tag{1.85}$$

where $C_p = (\partial H/\partial T)_{p,\xi}$, $h_T = (\partial H/\partial p)_{T,\xi} - V$, and $r_{T,p} = -(\partial H/\partial \xi)_{T,p}$ designate the specific heat at constant pressure, the heat compressibility and the *heat of reaction* at constant temperature and pressure, respectively. The heat of reaction is positive if the reaction is *exothermic* (which corresponds to delivered heat) and negative if the reaction is *endothermic* (which corresponds to absorbed heat). In terms of variation of affinity with temperature, the heat of reaction is given by

$$r_{T,p} = -T\left(\frac{\partial \mathcal{A}}{\partial T}\right)_{p,\xi_{\mathrm{eq}}}. \tag{1.86}$$

This is directly established by using the result

$$H = G + TS = G - T\left(\frac{\partial G}{\partial T}\right)_{p,\xi}. \tag{1.87}$$

From the definition of the heat of reaction, one has

$$r_{T,p} = -\left(\frac{\partial G}{\partial \xi}\right)_{T,p} + T\left[\frac{\partial}{\partial T}\left(\frac{\partial G}{\partial \xi}\right)_{T,p}\right]_{p,\xi}, \tag{1.88}$$

with $(\partial G/\partial \xi)_{T,p} = -\mathcal{A}$ in virtue of Gibbs' equation (1.75), substituting this result in (1.88) and recognizing that $\mathcal{A}_{\mathrm{eq}} = 0$, one obtains (1.86). By means of (1.78) of μ_k and (1.79) of $\mathcal{A}$, it is easily found from (1.86) that

$$r_{T,p} = -RT^2\frac{\partial}{\partial T}\ln K(T,p). \tag{1.89}$$

This is the van't Hoff relation, which is very important in chemical thermodynamics; it permits to determine the heat of reaction solely from the measurements of the equilibrium constant $K(T,p)$ at different temperatures.

1.7.3 Stability of Chemical Equilibrium and Le Chatelier's Principle

For the clarity of the presentation, let us recall the condition (1.65) for stability of diffusion which can be cast in the form:

$$\sum_{k,l}^{n}\mu_{kl}\mathrm{d}N_k\mathrm{d}N_l > 0, \quad \mu_{kl} = \frac{\partial \mu_k}{\partial N_l}. \tag{1.90}$$

Assuming that there are no simultaneous reactions and inserting (1.70), one obtains

$$\sum_{k,l}^{n} \mu_{kl}\nu_k\nu_l(\mathrm{d}\xi)^2 > 0. \tag{1.91}$$

From the other side, it follows from the definition of the affinity that

$$\frac{\partial \mathcal{A}}{\partial \xi} = -\sum_{k,l}^{n} \nu_k \frac{\partial N_k}{\partial N_l}\frac{\partial N_l}{\partial \xi} = -\sum_{k,l}^{n} \mu_{kl}\nu_k\nu_l, \tag{1.92}$$

so that the criterion of stability (1.90) can be cast in the simple form

$$-\frac{\partial \mathcal{A}}{\partial \xi}(\mathrm{d}\xi)^2 > 0 \quad \text{or} \quad \frac{\partial \mathcal{A}}{\partial \xi} < 0. \tag{1.93}$$

The above result is easily generalized for r simultaneous reactions:

$$\sum_{m,n}^{r} \frac{\partial \mathcal{A}_m}{\partial \xi_n}\mathrm{d}\xi_m \mathrm{d}\xi_n < 0 \quad (m, n = 1, 2, \ldots, r). \tag{1.94}$$

Perhaps one of the most interesting consequences of the stability criterion (1.93) is in the form of Le Chatelier's moderation principle. Let us first examine the effect on chemical equilibrium of a temperature change at constant pressure. The shift of equilibrium with temperature is measured by the quantity $(\partial\xi/\partial T)_{p,\mathcal{A}}$, index $\mathcal{A}$ is introduced here because in chemical equilibrium, $\mathcal{A}$ is constant, in fact zero. It is a mathematical exercise to prove that (see Problem 1.8)

$$\left(\frac{\partial \xi}{\partial T}\right)_{p,\mathcal{A}} = -\frac{(\partial \mathcal{A}/\partial T)_{p,\xi}}{(\partial \mathcal{A}/\partial \xi)_{p,T}}. \tag{1.95}$$

The numerator is related to the heat of reaction by (1.86) from which follows that:

$$\left(\frac{\partial \xi}{\partial T}\right)_{p,\mathcal{A}} = \frac{1}{T}\frac{r_{T,p}}{(\partial \mathcal{A}/\partial \xi)_{p,T}}. \tag{1.96}$$

According to the stability condition (1.93), the denominator is a negative quantity so that the sign of $(\partial\xi/\partial T)_{p,\mathcal{A}}$ is opposite to the sign of the heat of reaction $r_{T,p}$. Therefore, an increase of temperature at constant pressure will shift the reaction in the direction corresponding to endothermic reaction ($r_{T,p} < 0$). This is in the direction in which heat is absorbed, thus opposing the increase of temperature. Similar results are obtained by varying the pressure: an increase of pressure at constant temperature will cause the reaction to progress in the direction leading to a diminution of volume, thus weakening the action of the external effect. These are particular examples of the more general principle of Le Chatelier stating that any system in chemical equilibrium undergoes, under the effect of external stimuli, a compensating change which will be always in the opposite direction.

1.8 Final Comments

Equilibrium thermodynamics constitutes a unique and universal formalism whose foundations are well established and corroborated by experience. It has also been the subject of numerous applications. It should nevertheless be kept in mind that equilibrium thermodynamics is of limited range as it deals essentially with equilibrium situations and idealized reversible processes. It is therefore legitimate to ask to what extent equilibrium thermodynamics can be generalized to cover more general situations as non-homogeneous systems, far from equilibrium states and irreversible processes. Many efforts have been spent to meet such objectives and have resulted in the developments of various approaches coined under the generic name of non-equilibrium thermodynamics. It is our purpose in the forthcoming chapters to present, to discuss, and to compare the most recent and relevant – at least in our opinion – of these beyond of equilibrium theories.

There exists a multiplicity of excellent textbooks on equilibrium thermodynamics and it would be unrealistic to go through the complete list. Let us nevertheless mention the books by Callen (1985), Duhem (1911), Gibbs (1948), Kestin (1968), Prigogine (1947), Kondepudi and Prigogine (1998) and Zemansky (1968), which have been a source of inspiration for the present chapter.

1.9 Problems

1.1. *Mechanical work.* Starting from the mechanical definition of work $đW = \boldsymbol{F} \cdot \mathrm{d}\boldsymbol{x}$ (scalar product of force and displacement), show that the work done during the compression of a gas of volume V is $đW = -p\,\mathrm{d}V$, and that the same expression is valid for an expansion.

1.2. *Carnot cycle.* Show that the work performed by an engine during an irreversible cycle operating between two thermal reservoirs at temperatures T_1 and $T_2 < T_1$ is given by $W = W_{\max} - T_2 \Delta S$, where ΔS is the increase of entropy of the Universe, and $W_{\max}$ is the corresponding work performed in a reversible Carnot cycle.

1.3. *Fundamental relation.* In the entropy representation, the fundamental equation for a monatomic ideal gas is

$$S(U, V, N) = \frac{N}{N_0} S_0 + NR \ln \left[\left(\frac{U}{U_0} \right)^{3/2} \left(\frac{V}{V_0} \right) \left(\frac{N}{N_0} \right)^{-5/2} \right],$$

with R the ideal gas constant, and the subscript 0 standing for an arbitrary reference state. By using the formalism of equilibrium thermodynamics, show that the thermal and caloric equations of state for this system are $pV = NRT$ and $U = \frac{3}{2} NRT$, respectively.

1.4. *Maxwell's relations.* The equality of the second crossed derivatives of the thermodynamic potentials is a useful tool to relate thermodynamic quantities. Here, we will consider some consequences related to second derivatives of the entropy. (a) Show that $\mathrm{d}S(T,V)$ at constant N may be expressed as

$$\mathrm{d}S = \frac{1}{T}\left(\frac{\partial U}{\partial T}\right)_V \mathrm{d}T + \frac{1}{T}\left[\left(\frac{\partial U}{\partial V}\right)_T + p\right]\mathrm{d}V.$$

(b) By equating the second crossed derivatives of S in this expression, show that

$$p + T\left(\frac{\partial p}{\partial T}\right)_V = \left(\frac{\partial U}{\partial V}\right)_T.$$

(c) Using this equation, show that for ideal gases, for which $pV = NRT$, the internal energy does not depend on V, i.e. $(\partial U/\partial V)_T = 0$. (d) For electromagnetic radiation, $p = (1/3)(U/V)$. Using this result and the relation obtained in (b), show that for that system, one has $U = aVT^4$. *Remark*: This expression is closely related to the Stefan–Boltzmann law for the heat flux radiated by a black body, namely $q = \sigma T^4$, with σ the Stefan–Boltzmann constant, indeed it follows from the relation $q = 1/4c(U/V)$, with c the speed of light, and a given by $a = 4\sigma/c$.

1.5. *Maxwell's relations.* Prove that $c_p - c_v = Tv\alpha^2/\kappa_T$ by making use of Maxwell's relations.

1.6. *Van der Waals gases.* The thermal equation of state for real gases was approximated by van der Waals in the well-known expression

$$\left[p + a\left(\frac{N}{V}\right)^2\right](V - bN) = NRT,$$

where a and b are positive constants, fixed for each particular gas, which are, respectively, related to the attractive and repulsive intermolecular forces and are null for ideal gases. (a) Using the expressions derived in Problem 1.4, show that the caloric equation of state $U = U(T,V,N)$ has the form

$$U(T,V,N) = U_{\mathrm{id}}(T,N) - a\frac{N^2}{V},$$

with $U_{\mathrm{id}}(T,N)$ is the internal energy for ideal gases. (b) Calculate the change of temperature in an adiabatic expansion against the vacuum, i.e. the variation of T in terms of V at constant internal energy. (c) Find the curve separating the mechanically stable region from the mechanically unstable region in the plane $p - V$, the mechanical stability condition being given by (1.60). (d) The maximum of such a curve, defined by the additional condition $(\partial^2 p/\partial V^2)_{T,N} = 0$, is called the critical point. Show that the values of p, V, and T at the critical point are $p_{\mathrm{c}} = (1/27)(a/b^2)$, $V_{\mathrm{c}} = 3bN$, $T_{\mathrm{c}} = (8/27)(a/bR)$, respectively.

1.7. *Kelvin's statement of the second law.* Verify that in a reversible transformation at fixed temperature and volume, the maximum reversible work delivered by the system is equal to the decrease of the Helmholtz's free energy $-\mathrm{d}F$. As an aside result, show that it is not possible to obtain work from an engine operating in a cycle and in contact with one single source of heat. This result is known as Kelvin's statement of the second law.

1.8. *Mathematical relation.* Verify the general result

$$\left(\frac{\partial y}{\partial x}\right)_{f,z} = -\frac{(\partial f/\partial x)_{y,z}}{(\partial f/\partial y)_{x,z}}.$$

Hint: Consider $f = f(X,Y,Z) =$ constant as an implicit function of the variables X,Y,Z and write explicitly $\mathrm{d}f = 0$.

1.9. *Stability coefficient.* Prove that (see Box 1.5)

$$\frac{\partial^2 S}{\partial V^2} - \frac{(\partial^2 S/\partial U \partial V)^2}{\partial^2 S/\partial U^2} = \frac{1}{T}\left(\frac{\partial p}{\partial V}\right)_T.$$

Hint: Construct the Massieu function (i.e. Legendre transformation of entropy), namely $S[T^{-1}] = S - T^{-1}U$. From the differential of $S[T^{-1}]$, derive $(\partial^2 S[T^{-1}]/\partial V^2)_T$.

1.10. *Stability conditions.* Reformulate the stability analysis of Sect. 1.6.1 by considering that the total entropy and volume are kept constant and by expanding the total energy in Taylor's series around equilibrium.

1.11. *Le Chatelier's principle.* Show that an increase in pressure, at constant temperature, causes the chemical reaction to proceed in that direction which decreases the total volume.

1.12. *Le Chatelier's principle.* The reaction of dissociation of hydrogen iodide $2\mathrm{HI} \rightarrow \mathrm{H}_2 + \mathrm{I}_2$ is endothermic. Determine in which direction equilibrium will be shifted when (a) the temperature is decreased at constant pressure and (b) the pressure is decreased at constant temperature.

Chapter 2
Classical Irreversible Thermodynamics

Local Equilibrium Theory of Thermodynamics

Equilibrium thermodynamics is concerned with ideal processes taking place at infinitely slow rate, considered as a sequence of equilibrium states. For arbitrary processes, it may only compare the initial and final equilibrium states but the processes themselves cannot be described. To handle more realistic situations involving finite velocities and inhomogeneous effects, an extension of equilibrium thermodynamics is needed.

A first insight is provided by the so-called "classical theory of irreversible processes" also named "classical irreversible thermodynamics" (CIT). This borrows most of the concepts and tools from equilibrium thermodynamics but transposed at a local scale because non-equilibrium states are usually inhomogeneous. The objective is to cope with non-equilibrium situations where basic physical quantities like mass, temperature, pressure, etc. are not only allowed to change from place to place, but also in the course of time. As shown in the present and the forthcoming chapters, this theory has been very useful in dealing with a wide variety of practical problems. Pioneering works in this theory were accomplished by Onsager (1931) and Prigogine (1961); these authors were awarded the Nobel Prize in Chemistry in 1968 and 1977, respectively. Other important and influential contributions are also found in the works of Meixner and Reik (1959), de Groot and Mazur (1962), Gyarmati (1970), and many others, which have enlarged the theory to a wider number of applications and have clarified its foundations and its limits of validity.

The principal aims of classical irreversible thermodynamics are threefold. First to provide a thermodynamic support to the classical transport equations of heat, mass, momentum, and electrical charge, as the Fourier's law (1810) relating the heat flux to the temperature gradient, Fick's relation (1850) between the flux of matter and the mass concentration gradient, Ohm's equation (1855) between electrical current and potential, and Newton–Stokes' law (1687, 1851) relating viscous pressure to velocity gradient in fluids. A second objective is to propose a systematic description of the coupling between thermal, mechanical, chemical, and electromagnetic effects, as the Soret (1879) and Dufour (1872) effects, coupling heat, and mass transport, and the

Seebeck (1821), Peltier (1836) and Thomson effects, coupling thermal transport, and electric current. A third objective is the study of stationary non-equilibrium dissipative states, whose properties do not depend on time, but which are characterized by a non-homogeneous distribution of the variables and non-vanishing values of the fluxes.

The present chapter is divided in two parts: in the first one, we recall briefly the general statements underlying the classical theory of irreversible processes. The second part is devoted to the presentation of a few simple illustrations, as heat conduction in rigid bodies, matter transport, and hydrodynamics. Chemical reactions and coupled transport phenomena, like thermoelectricity, thermodiffusion, and diffusion through a membrane are dealt with separately in Chaps. 3 and 4.

2.1 Basic Concepts

The relevance of transport equations, which play a central role in non-equilibrium thermodynamics – comparable, in some way, to equations of state in equilibrium thermodynamics – justifies some preliminary considerations. Transport equations describe the amount of heat, mass, electrical charge, or other quantities which are transferred per unit time between different systems and different regions of a system as a response to a non-homogeneity in temperature T, molar concentration c, electric potential φ_e. Historically, the first incursions into this subject are allotted to Fourier, Fick, and Ohm, who proposed the nowadays well-known laws:

$$\boldsymbol{q} = -\lambda \nabla T \quad \text{(Fourier's law)}, \tag{2.1}$$

$$\boldsymbol{J} = -D \nabla c \quad \text{(Fick's law)}, \tag{2.2}$$

$$\boldsymbol{I} = \sigma_e \nabla \varphi_e \quad \text{(Ohm's law)}. \tag{2.3}$$

Here $\boldsymbol{q}$ is the heat flux (amount of internal energy per unit time and unit area transported by conduction), $\boldsymbol{J}$ is the diffusion flux (amount of matter, expressed in moles, transported per unit time and unit area), and $\boldsymbol{I}$ is the flux of electric current (electric charge transported per unit time and area). The coefficients λ, D, and σ_e are the thermal conductivity, diffusion coefficient, and electric conductivity, respectively. The knowledge of these various transport coefficients in terms of temperature, pressure, and mass concentration has important consequences in material sciences and more generally on our everyday life. For instance, a low value of thermal conductivity is needed for a better isolation of buildings; in contrast, large values are preferred to avoid excessive heating of computers; the diffusion coefficient is a fundamental parameter in biology and in pollution dispersal problems, while electrical conductivity has a deep influence on the development and management of electrical plants, networks, and microelectronic devices. The value of some transport coefficients, as for instance electrical conductivity, may present huge discontinuities as, for example, in superconductors, where conductivity diverges, thus implying a vanishing electrical resistivity, or in insulators, where conductivity vanishes.

The physical content of the above transport laws is rather intuitive: heat will flow from regions with higher temperature to regions at lower temperature and the larger is the temperature gradient the larger is the heat flow. Analogously, matter diffuses from regions with higher mass concentration to regions with lower concentration, and electric positive charges move from regions with higher electrical potential to regions with lower potential.

The evolution of a system in the course of time and space requires the knowledge of the balance between the ingoing and the outgoing fluxes. If the outgoing flow is larger than the incoming one, the amount of internal energy, number of moles, or electric charge in the system will increase, and it will decrease if the situation is reversed. When ingoing and outgoing fluxes are equal, the properties of the system will not change in the course of time, and the system is in a non-equilibrium steady state. In equilibrium, all fluxes vanish. It should be noticed that the above considerations apply only for so-called conserved quantities, which means absence of production or consumption inside the system. When source terms are present, as for instance in chemical reactions, one should add new contributions expressing the amount of moles, which is produced or destroyed.

Expressions (2.1)–(2.3) of the classical transport laws were originally proposed either from theoretical considerations or on experimental grounds. As stated before, non-equilibrium thermodynamics aims to propose a general scheme for the derivation of the transport laws by ensuring that they are compatible with the laws of thermodynamics (for instance, thermal conductivity must be positive because, otherwise, heat would spontaneously flow from lower to higher temperature, in conflict with the second law). Indeed, when λ, D, and σ_e are scalar quantities it is relatively evident that they must be positive; however, when they are tensors (as in anisotropic systems or in the presence of magnetic fields) the thermodynamic restrictions on the values of their components are not obvious and to obtain them a careful and systematic analysis is required. It opens also the way to the study of coupled situations where there are simultaneously non-homogeneities in temperature, concentration, and electric potential, instead of considering these effects separately as in (2.1)–(2.3).

2.2 Local Equilibrium Hypothesis

The most important hypothesis underlying CIT is undoubtedly the local equilibrium hypothesis. According to it, *the local and instantaneous relations between thermodynamic quantities in a system out of equilibrium are the same as for a uniform system in equilibrium.* To be more explicit, consider a system split mentally in a series of cells, which are sufficiently large for microscopic fluctuations to be negligible but sufficiently small so that equilibrium is realized to a good approximation in each individual cell. The size of such cells has been a subject of debate, on which a good analysis can be found in Kreuzer (1981) and Hafskjold and Kjelstrup (1995).

The local equilibrium hypothesis states that at a given instant of time, equilibrium is achieved in each individual cell or, using the vocabulary of continuum physics, at each material point. It should, however, be realized that the state of equilibrium is different from one cell to the other so that, for example, exchanges of mass and energy are allowed between neighbouring cells. Moreover, in each individual cell the equilibrium state is not frozen but changes in the course of time. A better description of this situation is achieved in terms of two timescales: the first, τ_{m}, denotes the equilibration time inside one cell and it is of the order of the time interval between two successive collisions between particles, i.e. 10^{-12} s, at normal pressure and temperature. The second characteristic time τ_{M} is a macroscopic one whose order of magnitude is related to the duration of an experiment, say about 1 s. The ratio between both reference times is called the Deborah number $De = \tau_{\mathrm{m}}/\tau_{\mathrm{M}}$. For $De \ll 1$, the local equilibrium hypothesis is fully justified because the relevant variables evolve on a large timescale τ_{M} and do practically not change over the time τ_{m}, but the hypothesis is not appropriate to describe situations characterized by $De \gg 1$. Large values of De are typical of systems with long relaxation times, like polymers, for which τ_{m} may be of the order of 100 s, or of high-frequency or very fast phenomena, such as ultrasound propagation, shock waves, nuclear collisions, for which τ_{M} is very short, say between 10^{-5} and 10^{-10} s.

A first consequence of the local equilibrium assumption is that all the variables defined in equilibrium as entropy, temperature, chemical potential, etc. are univocally defined outside equilibrium, but they are allowed to vary with time and space. Another consequence is that the local state variables are related by the same state equations as in equilibrium. This means, in particular, that the Gibbs' relation between entropy and the state variables remains locally valid for each value of the time t and the position vector $\boldsymbol{r}$. For example, in the case of a n-component fluid of total mass m, the local Gibbs' equation will be written as

$$\mathrm{d}s = T^{-1}\,\mathrm{d}u + pT^{-1}\,\mathrm{d}v - T^{-1}\sum_{k=1}^{n}\mu_k\,\mathrm{d}c_k, \tag{2.4}$$

where s is the specific entropy (per unit mass), u is the specific internal energy, related to the specific total energy e by $u = e - \frac{1}{2}\boldsymbol{v}\cdot\boldsymbol{v}$, with $\boldsymbol{v}$ the velocity field of the centre of mass of the cell, T is the absolute temperature, p is the hydrostatic pressure, v is the specific volume related to the mass density ρ by $v = 1/\rho$, $c_k = m_k/m$, the mass fraction of substance k, and μ_k its chemical potential.

A third consequence follows from the property that, locally, the system is stable. Therefore, in analogy with equilibrium situations, such quantities as heat capacity, isothermal compressibility or the Lamé coefficients in the theory of elasticity are positive definite.

More generally, Gibbs' equation will take the form

$$\mathrm{d}s(\boldsymbol{r},t) = \sum_i \Gamma_i(\boldsymbol{r},t)\,\mathrm{d}a_i(\boldsymbol{r},t), \tag{2.5}$$

where $a_i(\boldsymbol{r},t)$ is an extensive state variable, like u, v, c_k, while $\Gamma_i(\boldsymbol{r},t)$ is the corresponding conjugate intensive state variable, for instance T, p or μ_k. The above relation is assumed to remain valid when expressed in terms of the material (or Lagrangian) time derivative $\mathrm{d}/\mathrm{d}t = \partial/\partial t + \boldsymbol{v}\cdot\nabla$, i.e. by following a small cell moving with velocity $\boldsymbol{v}$:

$$\frac{\mathrm{d}s(\boldsymbol{r},t)}{\mathrm{d}t} = \sum_i \Gamma_i(\boldsymbol{r},t)\frac{\mathrm{d}a_i(\boldsymbol{r},t)}{\mathrm{d}t}. \tag{2.6}$$

From the kinetic theory point of view, the local equilibrium hypothesis is justified only for conditions where the Maxwellian distribution is approximately maintained. Otherwise, it should be generalized, as indicated in the second part of this book.

2.3 Entropy Balance

An important question is whether a precise definition can be attached to the notion of entropy when the system is driven far from equilibrium. In equilibrium thermodynamics, entropy is a well-defined function of state only in equilibrium states or during reversible processes. However, thanks to the local equilibrium hypothesis, entropy remains a valuable state function even in non-equilibrium situations. The problem of the definition of entropy and corollary of intensive variables as temperature will be raised as soon as the local equilibrium hypothesis is given up.

By material body (or system) is meant a continuum medium of total mass m and volume V bounded by a surface Σ. Consider an arbitrary body, outside equilibrium, whose total entropy at time t is S. The rate of variation of this extensive quantity may be written as the sum of the rate of exchange with the exterior $\mathrm{d}^{\mathrm{e}}S/\mathrm{d}t$ and the rate of internal production, $\mathrm{d}^{\mathrm{i}}S/\mathrm{d}t$:

$$\frac{\mathrm{d}S}{\mathrm{d}t} = \frac{\mathrm{d}^{\mathrm{e}}S}{\mathrm{d}t} + \frac{\mathrm{d}^{\mathrm{i}}S}{\mathrm{d}t}. \tag{2.7}$$

As in Chap. 1, we adopt the convention that any quantity (like mass, energy, entropy) is counted positive if supplied to the system and negative if transferred from it to the surroundings. The quantity $T\mathrm{d}^{\mathrm{i}}S/\mathrm{d}t$ is sometimes called the uncompensated heat or the rate of dissipation. For further purpose, it is convenient to introduce the notion of entropy flux $\boldsymbol{J}^s$, i.e. the entropy crossing the boundary surface per unit area and unit time, and the rate of entropy production σ^s, i.e. the entropy produced per unit volume and unit time inside the system. In terms of these quantities, $\mathrm{d}^{\mathrm{e}}S/\mathrm{d}t$ and $\mathrm{d}^{\mathrm{i}}S/\mathrm{d}t$ may be written as

$$\frac{\mathrm{d}^{\mathrm{e}}S}{\mathrm{d}t} = -\int_{\Sigma} \boldsymbol{J}^s \cdot \boldsymbol{n}\, \mathrm{d}\Sigma, \tag{2.8}$$

$$\frac{\mathrm{d}^{\mathrm{i}}S}{\mathrm{d}t} = \int_{V} \sigma^s\, \mathrm{d}V, \tag{2.9}$$

in which $\boldsymbol{n}$ is the unit normal pointing outwards the volume of the body.

Once entropy is defined, it is necessary to formulate the second law, i.e. to specify which kinds of behaviours are admissible in terms of the entropy behaviour. The classical formulation of the second law due to Clausius states that, in isolated systems, the possible processes are those in which the entropy of the final equilibrium state is higher or equal (but not lower) than the entropy of the initial equilibrium state. In the classical theory of irreversible processes, one introduces an even stronger restriction by requiring that the entropy of an isolated system must increase everywhere and at any time, i.e. $\mathrm{d}S/\mathrm{d}t \geq 0$. In non-isolated systems, the second law will take the more general form

$$\mathrm{d}^{\mathrm{i}}S/\mathrm{d}t > 0 \text{ (for irreversible processes)}, \tag{2.10a}$$

$$\mathrm{d}^{\mathrm{i}}S/\mathrm{d}t = 0 \text{ (for reversible processes or at equilibrium)}. \tag{2.10b}$$

It is important to realize that inequality (2.10a) does nor prevent that open or closed systems driven out of equilibrium may be characterized by $\mathrm{d}S/\mathrm{d}t < 0$; this occurs for processes for which $\mathrm{d}^{\mathrm{e}}S/\mathrm{d}t < 0$ and larger in absolute value than $\mathrm{d}^{\mathrm{i}}S/\mathrm{d}t$. Several examples are discussed in Chap. 6.

After introducing the specific entropy s in such a way that $S = \int \rho s\, \mathrm{d}V$ and using the definitions (2.8) and (2.9), the entropy balance (2.7) reads as

$$\frac{\mathrm{d}}{\mathrm{d}t}\int \rho s\, \mathrm{d}V = -\int_{\Sigma} \boldsymbol{J}^s \cdot \boldsymbol{n}\, \mathrm{d}\Sigma + \int_{V} \sigma^s\, \mathrm{d}V. \tag{2.11}$$

In virtue of the Gauss and Reynolds theorems, the above equation takes the form

$$\int \rho \frac{\mathrm{d}s}{\mathrm{d}t} \mathrm{d}V = -\int_{V} \nabla \cdot \boldsymbol{J}^s\, \mathrm{d}V + \int_{V} \sigma^s\, \mathrm{d}V, \tag{2.12}$$

where $\nabla \equiv (\partial/\partial x, \partial/\partial y, \partial/\partial z)$ designates the nabla operator whose components are the partial space derivatives in Cartesian coordinates. Assuming that (2.12) is valid for any volume V and that the integrands are continuous functions of position, one can write the following local balance relation

$$\rho \frac{\mathrm{d}s}{\mathrm{d}t} = -\nabla \cdot \boldsymbol{J}^s + \sigma^s, \tag{2.13}$$

with, in virtue of the second law as stated by ((2.10a) and (2.10b)),

$$\sigma^s \geq 0, \tag{2.14}$$

where the equality sign refers to reversible processes. This quantity is important in engineering because the product $T\sigma^s$ is a measure of the degradation or dissipation of energy in engines, and its minimization may be useful to enhance their efficiency.

Finally, it is worth to mention that the second law introduces an asymmetry in time for irreversible processes, which is often known as an arrow of time, an interesting physical and even philosophical concept that is briefly presented in Box 2.1.

Box 2.1 Irreversibility and the Arrow of Time
An interesting aspect of the principle of entropy increase is the special light it sheds on the concept of time, a view that departs radically from classical mechanics. The equations of Newtonian mechanics are deterministic and reversible with respect to time reversal. Time is considered as an external parameter, which describes the chronology of a succession of events; it is the time given by our watch. Reversible processes do not distinguish between the future and the past. In contrast, the principle of increase of entropy makes possible the distinction between future and past as it implies an arrow of time, which imposes that irreversible processes proceed spontaneously within a given direction in time. It establishes a fundamental anisotropy in Nature and provides a criterion allowing to decide whether a process is going forwards or backwards.

In relation with the notions of time, reversibility and irreversibility was raised the following problem. At the microscopic level, the motion of the individual particles composing the macroscopic systems is described by Newton's equation and is therefore of reversible nature, whereas at the macroscopic level, the systems behave irreversibly. The antagonism between these two behaviours has been a major source of debate since the microscopic interpretation of entropy by Boltzmann, in the years 1870. A widespread view is that the behaviour of the systems is intrinsically reversible but that the large number of particles makes that actually, it evolves towards an irreversible dynamics due to our inability to follow each individual particle. In other words, irreversibility is an illusion raised up by our ignorance. Following this attitude, irreversibility should not be related to the system itself but to the observer.

Irreversibility is also associated with loss of information, a view reinforced by the information theory. Accordingly, the entropy is interpreted as a lack of information about the microscopic state of the system, and loss of information means higher entropy. When entropy production is large, much information is lost per unit time, whereas in reversible processes no information is lost. The strong interrelations between information theory and thermodynamics have been underlined in many books and papers (e.g. Shannon 1948; Jaynes 1963; Keizer 1987).

In the 1970s, the study of deterministic chaos has opened new perspectives. Accordingly, though the equations of mechanics are reversible and deterministic, the sensibility of their solutions with respect to slight changes of the initial conditions or by varying some external parameters as temperature gradients (see Chap. 6) makes that the predictability is lost. Interesting conceptual discussions on this subject have been put forward by Prigogine (e.g. Prigogine and Stengers 1979; Prigogine 1996), who argues that irreversibility is present at all levels of description, microscopic and macroscopic, and merges out, as a source of order, from the instabilities present in the system. The breaking of time symmetry is introduced by constructing an appropriate operator playing the role of entropy and monotonically increasing in time. The formalism is coined under the name of subdynamics. Prigogine's line of thought is not unanimously accepted and target of keen discussions (see, e.g. Thom 1972; Bricmont 1995; Van Kampen 2000).

The above considerations show that the problem of evolution of reversibility towards irreversibility has not received a definitive answer yet (Lebowitz 1999). A last remark will concern ageing, which is a sensible problem to everybody and assuredly the most visible irreversible face of life. Experiences of our daily life tell us that youth will not return. There are many theories of biological ageing but these topics are outside the scope of the present book. It seems rather evident that ageing is related to the entropy production, or the amount of irreversibility, inside the metabolic processes, and that small entropy production should produce slow ageing. More considerations about the convergence of non-equilibrium thermodynamics and biology may be found in Chap. 4.

2.4 General Theory

Our main objective is to present a macroscopic description of irreversible processes and our major task is to determine the evolution equations of the relevant local variables, in accordance with the fundamental laws of thermodynamics. To give a general idea of the contents and assumptions underlying classical irreversible thermodynamics, we consider a system outside equilibrium and characterized by a finite set of variables to be specified later on. To make clear the general structure, we will proceed in several steps. The same steps, as introduced in this section, will be systematically repeated in the future.

Step 1. *Space of state variables*
The choice of state variables is determined by three arguments:

1. It depends on the system under consideration and the degree of accuracy that one wishes to achieve.
2. The variables must be "dynamically admissible", which means that their timescale is sufficiently different from the timescales corresponding to the more microscopic levels of description.
3. Theoretical predictions should be in agreement with experimental observations, and in addition, the results should be reproducible. This does not mean that the state variables are directly measurable but it is required to express them in terms of measured quantities.

According to the local equilibrium hypothesis, the space of the state variables is the ensemble $a(\boldsymbol{r},t) = [a_1(\boldsymbol{r},t), a_2(\boldsymbol{r},t), \ldots, a_N(\boldsymbol{r},t)]$ of the extensive thermodynamic variables appearing in Gibbs' equation, plus the velocity field $\boldsymbol{v}(\boldsymbol{r},t)$, i.e. the union of a mechanical variable (the barycentric velocity of each subcell) and thermodynamic variables (i.e. each subcell is considered as a thermodynamic system described by the usual set of thermodynamic quantities as in equilibrium). In the case of an n-component fluid, the thermodynamic variables are simply the specific volume $v(\boldsymbol{r},t)$, the mass fraction $c_k(\boldsymbol{r},t)(k = 1, 2, \ldots, n)$ of the n constituents and the internal energy $u(\boldsymbol{r},t)$.
Step 2. *Evolution equations*
They take the form of balance equations written in whole generality as

$$\rho \frac{\mathrm{d}a}{\mathrm{d}t} = -\nabla \cdot \boldsymbol{J}^a + \sigma^a, \tag{2.15}$$

where $\boldsymbol{J}^a$ is the flux term expressing the exchange with the surroundings and σ^a is the corresponding source term; when this term is zero, a is conserved. To be explicit, and to identify the flux and source terms, let us write down the balance equations of total mass, mass concentrations, energy and momentum for a n-constituent mixture (de Groot and Mazur 1962):

$$\rho \frac{\mathrm{d}v}{\mathrm{d}t} = \nabla \cdot \boldsymbol{v}, \tag{2.16}$$

$$\rho \frac{\mathrm{d}c_k}{\mathrm{d}t} = -\nabla \cdot \boldsymbol{J}^k + \sigma^k, \tag{2.17}$$

$$\rho \frac{\mathrm{d}u}{\mathrm{d}t} = -\nabla \cdot \boldsymbol{q} - \mathbf{P}^{\mathrm{T}} : \nabla \boldsymbol{v} + \rho r, \tag{2.18}$$

$$\rho \frac{\mathrm{d}\boldsymbol{v}}{\mathrm{d}t} = -\nabla \cdot \mathbf{P} + \rho \boldsymbol{F}, \tag{2.19}$$

where the superscript T means transposition. In (2.16), the flux term is simply the velocity and there is no source because of conservation of mass. In (2.17)–(2.19), the fluxes are the diffusion flux $\boldsymbol{J}^k$ of substance k, the heat flux $\boldsymbol{q}$ and the pressure tensor $\mathbf{P}$, respectively; the corresponding sources are the rate of production σ^k of substance k by chemical reactions, the mechanical work $-\mathbf{P}^{\mathrm{T}} : (\nabla \boldsymbol{v})$ plus some possible energy source r per unit mass and the external body force $\boldsymbol{F}$ per unit mass. The main difference between $\boldsymbol{q}$ and r is that $\boldsymbol{q}$ measures the transfer of heat across the bounding surface while r is the energy supply distributed within the volume. We refer to $\mathbf{P}$ as the pressure tensor because pure hydrostatic pressure corresponds to $\mathbf{P} = p\mathbf{I}$, where $\mathbf{I}$ is the identity tensor with components δ_{ij} ($\delta_{ij} = 1$ for $i = j$ and $\delta_{ij} = 0$ for $i \neq j$); the colon in (2.18) stands for the double scalar product $\mathbf{A} : \mathbf{B} = A_{ij}B_{ji}$. Einstein's summation convention on repeated indices, when they refer to Cartesian coordinates, will be used throughout the book.

The set (2.16)–(2.19) describes the evolution of the system, i.e. how the basic properties change with time at every point in space. Note that it contains more unknown quantities than equations, because the fluxes $\boldsymbol{q}$ and $\mathbf{P}$ are generally not known and have to be expressed in terms of the basic variables to close the system. In that respect the second law is very useful as shown by the next step.

Step 3. *Entropy production and second law of thermodynamics*
It is the purpose to examine the consequences resulting from the entropy inequality $\sigma^s > 0$. Substitution of the evolution equations (2.15) of the basic variables in the Gibbs' relation (2.6) leads to an explicit equation for the entropy evolution which, after comparison with the general expression (2.13) of the entropy balance, allows us to identify $\boldsymbol{J}^s$ and σ^s. The latter is found to consist of a sum of products of so-called thermodynamic fluxes J_α and thermodynamic forces X_α:

$$\sigma^s = \sum_\alpha J_\alpha X_\alpha. \tag{2.20}$$

Note that the thermodynamic forces are not forces in the mechanical sense, but they are quantities generally related to the gradients of the intensive variables whereas the fluxes J_α can be identified with the fluxes of energy, mass, momentum, etc. In (2.20), the fluxes and forces may be scalars, vectors, or tensors, and the product $J_\alpha X_\alpha$ stands indifferently for the usual product between two scalars, the scalar product between two vectors and the double scalar product between two tensors. Explicit expressions for J_α and X_α will be given later on while treating applications. It is important to point out that at equilibrium (or for reversible processes), the thermodynamic fluxes and forces vanish identically so that entropy production is zero in such situations, as it should.

Step 4. *Linear flux–force relations*
Experience indicates that the thermodynamic fluxes and forces are not independent but that there exists a relationship between them. Moreover, it has been observed that for a large class of irreversible processes, the fluxes are linear functions of the forces, to a good approximation. This is furthermore the simplest way to ensure that the rate of entropy production is a positive quantity. Within the hypothesis of linearity, one has

$$J_\alpha = \sum_\beta L_{\alpha\beta} X_\beta. \tag{2.21}$$

These flux–force relations are named phenomenological relations, constitutive or transport equations, they express the relation between causes (the forces) and effects (the fluxes) and the specific properties of the materials under study. At equilibrium, both members of (2.21) vanish identically. The phenomenological coefficients $L_{\alpha\beta}$ are generally depending on the intensive variables T, p, and c_k; the coefficient $L_{\alpha\alpha}$ connects a flow J_α to its conjugate force X_α while $L_{\alpha\beta}$ describes the coupling between two irreversible processes α and β. In thermoelectricity for instance, $L_{\alpha\alpha}$ is related to the electrical resistance and $L_{\alpha\beta}$ to the coupling between the electric current and the heat flow. Simple examples of linear phenomenological relations are the Fourier's, Fick's, and Ohm's equations (2.1)–(2.3).

Step 5. *Restrictions due to material symmetry: Curie's law*
According to relation (2.21), one should in principle be allowed to couple any flux to any force. However, material symmetry reduces the number of couplings between fluxes and forces. This property is known in CIT as Curie's law. It reflects the property that macroscopic causes cannot have more elements of symmetry than the effects they produce. This restriction plays an important role, especially in isotropic systems, for which the properties at equilibrium are the same in all directions.

Truly, as shown in Box 2.2, it is evident that Curie's law is nothing more than an application of the representation theorems of isotropic tensors (Spencer and Rivlin 1959; Truesdell and Noll 1965). To illustrate this statement, consider an isotropic body subject to a hypothetical irreversible process described by the fluxes j (a scalar), $\boldsymbol{J}$ (a vector), $\mathbf{T}$ (a symmetric tensor of order 2). In addition, the fluxes j and $\boldsymbol{J}$ are assumed to depend on the thermodynamic forces $\boldsymbol{x}$ (a vector) and $\mathbf{X}$ (a tensor of order 2) but $\mathbf{T}$ depends only on $\mathbf{X}$.

As a consequence of isotropy, and under the hypothesis of linear flux–force relations (see Box 2.2), one obtains the following phenomenological relations:

$$j = l \operatorname{tr} \mathbf{X}, \tag{2.22a}$$

$$\boldsymbol{J} = A_1 \boldsymbol{x}, \tag{2.22b}$$

$$\mathbf{T} = B_1 (\operatorname{tr} \mathbf{X}) \mathbf{I} + B_2 \mathbf{X}, \tag{2.22c}$$

where the phenomenological scalar coefficients l, A_1, B_1, and B_2 are independent of $\boldsymbol{x}$ and $\mathbf{X}$, and tr $\mathbf{X}$ denotes the trace of tensor $\mathbf{X}$. The results ((2.22a), (2.22b), and (2.22c)) exhibit the property that, in isotropic systems and within the linear regime, it is forbidden to couple fluxes and forces of different tensorial character. For instance a chemical affinity (a scalar) cannot give raise to a heat flux (a vector), similarly a temperature gradient (a vector) is unable to induce a mechanical stress (a tensor of order 2).

Box 2.2 Curie's Law

The existence of spatial symmetries in a material system contributes to simplify the scheme of the phenomenological relations. Because of invariance of the phenomenological equations under special orthogonal transformations, some couplings between fluxes and forces are not authorized in isotropic systems. In the examples treated in this chapter, no fluxes and forces of tensorial order higher than the second will occur. We shall therefore consider an isotropic material characterized by the three following phenomenological relations

$$j = j(\boldsymbol{x}, \mathbf{X}), \quad \boldsymbol{J} = \boldsymbol{J}(\boldsymbol{x}, \mathbf{X}), \quad \mathbf{T} = \mathbf{T}(\mathbf{X}), \tag{2.2.1}$$

which contain fluxes and forces of tensorial order 0 (the scalar j), order 1 (the vectors $\boldsymbol{J}$ and $\boldsymbol{x}$) and order 2 (the symmetric tensors $\mathbf{T}$ and $\mathbf{X}$). Isotropy imposes that the above relations transform as follows under an orthogonal transformation $\boldsymbol{Q}$:

$$j(\mathbf{Q}\cdot\boldsymbol{x}; \mathbf{Q}\cdot\mathbf{X}\cdot\mathbf{Q}^{\mathrm{T}}) = j(\boldsymbol{x}, \mathbf{X}), \tag{2.2.2}$$

$$\boldsymbol{J}(\mathbf{Q}\cdot\boldsymbol{x}; \mathbf{Q}\cdot\mathbf{X}\cdot\mathbf{Q}^{\mathrm{T}}) = \mathbf{Q}\cdot\boldsymbol{J}(\boldsymbol{x}, \mathbf{X}), \tag{2.2.3}$$

$$\mathbf{T}(\mathbf{Q}\cdot\mathbf{X}\cdot\mathbf{Q}^{\mathrm{T}}) = \mathbf{Q}\cdot\mathbf{T}(\mathbf{X})\cdot\mathbf{Q}^{\mathrm{T}}. \tag{2.2.4}$$

According to the theorems of representation of isotropic tensors (e.g. Spencer and Rivlin 1959), the functions j, $\boldsymbol{J}$, and $\mathbf{T}$ are isotropic if and only if

$$j(\boldsymbol{x}, \mathbf{X}) = j[I_X, II_X, III_X, \boldsymbol{x}\cdot\boldsymbol{x}, \boldsymbol{x}\cdot(\mathbf{X}\cdot\boldsymbol{x}), \boldsymbol{x}\cdot(\mathbf{X}\cdot\mathbf{X}\cdot\boldsymbol{x})], \tag{2.2.5}$$

$$\boldsymbol{J}(\boldsymbol{x}, \mathbf{X}) = (A_1\mathbf{I} + A_2\mathbf{X} + A_3\mathbf{X}\cdot\mathbf{X})\cdot\boldsymbol{x}, \tag{2.2.6}$$

$$\mathbf{T}(\mathbf{X}) = B_1\mathbf{I} + B_2\mathbf{X} + B_3\mathbf{X}\cdot\mathbf{X}, \tag{2.2.7}$$

the flux j and the coefficients A_i $(i = 1, 2, 3)$ are isotropic scalar functions of $\boldsymbol{x}$ and $\mathbf{X}$ and the B_i are isotropic scalars of $\mathbf{X}$ alone; I_X, II_X, and III_X are the principal invariants of the tensor $\mathbf{X}$, namely

$$I_X = \operatorname{tr}\mathbf{X}, \quad II_X = \tfrac{1}{2}[I_X^2 - \operatorname{tr}(\mathbf{X}\cdot\mathbf{X})], \quad III_X = \det\ \mathbf{X}. \tag{2.2.8}$$

By restricting the analysis to *linear* laws, (2.2.5)–(2.2.7) will take the simple form

$$j = l(\operatorname{tr}\mathbf{X}), \boldsymbol{J} = A_1\boldsymbol{x},\ \mathbf{T} = B_1(\operatorname{tr}\mathbf{X})\mathbf{I} + B_2\ \mathbf{X}, \tag{2.2.9}$$

wherein l, A_1, B_1, and B_2 are now scalars independent of $\boldsymbol{x}$ and $\mathbf{X}$. Relations (2.2.9) exhibit the property that in linear constitutive equations, spatial symmetry allows exclusively the coupling between fluxes and forces of the same tensorial order, as concluded from Curie's law.

Step 6. *Restrictions on the sign of phenomenological coefficients*
A direct restriction on the sign of the phenomenological coefficients arises as a consequence of the second law. Substitution of the linear flux–force relations (2.21) into (2.20) of the rate of entropy production yields the quadratic form

$$\sigma^s = \sum_{\alpha\beta} L_{\alpha\beta} X_\alpha X_\beta \geq 0. \tag{2.23}$$

According to standard results in algebra, the necessary and sufficient conditions for $\sigma^s \geq 0$ are that the determinant $|L_{\alpha\beta} + L_{\beta\alpha}|$ and all its principal minors are non-negative. It follows that

$$L_{\alpha\alpha} \geq 0, \tag{2.24}$$

while the cross-coefficients $L_{\alpha\beta}$ must satisfy

$$L_{\alpha\alpha} L_{\beta\beta} \geq \tfrac{1}{4}(L_{\alpha\beta} + L_{\beta\alpha})^2. \tag{2.25}$$

In virtue of inequality (2.24), all the transport coefficients like the heat conductivity, the diffusion coefficient, and the electrical resistance are positive, meaning that heat flows from high to low temperature, electrical current from high to low electric potential, and neutral solutes from higher to lower concentrations.

Step 7. *Restrictions on $L_{\alpha\beta}$ due to time reversal: Onsager–Casimir's reciprocal relations*
It was established by Onsager (1931) that, besides the restrictions on the sign, the phenomenological coefficients verify symmetry properties. The latter were presented by Onsager as a consequence of "microscopic reversibility", which is the invariance of the microscopic equations of motion with respect to time reversal $t \to -t$. Accordingly, by reversing the time, the particles retrace their former paths or, otherwise stated, there is a symmetry property between the past and the future. Invoking the principle of microscopic reversibility and using the theory of fluctuations, Onsager was able to demonstrate the symmetry property

$$L_{\alpha\beta} = L_{\beta\alpha}. \tag{2.26}$$

In Chaps. 4 and 11, we will present detailed derivations of (2.26).

It should, however, be stressed that the above result holds true only for fluctuations $a_\alpha(t) = A_\alpha(t) - A_\alpha^{\mathrm{eq}}$ of extensive state variables A_α with respect to their equilibrium values, which are even functions of time $a_\alpha(t) = a_\alpha(-t)$. In the case of odd parity of one of the variables, α or β, for which $a_\alpha(t) = -a_\alpha(-t)$, the coefficients $L_{\alpha\beta}$ are skew-symmetric instead of symmetric, as shown by Casimir (1945), i.e.

$$L_{\alpha\beta} = -L_{\beta\alpha}. \tag{2.27}$$

In a reference frame rotating with angular velocity $\boldsymbol{\omega}$ and in presence of an external magnetic field $\boldsymbol{\mathcal{H}}$, Onsager–Casimir's reciprocal relations take the form

$$L_{\alpha\beta}(\boldsymbol{\omega}, \boldsymbol{\mathcal{H}}) = \pm L_{\beta\alpha}(-\boldsymbol{\omega}, -\boldsymbol{\mathcal{H}}), \tag{2.28}$$

as it will be shown on dynamical bases in Chap. 11. The validity of the Onsager–Casimir's reciprocal relations is not limited to phenomenological transport coefficients that are scalar quantities as discussed earlier. Consider for example an irreversible process taking place in an anisotropic crystal, such that

$$\boldsymbol{J}_\alpha = \sum_\beta \mathbf{L}_{\alpha\beta} \cdot \boldsymbol{X}_\beta, \tag{2.29}$$

where fluxes and forces are vectors and $\mathbf{L}_{\alpha\beta}$ is a tensor of order 2. The Onsager–Casimir's reciprocal relations write now as

$$\mathbf{L}_{\alpha\beta} = \pm(\mathbf{L}_{\beta\alpha})^{\mathrm{T}}. \tag{2.30}$$

Transformation properties of the reciprocal relations have been discussed by Meixner (1943) and Coleman and Truesdell (1960).

At first sight, Onsager–Casimir's reciprocal relations may appear as a rather modest result. Their main merit is to have evidenced symmetry properties in coupled irreversible processes. As illustration, consider heat conduction in an anisotropic crystal. The reciprocity relations imply that a temperature gradient of $1°\mathrm{C\,m^{-1}}$ along the x-direction will give raise to a heat flux in the normal y-direction, which is the same as the heat flow generated along the x-axis by a temperature gradient of $1°\mathrm{C\,m^{-1}}$ along y. Another advantage of the Onsager–Casimir's reciprocal relations is that the measurement (or the calculation) of a coefficient $L_{\alpha\beta}$ alleviates the repetition of the same operation for the reciprocal coefficient $L_{\beta\alpha}$; this is important in practice as the cross-coefficients are usually much smaller (of the order of 10^{-3} to 10^{-4}) than the direct coupling coefficients, and therefore difficult to measure or even to detect.

Although the proof of the Onsager–Casimir's reciprocal relations was achieved at the microscopic level of description and for small deviations of fluctuations from equilibrium, these symmetry properties have been widely applied in the treatment of coupled irreversible processes taking place at the macroscopic scale even very far from equilibrium. It should also be kept in mind that the validity of the reciprocity properties is secured as far as the flux–force relations are linear, but that they are not of application in the non-linear regime.

2.5 Stationary States

Stationary states play an important role in continuum physics; they are defined by the property that the state variables, including the velocity, remain unchanged in the course of time. For instance, if heat is supplied at one end of a system and removed at the other end at the same rate, the temperature at each point will not vary in time but will change from one position to the other. Such a state cannot be confused with an equilibrium state, which is characterized by a uniform temperature field, no heat flow, and a zero entropy production. It is to be emphasized that the evolution of a system towards an equilibrium state or a steady state is conditioned by the nature of the boundary conditions. Since in a stationary state, entropy does not change in the course of time, we can write in virtue of (2.7) that

$$-\left|\frac{\mathrm{d}S}{\mathrm{d}t}\right|_{\text{out}} + \left|\frac{\mathrm{d}S}{\mathrm{d}t}\right|_{\text{in}} + \frac{\mathrm{d}^{\mathrm{i}}S}{\mathrm{d}t} = 0, \tag{2.31}$$

since the rate of entropy production is positive, it is clear that the entropy delivered by the system to the external environment is larger than the entropy that is entering. Using the vocabulary of engineers, the system degrades the energy that it receives and this degradation is the price paid to maintain a stationary state.

Stationary states are also characterized by interesting extremum principles as demonstrated by Prigogine (1961): the most important is the principle of minimum entropy production, which is discussed further. The importance of variational principles has been recognized since the formulation of Hamilton's least action principle in mechanics stating that the average kinetic energy less the average potential energy is minimum along the path of a particle moving from one point to another. Quoting Euler, "since the construction of the universe is the most perfect possible, being the handy work of an all-wise Maker, nothing can be met in the world in which some minimum or maximum property is not displayed". It is indeed very attractive to believe that a whole class of processes is governed by a single law of minimum or maximum. However, Euler's enthusiasm has to be moderated, as most of the physical phenomena cannot be interpreted in terms of minima or maxima. In equilibrium thermodynamics, maximization of entropy for isolated systems or minimization of Gibbs' free energy for systems at constant temperature and pressure, for instance, provide important examples of variational principles. Out of equilibrium, such variational formulations are much more limited. For this reason, we pay here a special attention to the minimum entropy production theorem, which is the best known among the few examples of variational principles in non-equilibrium thermodynamics.

2.5.1 Minimum Entropy Production Principle

Consider a non-equilibrium process, for instance heat conduction or thermodiffusion taking place in a volume V at rest subjected to time-independent constraints at its surface. The state variables $a_1, a_2, \ldots, a_n$ are assumed to obey conservation laws of the form

$$\rho\frac{\partial a_\alpha}{\partial t} = -\nabla\cdot\boldsymbol{J}_\alpha \quad (\alpha = 1, 2, \ldots, n), \tag{2.32}$$

where $\partial/\partial t$ is the partial or Eulerian time derivative; processes of this kind, characterized by absence of global velocity, are called purely dissipative. We have seen in Sect. 2.4 that the total entropy P produced inside the system can be written as

$$P = \int \sum_{\alpha,\beta} L_{\alpha\beta} X_\alpha X_\beta \, \mathrm{d}V. \tag{2.33}$$

Since the thermodynamic forces take usually the form of gradients of intensive variables

$$\boldsymbol{X}_\alpha = \nabla \Gamma_\alpha, \tag{2.34}$$

where Γ_α designates an intensive scalar variable, (2.33) becomes

$$P = \int \sum_{\alpha,\beta} L_{\alpha\beta} \nabla \Gamma_\alpha \cdot \nabla \Gamma_\beta \, \mathrm{d}V. \tag{2.35}$$

We now wish to show that the entropy production is minimum in the stationary state. Taking the time derivative of (2.35) and supposing that the phenomenological coefficients are constant and symmetric, one obtains

$$\frac{\mathrm{d}P}{\mathrm{d}t} = 2 \int \sum_{\alpha,\beta} L_{\alpha\beta} \nabla \Gamma_\alpha \cdot \nabla \left(\frac{\partial \Gamma_\beta}{\partial t} \right) \mathrm{d}V. \tag{2.36}$$

After integration by parts and recalling that the boundary conditions are time independent, it is found that

$$\begin{aligned} \frac{\mathrm{d}P}{\mathrm{d}t} &= -2 \int \sum_\beta \frac{\partial \Gamma_\beta}{\partial t} \nabla \cdot \left(\sum_\alpha L_{\alpha\beta} \nabla \Gamma_\alpha \right) \mathrm{d}V \\ &= -2 \int \sum_\beta \frac{\partial \Gamma_\beta}{\partial t} \nabla \cdot \boldsymbol{J}_\beta \, \mathrm{d}V = 2 \int \rho \sum_\beta \frac{\partial \Gamma_\beta}{\partial t} \frac{\partial a_\beta}{\partial t} \, \mathrm{d}V, \end{aligned} \tag{2.37}$$

wherein use has been made successively of the linear flux–force relations (2.21) and the conservation law (2.32). In the stationary state, for which $\partial \alpha_\beta / \partial t = 0$, one has

$$\frac{\mathrm{d}P}{\mathrm{d}t} = 0. \tag{2.38}$$

During the transient regime, (2.37) can be written as

$$\frac{\mathrm{d}P}{\mathrm{d}t} = \int \sum_{\alpha,\beta} \frac{\partial a_\alpha}{\partial \Gamma_\beta} \frac{\partial \Gamma_\alpha}{\partial t} \frac{\partial \Gamma_\beta}{\partial t} \, \mathrm{d}V, \tag{2.39}$$

and since the $\partial \alpha_\alpha / \partial \Gamma_\beta$ terms (which represent, for instance, minus the heat capacity or minus the coefficient of isothermal compressibility) are negative quantities because of stability of equilibrium, one may conclude that

$$\frac{\mathrm{d}P}{\mathrm{d}t} \leq 0. \tag{2.40}$$

This result proves that the total entropy production P decreases in the course of time and that it reaches its minimum value in the stationary state. An important aside result is that stationary states with a minimum entropy production are unconditionally stable. Indeed, after application of an arbitrary disturbance in the stationary state, the system will move towards a transitory regime with a greater entropy production. But as the latter can only decrease, the system will go back to its stationary state, which is therefore referred to as stable. It is also worth to mention that P, a positive definite functional with a negative time derivative, provides an example of Lyapounov's function (Lyapounov 1966), whose occurrence is synonymous of stability, as discussed in Chap. 6.

It should, however, be emphasized that the above conclusions are far from being general, as their validity is subordinated to the observance of the following requirements:

1. Time-independent boundary conditions
2. Linear phenomenological laws
3. Constant phenomenological coefficients
4. Symmetry of the phenomenological coefficients

In practical situations, it is frequent that at least one of the above restrictions is not satisfied, so that the criterion of minimum entropy production is of weak bearing. It follows also that most of the stationary states met in the nature are not necessarily stable as confirmed by our everyday experience.

The result (2.38) can still be cast in the form of a variational principle

$$\delta P = \delta \int \sigma^s(\Gamma_\alpha, \nabla \Gamma_\alpha, \ldots)\,\mathrm{d}V = 0, \tag{2.41}$$

where the time derivative symbol $\mathrm{d}/\mathrm{d}t$ has been replaced by the variational symbol δ. Since the corresponding Euler–Lagrange equations are shown to be the stationary balance relations, it turns out that the stationary state is characterized by an extremum of the entropy production, truly a minimum, as it can be proved that the second variation is positive definite $\delta^2 P > 0$.

It should also be realized that the minimum entropy principle is not an extra law coming in complement of the classical balance equations of mass, momentum, and energy, but nothing else than a reformulation of these laws in a condensed form, just like in classical mechanics, Hamilton's principle is a reformulation of Newton's equations.

The search for variational principles in continuum physics has been a subject of continuous and intense activity (Glansdorff and Prigogine 1964, 1971; Finlayson 1972; Lebon 1980). A wide spectrum of applications in macroscopic physics, chemistry, engineering, ecology, and econophysics is discussed in Sieniutycz and Farkas (2004). It should, however, be stressed that it is only in exceptional cases that there exists a "true" variational principle for processes that dissipate energy. Most of the principles that have been proposed refer either to equilibrium situations, as the maximum entropy principle in equilibrium thermodynamics, the principle of virtual work in statics, the minimum energy principle in elasticity, or to ideal reversible motions as the principle of least action in rational mechanics or the minimum energy principle for Eulerian fluids.

2.6 Applications to Heat Conduction, Mass Transport, and Fluid Flows

To better understand and illustrate the general theory, we shall deal with some applications, like heat conduction in a rigid body and matter diffusion involving no coupling of different thermodynamic forces and fluid flow. More complex processes involving coupling, like thermoelectricity, thermodiffusion and diffusion through membranes are treated in Chap. 3. The selection of these problems has been motivated by the desire to propose a pedagogical approach and to cover situations frequently met in practical problems by physicists, chemists and engineers. Chemistry will receive a special treatment in Chap. 4 where we deal at length with chemical reactions and their coupling with mass transport, a subject of utmost importance in biology. Despite its success, CIT has been the subject of several limitations and criticisms, which are discussed in Sect. 2.7 of the present chapter.

2.6.1 Heat Conduction in a Rigid Body

The problem consists in finding the temperature distribution in a rigid body at rest, subject to arbitrary time-dependent boundary conditions on temperature, or on the heat flux. Depending on the geometry and the physical properties of the system and on the nature of the boundary conditions, a wide variety of situations may arise, some of them being submitted as problems at the end of the chapter. For the sake of pedagogy, we follow the same presentation as in Sect. 2.4.

Step 1. *State variable(s)*
Here we may select indifferently the specific internal energy $u(\boldsymbol{r}, t)$ or the temperature field $T(\boldsymbol{r}, t)$, which should be preferred in practical applications.

Step 2. *Evolution equation*
In absence of source term, the evolution equation (2.18) for $u(\boldsymbol{r}, t)$ is simply

$$\rho \frac{\mathrm{d}u}{\mathrm{d}t} = -\nabla \cdot \boldsymbol{q}, \tag{2.42}$$

here $\mathrm{d}/\mathrm{d}t$ reduces to the partial time derivative $\partial/\partial t$ as $\boldsymbol{v} = 0$. Equation (2.42) contains two unknown quantities, the heat flux $\boldsymbol{q}$ to be given by a constitutive relation and the internal energy $u(T)$ to be expressed by means of an equation of state.

Step 3. *Entropy production and second law*
According to the second law, the rate of entropy production defined by

$$\sigma^s = \rho \frac{\mathrm{d}s}{\mathrm{d}t} + \nabla \cdot \boldsymbol{J}^s \geq 0, \tag{2.43}$$

is positive definite. The expression of $\mathrm{d}s/\mathrm{d}t$ is obtained from Gibbs' equation

$$\frac{\mathrm{d}s}{\mathrm{d}t} = T^{-1} \frac{\mathrm{d}u}{\mathrm{d}t}, \tag{2.44}$$

where $\mathrm{d}u/\mathrm{d}t$ is given by the energy balance equation (2.42). Substituting (2.44) in (2.43) results in

$$\sigma^s = \boldsymbol{q} \cdot \nabla T^{-1} + \nabla \cdot (\boldsymbol{J}^s - T^{-1}\boldsymbol{q}). \tag{2.45}$$

Since σ^s represents the rate of entropy production inside the body, its expression cannot contain a flux term like $\nabla \cdot (\boldsymbol{J}^s - T^{-1}\boldsymbol{q})$, which describes the rate of exchange with the outside, as a consequence this term must be set equal to zero so that

$$\boldsymbol{J}^s = T^{-1}\boldsymbol{q}, \tag{2.46}$$

whereas (2.45) of σ^s reduces to

$$\sigma^s = \boldsymbol{q} \cdot \nabla T^{-1}. \tag{2.47}$$

This illustrates the general statement (2.20) that the entropy production is a bilinear form in the force ∇T^{-1} (the cause) and the flux of energy $\boldsymbol{q}$ (the effect).

Step 4. *Linear flux–force relation*
The simplest way to ensure that $\sigma^s \geq 0$ is to assume a linear relationship between the heat flux and the temperature gradient; for isotropic media,

$$\boldsymbol{q} = L_{qq}(T)\nabla T^{-1}, \tag{2.48}$$

where $L_{qq}(T)$ is a scalar phenomenological coefficient depending generally on the temperature. Defining the heat conductivity by $\lambda(T) = L_{qq}(T)/T^2$, the flux–force relation (2.48) takes the more familiar form

$$\boldsymbol{q} = -\lambda \nabla T, \tag{2.49}$$

which is nothing else than the *Fourier's law* stating that the heat flux is proportional to the temperature gradient. We observe in passing that Curie's principle is satisfied as (2.49) is a relationship between flux and force of the same tensor character, namely vectors. In an anisotropic crystal, Fourier's relation reads as

$$\boldsymbol{q} = -\boldsymbol{\lambda} \cdot \nabla \mathrm{T}, \tag{2.50}$$

where the heat conductivity $\boldsymbol{\lambda}$ is now a tensor of order 2.

Step 5. *Restriction on the sign of the transport coefficients*
Substitution of (2.49) in (2.48) of the rate of entropy production yields

$$\sigma^s = \frac{1}{\lambda T^2} \boldsymbol{q} \cdot \boldsymbol{q}, \tag{2.51}$$

and from the requirement that $\sigma^s \geq 0$, it is inferred that $\lambda \geq 0$. Roughly speaking this means that in an isotropic medium, the heat flux takes place in a direction opposite to the temperature gradient; therefore, heat will flow spontaneously from high to low temperature, in agreement with our everyday experience. In an anisotropic system, flux and force will generally be oriented in different directions but the positiveness of tensor $\boldsymbol{\lambda}$ requires that the angle between them cannot be smaller than $\pi/2$.

Step 6. *Reciprocal relations*
In the general case of an anisotropic medium, the flux–force relation is the Fourier's law expressed in the form (2.50). According to Onsager's reciprocal relations, the second-order tensor $\boldsymbol{\lambda}$ is symmetric so that, in Cartesian coordinates,

$$\lambda_{ij} = \lambda_{ji}, \tag{2.52}$$

a result found to be experimentally satisfied in crystals wherein, however, spatial symmetry may impose further symmetry relations. For instance, in crystals pertaining to the hexagonal or the tetragonal class, spatial symmetry requires that the conductivity tensor is skew-symmetric. By combining this result with the symmetry property (2.52), it turns out that the elements $\lambda_{ij}(i \neq j)$ are zero; it follows that for these classes of crystal, the application of a temperature gradient in the x-direction cannot produce a heat flow in the perpendicular y-direction. Very old experiences by Soret (1893) and Voigt (1903) confirmed this result, which is presented as one of the confirmations of the Onsager–Casimir's reciprocal relations.

Step 7. *The temperature equation*
We now wish to calculate the temperature distribution in an isotropic rigid solid as a function of time and space. The corresponding differential equation is easily obtained by introducing Fourier's law (2.49) in the energy balance equation (2.42) and the result is

$$\rho c_v \frac{\partial T}{\partial t} = \nabla \cdot (\lambda \nabla T), \tag{2.53}$$

where use is made of the definition of the heat capacity $c_v = \partial u / \partial T$.

In the case of constant heat conductivity and heat capacity, and introducing the heat diffusivity defined by $\chi = \lambda / \rho c_v$, (2.53) reads as

$$\frac{\partial T}{\partial t} = \chi \frac{\partial^2 T}{\partial x^2}. \tag{2.54}$$

Relation (2.53) is classified as a parabolic partial differential equation. In Box 2.3 is presented the mathematical method of solution of this important equation under given typical initial and boundary conditions for an infinite one-dimensional rod. It is the same kind of equation that governs matter diffusion, as shown in Sect. 2.6.2.

Box 2.3 Method of Solution of the Heat Diffusion Equation

A convenient method to solve (2.54) is to work in the Fourier space (k, t) with k designating the wave number. It is interesting to recall that Fourier devised originally the transform bearing his name to solve the heat diffusion equation. Let us write $T(x, t)$ as a Fourier integral of the form

$$T(x,t) = \int_{-\infty}^{+\infty} T(k,t)\,\exp(\mathrm{i}kx)\,\mathrm{d}k, \tag{2.3.1}$$

with $T(k, t)$ the Fourier transform. The initial and boundary conditions are assumed to be given by $T(x, 0) = g(x)$, $T(\pm\infty, t) = 0$; introduction of (2.3.1) in (2.54) leads to the ordinary differential equation

$$\frac{\mathrm{d}T(k,t)}{\mathrm{d}t} = -k^2\chi T(k,t), \tag{2.3.2}$$

whose solution is directly given by

$$T(k,t) = T(k,0)\,\exp(-k^2\chi t), \tag{2.3.3}$$

where $T(k, 0)$ is the Fourier transform of the initial temperature profile

$$T(k,0) = \tfrac{1}{2\pi}\int_{-\infty}^{+\infty} g(x')\,\exp(-\mathrm{i}kx')\,\mathrm{d}x'. \tag{2.3.4}$$

Substitution of (2.3.3) and (2.3.4) in (2.3.1) yields $T(x, t)$ in terms of the initial distribution $g(x)$:

$$T(x,t) = \tfrac{1}{2\pi}\int_{-\infty}^{+\infty}\mathrm{d}k\int_{-\infty}^{+\infty} g(x')\,\exp(-k^2\chi t)\,\exp[\mathrm{i}k(x-x')]\,\mathrm{d}x'. \tag{2.3.5}$$

By carrying the integration with respect to k, we obtain the final solution in the form

$$T(x,t) = \tfrac{1}{(4\pi\chi t)^{1/2}}\int_{-\infty}^{+\infty} g(x')\,\exp[-(x-x')^2/4\chi t]\,\mathrm{d}x'. \tag{2.3.6}$$

When the initial temperature dependence corresponds to a local heating at one particular point x_0 of the solid, namely

$$g(x) = g_0\delta(x - x_0), \tag{2.3.7}$$

where g_0 is an arbitrary constant proportional to the energy input at the initial time, and $\delta(x - x_0)$ the Dirac function, (2.3.6) will be given by

$$T(x,t) = \frac{g_0}{(4\pi\chi t)^{1/2}}\exp[-(x-x_0)^2/4\chi t]. \tag{2.3.8}$$

We deduce from (2.3.8) that the temperature diffuses in the whole rod from the point x_0 over a distance proportional to $(\chi t)^{1/2}$; as a consequence, large values of the thermal diffusivity imply a rapid diffusion of temperature. Solution (2.3.8) exhibits also the property that, after application of a temperature disturbance at a given point of the system, it will be experienced instantaneously everywhere in the whole body. Such a result is in contradiction with the principle of causality, which demands that response will be felt after the application of a cause. In the above example, cause and effect occur simultaneously. This property is typical of classical irreversible thermodynamics but is not conceptually acceptable. This failure was one of the main motivations to propose another thermodynamic formalism currently known under the name of Extended Irreversible Thermodynamics (see Chap. 7).

Box 2.4 Stationary States and Minimum Entropy Production Principle

In the case of thermal conduction described by the phenomenological equation $\boldsymbol{q} = L_{qq}\nabla T^{-1}$, the total entropy production P in the system, say a one-dimensional rod of length l, is according to (2.47),

$$P = \int_0^l \sigma^s \, \mathrm{d}x = \int_0^l q(\partial T^{-1}/\partial x) \, \mathrm{d}x = \int_0^l L_{qq}(\partial T^{-1}/\partial x)^2 \, \mathrm{d}x. \quad (2.4.1)$$

For a constant phenomenological coefficient L_{qq}, its time derivative is

$$\frac{\mathrm{d}P}{\mathrm{d}t} = 2\int_0^l L_{qq} \frac{\partial T^{-1}}{\partial x} \frac{\partial}{\partial t} \frac{\partial T^{-1}}{\partial x} \, \mathrm{d}x. \quad (2.4.2)$$

After integration by parts and imposing time-independent boundary conditions, one has

$$\frac{\mathrm{d}P}{\mathrm{d}t} = -2\int_0^l \frac{\partial T^{-1}}{\partial t} \frac{\partial}{\partial x}\left(L_{qq}\frac{\partial}{\partial x}T^{-1}\right) \mathrm{d}x = 2\int_0^l \rho \frac{\partial T^{-1}}{\partial t}\frac{\partial u}{\partial t} \, \mathrm{d}x, \quad (2.4.3)$$

where use is made of the energy balance equation (2.42). Since $\partial u/\partial t = c_v \, \partial T/\partial t$ with $c_v > 0$, it is finally found that

$$\frac{\mathrm{d}P}{\mathrm{d}t} = -2\int_0^l \frac{\rho c_v}{T^2}\left(\frac{\partial T}{\partial t}\right)^2 \mathrm{d}x \leq 0 \quad (2.4.4)$$

is negative outside the stationary state and zero in the stationary state for which $\partial T/\partial t = 0$.

It is directly recognized that the total entropy production P is a Lyapounov function, as $P > 0$ together with $\mathrm{d}P/\mathrm{d}t \leq 0$, and this ensures that the stationary state is stable. It is, however, important to realize that the phenomenological coefficient L_{qq} is related to the heat conductivity by $L_{qq} = \lambda T^2$ and therefore the validity of the minimum entropy production principle is conditioned by the condition that λ varies like T^{-2}; the principle is therefore not applicable to systems with a constant heat conductivity or with an arbitrary dependence of λ with respect to the temperature. This is the reason why the criterion of minimum entropy production remains an exception.

2.6.2 Matter Diffusion Under Isothermal and Isobaric Conditions

In this section, we briefly discuss the problem of isothermal and isobaric diffusion of two non-viscous isotropic non-reacting fluids, in the absence of external forces. There is no difficulty to generalize the forthcoming considerations and results to the more general case of a n-component mixture of viscous fluids.

Let ρ_1 and ρ_2 denote the densities (mass per unit volume) and $c_1(=\rho_1/\rho)$, $c_2(=\rho_2/\rho)$ the mass fractions of the two constituents ($c_1 + c_2 = 1$), $\rho = \rho_1 + \rho_2$ is the total mass density. Designating by $\boldsymbol{v}_k(\boldsymbol{r}, t)$ the local macroscopic velocity of substance k, the centre of mass or barycentric velocity is given by

$$\rho \boldsymbol{v} = \rho_1 \boldsymbol{v}_1 + \rho_2 \boldsymbol{v}_2. \tag{2.55}$$

The diffusion flux $\boldsymbol{J}_k$ of substance k with respect to the centre of mass is defined by

$$\boldsymbol{J}_k = \rho_k(\boldsymbol{v}_k - \boldsymbol{v}), \tag{2.56}$$

and is expressed in kg m^{-2} s^{-1}; in some circumstances, it may be convenient to replace in (2.56) the barycentric velocity by an arbitrary reference velocity (de Groot and Mazur 1962). In virtue of the definition of the barycentric velocity, it is directly seen that $\boldsymbol{J}_1 + \boldsymbol{J}_2 = 0$.

The analysis carried out in Sect. 2.6.1 for heat transport is easily repeated for matter diffusion. In the present problem, the Gibbs' equation takes the form

$$T \frac{\mathrm{d}s}{\mathrm{d}t} = \frac{\mathrm{d}u}{\mathrm{d}t} + p \frac{\mathrm{d}v}{\mathrm{d}t} - (\bar{\mu}_1 - \bar{\mu}_2) \frac{\mathrm{d}c_1}{\mathrm{d}t}, \tag{2.57}$$

wherein $\bar{\mu}_1$ and $\bar{\mu}_2$ are the respective chemical potentials and where use has been made of $\mathrm{d}c_2 = -\mathrm{d}c_1$. Combining (2.57) with the balance equations (2.16)–(2.18) written without source terms, it is easily shown (see Problem 2.6) that the entropy production is given by

$$T\sigma^s = -\boldsymbol{J}_1 \cdot \nabla(\bar{\mu}_1 - \bar{\mu}_2) \geq 0. \tag{2.58}$$

Within the hypothesis of linear flux–force relations, one obtains

$$\boldsymbol{J}_1 = -L\nabla(\bar{\mu}_1 - \bar{\mu}_2) \quad (L > 0), \tag{2.59}$$

or, in virtue of Gibbs–Duhem's relation $c_1\,\mathrm{d}\bar{\mu}_1 + c_2\,\mathrm{d}\bar{\mu}_2 = c_1\nabla\bar{\mu}_1 + c_2\nabla\bar{\mu}_2 = 0$,

$$\boldsymbol{J}_1 = -\frac{L}{c_2}\nabla\bar{\mu}_1 = -\frac{L(\partial\bar{\mu}_1/\partial c_1)_{T,p}}{c_2}\nabla c_1, \tag{2.60}$$

wherein $(\partial\bar{\mu}_1/\partial c_1)_{T,p} \geq 0$ to satisfy the requirement of stability of equilibrium. After identifying the coefficient of ∇c_1 with ρD, where D (in $\mathrm{m}^2\,\mathrm{s}^{-1}$) is the positive diffusion coefficient, one finds back the celebrated Fick's law

$$\boldsymbol{J}_1 = -\rho D\nabla c_1 \quad (D > 0). \tag{2.61}$$

Elimination of $\boldsymbol{J}_1$ between Fick's law and the balance equation $\rho\,\mathrm{d}c_1/\mathrm{d}t = -\nabla\cdot\boldsymbol{J}_1$ leads to a parabolic differential equation in $c_1(\boldsymbol{r},t)$ similar to (2.54). In the particular case of diffusion of N material particles located at a point $\boldsymbol{r} = \boldsymbol{r}_0$ at time t_0, the distribution of the mass fraction $c_1(\boldsymbol{r},t)$ obeys the same law as (2.3.8), namely

$$c_1(\boldsymbol{r},t) = \frac{N}{8(\pi Dt)^{3/2}}\exp[-(\boldsymbol{r}-\boldsymbol{r}_0)\cdot(\boldsymbol{r}-\boldsymbol{r}_0)/(4Dt)]. \tag{2.62}$$

The characteristic displacement length is now given by $l \propto \sqrt{Dt}$, and this result constitutes the main characteristic of diffusion phenomena, namely that the mean distance of diffusion is proportional to the square root of the time. Note that it is also usual to express Fick's law in terms of the molar concentration rather than the mass fraction; in such a case, Fick's law takes the form (2.2) instead of (2.61).

2.6.3 Hydrodynamics

As a further application, we will show that CIT provides a general framework for the macroscopic description of hydrodynamics. The problem that we wish to solve is the establishment of the complete set of partial differential equations governing the motion of a fluid. To be explicit, let us consider the laminar motion (turbulence is excluded) of a one-constituent isotropic, compressible and viscous fluid in presence of a temperature gradient, without internal energy supply.

The set of basic variables giving a complete knowledge of the system in space and time are the mass density $\rho(\boldsymbol{r},t)$, the velocity $\boldsymbol{v}(\boldsymbol{r},t)$ and the temperature $T(\boldsymbol{r},t)$ fields.

These variables obey the classical balance equations of mass, momentum, and energy already given in Sect. 2.4 but recalled here for the sake of clarity:

$$\rho\frac{\mathrm{d}\rho^{-1}}{\mathrm{d}t} = \nabla\cdot\boldsymbol{v}, \tag{2.63}$$

$$\rho\frac{\mathrm{d}\boldsymbol{v}}{\mathrm{d}t} = -\nabla\cdot\mathbf{P} + \rho\boldsymbol{F}, \tag{2.64}$$

$$\rho\frac{\mathrm{d}u}{\mathrm{d}t} = -\nabla\cdot\boldsymbol{q} - \mathbf{P}:\mathbf{V}. \tag{2.65}$$

The quantity $\mathbf{V}$ is the symmetric part of the velocity gradient tensor; in Cartesian coordinates, $V_{ij} = \frac{1}{2}(\partial v_i/\partial x_j + \partial v_j/\partial x_i)$. In fluid mechanics, the symmetric pressure tensor $\mathbf{P}$ is usually split into a reversible hydrostatic pressure $p\mathbf{I}$ and an irreversible viscous pressure $\mathbf{P}^{\mathrm{v}}$ in order that $\mathbf{P} = p\mathbf{I} + \mathbf{P}^{\mathrm{v}}$ with $\mathbf{I}$ the identity tensor. The symmetric viscous pressure tensor $\mathbf{P}^{\mathrm{v}}$ may further be decomposed into a bulk part $p^{\mathrm{v}}(= \frac{1}{3}\ \mathrm{tr}\ \mathbf{P}^{\mathrm{v}})$ and a traceless deviatoric part $\overset{0}{\mathbf{P}}{}^{\mathrm{v}}$ so that $\mathbf{P} = p^{\mathrm{v}}\mathbf{I} + \overset{0}{\mathbf{P}}{}^{\mathrm{v}}$. The possibility and physical meaning of an antisymmetric contribution to $\mathbf{P}^{\mathrm{v}}$ will be analysed in Problem 2.9. Gathering all these results, the pressure tensor can be written as

$$\mathbf{P} = (p + p^{\mathrm{v}})\mathbf{I} + \overset{0}{\mathbf{P}}{}^{\mathrm{v}}. \tag{2.66}$$

Instead of the pressure tensor $\mathbf{P}$, some authors prefer to use the stress tensor $\boldsymbol{\sigma}$, which is equal to minus the pressure tensor. All the results derived above and the forthcoming remain valid by working with the stress tensor. The evolution equations (2.63)–(2.65) constitute a set of five scalar relations with 16 unknown quantities namely, ρ, v, p, p^{v}, $\overset{0}{\mathbf{P}}{}^{\mathrm{v}}$, $\boldsymbol{q}$, u, and T. It is the aim of the classical theory of irreversible processes to provide the eleven missing equations. As usual, we start from Gibbs' relation

$$\frac{\mathrm{d}s}{\mathrm{d}t} = T^{-1}\frac{\mathrm{d}u}{\mathrm{d}t} - pT^{-1}\frac{\mathrm{d}\rho^{-1}}{\mathrm{d}t}, \tag{2.67}$$

wherein $\mathrm{d}u/\mathrm{d}t$ and $\mathrm{d}\rho^{-1}/\mathrm{d}t$ will be replaced, respectively, by expressions (2.65) and (2.63). This yields the following balance equation for the specific entropy s:

$$\rho\frac{\mathrm{d}s}{\mathrm{d}t} = -\nabla\cdot\frac{\boldsymbol{q}}{T} + \boldsymbol{q}\cdot\nabla T^{-1} - T^{-1}p^{\mathrm{v}}\nabla\cdot\boldsymbol{v} - T^{-1}\overset{0}{\mathbf{P}}{}^{\mathrm{v}} : \overset{0}{\mathbf{V}}, \tag{2.68}$$

where $\overset{0}{\mathbf{V}}$ is the traceless part of tensor $\mathbf{V}$. From (2.68), it is inferred that the expressions of the entropy flux and the entropy production are given by

$$\boldsymbol{J}^s = \frac{1}{T}\boldsymbol{q}, \tag{2.69}$$

$$\sigma^s = \boldsymbol{q}\cdot\nabla T^{-1} - T^{-1}p^{\mathrm{v}}\nabla\cdot\boldsymbol{v} - T^{-1}\overset{0}{\mathbf{P}}{}^{\mathrm{v}} : \overset{0}{\mathbf{V}}. \tag{2.70}$$

Expression (2.70) is bilinear in the fluxes $\boldsymbol{q}$, p^{v}, $\overset{0}{\mathbf{P}}{}^{\mathrm{v}}$, and the forces ∇T^{-1}, $T^{-1}(\nabla\cdot\boldsymbol{v})$, $T^{-1}\overset{0}{\mathbf{V}}$; by assuming linear relations between them and invoking Curie's law, one obtains the following set of phenomenological relations, also called the transport equations:

$$\boldsymbol{q} = L_{qq}\nabla T^{-1} \equiv -\lambda\nabla T, \tag{2.71}$$

$$p^{\mathrm{v}} = -l_{vv}T^{-1}\nabla\cdot\boldsymbol{v} \equiv -\zeta\nabla\cdot\boldsymbol{v}, \tag{2.72}$$

$$\overset{0}{\mathbf{P}}{}^{\mathrm{v}} = -L_{vv}T^{-1}\overset{0}{\mathbf{V}} \equiv -2\eta\overset{0}{\mathbf{V}}. \tag{2.73}$$

We recognize in (2.71) the Fourier's law by identifying the phenomenological coefficient L_{qq}/T^2 with the heat conductivity λ; (2.72) is the Stokes' relation if l_{vv}/T is identified with the bulk viscosity ζ, and finally (2.73) is the Newton's law of hydrodynamics when L_{vv}/T is put equal to 2η, with η the dynamic shear viscosity. It follows from (2.71)–(2.73) that the fluid will be set instantaneously in motion as soon as it is submitted to a force, in contrast with elastic solids, which may stay at rest even when subject to stresses. At equilibrium, both members of equations (2.71)–(2.73) vanish identically as it should. The phenomenological coefficients λ, ζ, and η depend generally on ρ and T but in most practical situations the dependence with respect to ρ is negligible. In the case of incompressible fluids for which $\nabla \cdot \boldsymbol{v} = 0$, the viscous pressure p^{v} is zero and the bulk viscosity does not play any role. However, in compressible fluids like dense gases and bubbly liquids, the bulk viscosity is by no means negligible. The bulk viscosity vanishes in the case of perfect gases; in real gases, it is far from being negligible and the ratio ζ/η between bulk and shear viscosity may even be of the order of hundred.

After substitution of the transport equations (2.71)–(2.73) in (2.70) of the entropy production, one obtains

$$\sigma^s = \frac{\lambda}{T^2}(\nabla T)^2 + \frac{\zeta}{T}(\nabla \cdot \boldsymbol{v})^2 + \frac{2\eta}{T}(\overset{0}{\mathbf{V}} : \overset{0}{\mathbf{V}}) \geq 0. \tag{2.74}$$

Positiveness of σ^s requires that

$$\lambda > 0, \quad \zeta > 0, \quad \eta > 0. \tag{2.75}$$

Negative values of these transport coefficients will therefore be in contradiction with the second law of thermodynamics.

By combining the nine transport equations (2.71)–(2.73) with the five evolution equations (2.63)–(2.65), one is faced with a set of 14 equations with 16 unknowns. The missing relations are provided by Gibbs' equation (2.67) from which are supplemented the two equations of state $T = T(u, \rho)$, $p = p(u, \rho)$, or solving with respect to u,

$$u = u(T, \rho), \quad p = p(T, \rho). \tag{2.76}$$

These equations, together with the evolution and transport relations describe completely the behaviour of the one-component isotropic fluid after that boundary and initial conditions have been specified. The state equations (2.76) have a status completely different from the phenomenological equations (2.71)–(2.73) because, in contrast with the latter, they do not vanish identically at equilibrium. By substitution of the phenomenological laws of Stokes (2.72) and Newton (2.73) in the momentum balance equation (2.64), one finds back the well-known Navier–Stokes' equation when it is assumed that the viscosity coefficients are constant:

$$\rho \frac{\mathrm{d}\boldsymbol{v}}{\mathrm{d}t} = -\nabla p + 2\eta \nabla^2 \mathbf{v} + \left(\frac{2}{3}\eta + \zeta\right) \nabla(\nabla \cdot \boldsymbol{v}) + \rho \boldsymbol{F}. \tag{2.77}$$

In the case of an isothermal and incompressible fluid, which is the most customary situation met in hydrodynamics, the behaviour of the fluid is completely described by (2.77) and $\nabla \cdot \boldsymbol{v} = 0$. For more general situations like multi-component mixtures with diffusion, the equations become much more complicated but the way to derive them follows the same systematic procedure as above. It should, however, be stressed that the results of this section are only applicable to the study of laminar flows of some particular class of fluids, namely these described by the linear flux–force relations of Fourier, Stokes, and Newton. The above formalism is not adequate for non-Newtonian fluids or polymeric solutions and cannot cope with turbulent motions. We note finally that the basic relations obtained so far, like for instance the Navier–Stokes' equation (2.77), are parabolic partial differential equations, which means that disturbances will propagate through the fluid with an infinite velocity.

2.7 Limitations of the Classical Theory of Irreversible Thermodynamics

Despite its numerous successes, CIT has raised several questions and some shortcomings have been pointed out. Let us comment briefly about the most frequent criticisms:

1. The theory is not applicable to irreversible processes described by *non-linear phenomenological equations.* Many actual processes, like chemical reactions or non-Newtonian flows, are indeed characterized by non-linear phenomenological relations and are therefore outside the scope of classical irreversible thermodynamics. It should, however, be noted that linearity between fluxes and forces does not require that the phenomenological coefficients, like the heat conductivity and the viscosity coefficients, are constant; they may indeed depend in particular on the temperature and the pressure with the consequence that the corresponding evolution equations, like Navier–Stokes' equation, may be non-linear.
2. The cornerstone of the theory is the *local equilibrium hypothesis*, accordingly the thermodynamic state variables are the same as in equilibrium and Gibbs' equation, which plays a fundamental role in CIT, remains locally valid, i.e. at each position in space and each instant of time. The advantage of this assumption is that entropy has a precise meaning, even in non-equilibrium states, and that it leads to an explicit expression for the entropy production from which are inferred appropriate forms for the constitutive relations. However, it is conceivable that other variables, not found at equilibrium, are able to influence the process. This is, for instance, the case of polymers of long molecular chains in which configuration influences considerably their behaviour. Other examples are superfluids and superconductors whose peculiar properties ask for the introduction of extra

variables. As the local equilibrium hypothesis implies large time and space scales, CIT is not appropriate for describing high-frequency phenomena as ultrasound propagation or nuclear collisions and short-wavelength systems and processes, like nano-structures, shock waves or light and neutron scattering.

Another important point to notice is that, when referred to generalized hydrodynamics, the phenomenological coefficients are frequency and wavelength dependent. This is at variance with the local equilibrium hypothesis, implying that these coefficients are frequency and wavelength independent.

3. By introduction of the Fourier's law with constant heat conductivity in the energy balance, one obtains an equation for the temperature, which is of the diffusion type; see for instance (2.54). From the mathematical point of view, this is a parabolic partial differential equation, which implies that after application of a disturbance, the latter will propagate at *infinite velocity* across the body so that it will be felt instantaneously and everywhere in the whole body. The same remark is applicable to the other equations of the theory and in particular to Navier–Stokes' equation. The problem of propagation of signals with an infinite velocity is conceptually a major inconvenient, which is not acceptable from a purely physical point of view. However it is not a tragedy in most practical situations because their characteristic time is much longer than the transit time of the signals transporting the thermodynamic information.
4. Another important subject of controversy concerns the validity of the *Onsager–Casimir's reciprocal relations*, which were derived from the condition of microscopic reversibility. Although the reciprocity property was originally shown by Onsager to be applicable at the microscopic level for particles in situations close to equilibrium, it is current to extrapolate this result at the macroscopic level for continuum media driven far from equilibrium. There is clearly no theoretical argument supporting such an extrapolation. Onsager's demonstration itself has been the subject of acrid criticisms because it is based on an important assumption with regard to the regression of the fluctuating thermodynamic variables. Indeed, it lies on the hypothesis that the rate of decay of the fluctuations takes place according to the same law as the macroscopic linear flux–force relations. Moreover, in the light of the demonstration given by Onsager, the thermodynamic fluxes are defined as time derivatives of extensive thermodynamic variables and the forces are the derivatives of the entropy with respect to the same state variables. Such conditions are certainly not respected by thermodynamic fluxes like the heat flux or the pressure tensor or thermodynamic forces as the temperature gradient or the velocity gradient tensor.
5. As shown explicitly in Box 2.2, the Curie law stating that in isotropic materials, fluxes couple only with forces of the same tensorial order, is directly derivable from algebra. It is also important to realize that this exclusion

law is not valid outside the linear regime even for isotropic bodies. As observed by Truesdell (1984), these results were well known even before Curie was born. Moreover, when one goes back to Curie's original papers on symmetry considerations, there is nothing, even vaguely mentioned, which is reminding of the law bearing his name in classical irreversible thermodynamics. In anisotropic systems, the coupling between terms of different tensorial order, as for instance chemical reactions (scalar) and mass transport (vectorial), may provide very important effects, as shown in Chap. 4.

6. Another point of debate is the question of *how the fluxes and forces are to be selected* from the expression of the entropy production. It is generally admitted that the thermodynamic fluxes are the physical fluxes that appear in the balance laws of mass, momentum, and energy, and that the forces are the conjugated terms in the bilinear expression of the entropy production. Another suggestion is to identify the forces as the gradient of intensive variables, as temperature and velocity, and to select as fluxes the conjugate terms in the entropy production. Unfortunately, such definitions are not applicable to chemical reactions. Another school of thought claims that it is indifferent how the fluxes and forces are selected, at the condition that non-singular linear transformations leave invariant the entropy production expression. However, as pointed out by Coleman and Truesdell (1960), one should be careful in the identification of fluxes and forces, as a wrong selection could destroy the reciprocal property of the phenomenological coefficients. Clearly, the controversy about how fluxes and forces are chosen is not closed but, in our opinion, this is not the more fundamental question about CIT, all the more as the discussions about this point have been sterile.

2.8 Problems

2.1. *Heat transport and entropy production.* Two heat reservoirs at temperature T_1 and T_2 are connected by a rigid heat-conducting one-dimensional rod. Find the longitudinal temperature profiles in the rod if the heat conductivity is (a) constant and (b) proportional to T^α, where α is a constant. Calculate the entropy produced along the rod per unit time.

2.2. *Fourier's and Newton's laws.* Consider a rigid cylindrical heat-conducting rod. The heat transfer along the rod is described by means of Fourier's law ($\boldsymbol{q} = -\lambda\nabla T$), whereas the heat exchange with the environment is described by Newton's cooling law, expressing that the heat flow is proportional to the temperature difference between the rod and the environment ($\boldsymbol{q}\cdot\boldsymbol{n} = -\alpha(T - T_{\text{env}})$), where $\boldsymbol{n}$ is the unit vector normal to the boundaries. (a) Derive the equation describing the temperature distribution along the rod. (b) Determine the temperature profile along a very long rod if its hotter boundary is kept at T_1 and the environment's temperature T_{env} is constant.

2.3. *Temperature profile.* A cylindrical tube of radius R and length L is immersed in an environment at temperature T_0. A fluid flows inside the tube at a flow rate Q; the temperature of the fluid when it enters the tube is T_1. Find the temperature T_2 of the outgoing fluid at the end of the tube as a function of the flow rate Q in terms of its specific heat per unit mass, its density and its thermal conductivity. The heat transfer coefficient across the lateral walls of the tube is a.

2.4. *Radiative heat transfer.* Assume that heat exchange between a spherical body of radius R and its environment occurs through electromagnetic radiation and obeys Stefan–Boltzmann's law. Accordingly, the quantity of heat emitted per unit area and unit time by a body at temperature T is $q = \varepsilon\sigma T^4$, where σ is the Stefan–Boltzmann constant and ε the emissivity of the system. (a) Establish the differential equation describing the cooling of the sphere when its surface is at temperature T and the environment at temperature T_{env}. (b) Find the corresponding entropy production per unit time. (c) Solve the problem when the difference $T - T_{\text{env}}$ is small compared to T_{env}. (d) How will the solution be modified if the radius of the sphere is 10 times larger?

2.5. *The age of the Earth.* A semi-infinite rigid solid of heat diffusivity χ, initially at temperature T_{10}, is suddenly put into contact with another semi-infinite solid at fixed temperature T_2. Assume that heat exchange is described by Fourier's law. (a) Show that the temperature profile as a function of time and position is

$$\theta(x,t) = \frac{\theta_0}{\sqrt{\pi\chi t}} \int_{-\infty}^{x} \mathrm{d}x' \exp\left(-\frac{(x-x')^2}{4\chi t}\right),$$

where $\theta = T(x,t) - T_2$ and $\theta_0 \equiv T_{10} - T_2$. (b) Verify that the temperature gradient at the surface is

$$\left(\frac{\partial\theta}{\partial x}\right)_{x=0} = -\frac{\theta_0}{\sqrt{\pi\chi t}}.$$

(c) At the end of the nineteenth century, lord Kelvin evaluated the age of the Earth from an analysis of the terrestrial temperature gradient near the surface of the Earth, found to be equal to $-3\times10^{-2}\,^{\circ}\mathrm{C\,m^{-1}}$. By taking $\chi \approx 3\times10^{5}\,\mathrm{m^2\,s^{-1}}$ and $\theta_0 \approx 3{,}800^{\circ}\mathrm{C}$, estimate the age of the Earth according to this model. The result found is much less than the actual estimation of 4.5 billion years (radioactivity, which was not known at Kelvin's time, will considerably influence the result).

2.6. *Matter diffusion.* Derive the expressions of the entropy flux and the entropy production for the problem of isothermal and isobaric diffusion in a binary mixture of non-viscous fluids, in absence of external body forces (see Sect. 2.6.2).

2.7. *Concentration profile.* A perfectly absorbing sphere of radius R is immersed in a solvent–solute solution with homogeneous concentration c_0. The diffusion coefficient of the solute is D. Since the sphere is perfectly absorbing, the concentration of solute on its surface is always zero. (a) Determine the concentration profile as a function of the radial distance. (b) Show that the amount of solute crossing the surface of the sphere per unit time is given by $I = 4\pi DRc_0$. *Hint*: In spherical symmetry, the diffusion equation has the form

$$\frac{\partial c}{\partial t} = \frac{D}{r^2}\frac{\partial}{\partial r}\left(r^2\frac{\partial c}{\partial r}\right).$$

2.8. *The Einstein relation.* A dilute suspension of small particles in a viscous fluid at homogeneous temperature T is under the action of the gravitational field. The friction coefficient of the particles with respect to the fluid is a ($a = 6\pi\eta r$ for spherical particles of radius r in a solvent with viscosity η). Owing to gravity, the particles have a sedimentation velocity $v_{\text{sed}} = m'g/a$, with m' the mass of one particle minus the mass of the fluid displaced by it (Archimede's principle); the corresponding sedimentation flux is $J_{\text{sed}} = nv_{\text{sed}}$, with n the number of particles per unit volume. Against the sedimentation flux is acting a diffusion flux $J_{\text{dif}} = -D\,\partial n/\partial z$, D being the diffusion coefficient. (a) Find the vertical profile of $n(z)$ in equilibrium. (b) Compare this expression with Boltzmann's result

$$n(z) = n(0)\,\exp[-(m'gz/k_{\text{B}}T)],$$

where k_{B} is the Boltzmann constant, and demonstrate Einstein's relation

$$D = k_{\text{B}}T/a.$$

2.9. *Micropolar fluids.* In some fluids (composed of elongated particles or rough spheres) the pressure tensor is non-symmetric. Its antisymmetric part is related to the rate of variation of an intrinsic angular momentum, and therefore, it contributes to the balance equation of angular momentum. The antisymmetric part of the tensor is usually related to an axial vector $\boldsymbol{P}^{\text{va}}$, whose components are defined as $P_1^{\text{va}} = P_{23}^{\text{va}}$, $P_2^{\text{va}} = P_{31}^{\text{va}}$, $P_3^{\text{va}} = P_{12}^{\text{va}}$. The balance equation for the internal angular momentum is

$$\rho j\dot{\boldsymbol{\omega}} + \nabla\cdot\mathbf{Q} = -2\boldsymbol{P}^{\text{va}},$$

with j the microinertia per unit mass of the fluid, $\boldsymbol{w}$ the angular velocity, and $\mathbf{Q}$ the flux of the intrinsic angular momentum, which is usually neglected. (a) Prove that the corresponding entropy production is

$$\sigma^s = -T^{-1}p^{\text{v}}\nabla\cdot\boldsymbol{v} - T^{-1}\overset{0}{\mathbf{P}}{}^{\text{vs}} : \overset{0}{(\nabla\boldsymbol{v})}{}^s - T^{-1}\boldsymbol{P}^{\text{va}}\cdot(\nabla\times\boldsymbol{v} - 2\boldsymbol{\omega}) + \boldsymbol{q}\cdot\nabla T^{-1},$$

where $\overset{0}{\mathbf{P}}{}^{\text{vs}}$ is the symmetric part of $\mathbf{P}^{\text{v}}$. *Hint*: Note that $\partial s/\partial\boldsymbol{\omega} = -\rho jT^{-1}\boldsymbol{\omega}$. (b) Show that the constitutive equation for $\boldsymbol{P}^{\text{va}}$ may be given by

$$\boldsymbol{P}^{\mathrm{va}} = -\eta_{\mathrm{r}}(\nabla \times \boldsymbol{v} - 2\boldsymbol{\omega}),$$

where η_{r} is the so-called rotational viscosity; explain why there is no coupling between $\boldsymbol{P}^{\mathrm{va}}$ and $\boldsymbol{q}$ (see, for instance, Snider and Lewchuk 1967; Rubí and Casas-Vázquez 1980).

2.10. *Flux–force relations.* It is known that by writing a linear relation of the form $J_\alpha = \sum_\beta L_{\alpha\beta} X_\beta$ between n-independent fluxes $J_\alpha (\alpha = 1, \ldots, n)$ and n-independent thermodynamic forces X_β, the matrix $L_{\alpha\beta}$ of phenomenological coefficients is symmetric, according to Onsager. (a) Show that the matrix $L_{\alpha\beta}$ is still symmetric when not all the fluxes are independent but one of them is a linear combination of the other ones, i.e. $\sum_{\alpha=1}^{n} a_\alpha J_\alpha = 0$. (b) Does this conclusion remain fulfilled when instead of a linear relation between the fluxes, there exists a linear relation $\sum_{\alpha=1}^{n} a_\alpha X_\alpha = 0$ between the forces?

2.11. *Minimum entropy production principle.* Consider a one-dimensional rigid heat conductor with fixed temperatures at its ends. Show that the Euler–Lagrange equation corresponding to the principle of minimum entropy production

$$\delta P(T^{-1}) = \delta \int q(\partial T^{-1}/\partial x)\, \mathrm{d}x = 0$$

is the steady energy balance $\partial q/\partial x = 0$ at the condition to write Fourier's law in the form $q = L\partial T^{-1}/\partial x$ with $L =$ constant. Show that $\delta^2 P \geq 0$; this result expresses that the Euler–Lagrange equation is a necessary condition for P to be a minimum. Is the principle still fulfilled when Fourier's law is written in the more classical form $q = -\lambda\, \partial T/\partial x$ with $\lambda =$ constant?

2.12. *Variational principles.* Prigogine's minimum entropy production is not the only variational principle in linear non-equilibrium thermodynamics. Other principles have been proposed, amongst others, by Onsager (1931) and Gyarmati (1970). For instance, consider the entropy production $P(X, J) \equiv \sum_i X_i J_i$ and the dissipation functions

$$\Phi(J, J) \equiv \frac{1}{2} \sum_{i,j} R_{ij} J_i J_j \quad \text{and} \quad \Psi(X, X) \equiv \frac{1}{2} \sum_{i,j} L_{ij} X_i X_j,$$

$\boldsymbol{X}$ being the thermodynamic forces and $\boldsymbol{J}$ the conjugated fluxes. Show that if the matrices L_{ij}, R_{ij}, and L_{ij} are constant and symmetric, the two following variational principles are valid (1) for prescribed forces $\boldsymbol{X}$ and varying fluxes $\boldsymbol{J}$, $P(X, J) - \Phi(J, J)$ is maximum implies that $X_k = \sum_i R_{ki} J_i$ and (2) for prescribed fluxes $\boldsymbol{J}$ and varying forces $\boldsymbol{X}$, $P(X, J) - \Psi(X, X)$ is maximum implies that $J_k = \sum_i L_{ki} X_i$. (The reader interested in these topics is referred to Ichiyanagi 1994.)

Chapter 3
Coupled Transport Phenomena

Thermoelectricity, Thermodiffusion, Membranes

The analysis of coupled processes is one of the most outstanding aspects of the classical theory of non-equilibrium thermodynamics. To emphasize this feature, this chapter is dedicated to the study of coupled transport processes, involving two fluxes and two thermodynamic forces. Emphasis is put on some practical applications in thermoelectricity (coupling of heat and electricity), thermodiffusion (coupling of heat and mass motion) and transport of matter through membranes. In the precedent chapter, we have already outlined, in general terms, the importance of the Onsager's reciprocal relations ruling the coefficients that describe such couplings. Here, we will discuss direct applications.

Coupled transport processes find a natural theoretical framework in non-equilibrium thermodynamics. However, they were discovered well before the developments of these theories during the early nineteenth century. Indeed, thermoelectric effects were observed by Seebeck and Peltier, respectively, in 1821 and 1835, even earlier than the quantitative formulation of Ohm's law in 1855. The first results on thermodiffusion were obtained by Ludwig in 1856, Dufour's effect was discovered in 1872, and Soret carried out his first systematic researches in 1879. Although these topics are rather old and classical, from the practical point of view they have entered recently in a completely new perspective, because of the recent developments in materials sciences, high-power lasers and renewed interest in optimization of energy generation. We will outline here these new features.

Transport properties play a central role in materials sciences. For instance, a problem of acute interest is to obtain superconductor materials with high critical temperatures. Another domain of intense activity is the optimal application of thermoelectricity to convert waste heat to usable electricity or to develop cooling devices; this requires materials with high Seebeck coefficient, high electrical conductivity and low thermal conductivity. This is not an easy task because usually a high electrical conductivity implies also a high thermal conductivity. Recently, research on thermoelectric materials has known an extraordinary burst: it is frequent to find semiconductors and superlattice devices whose figure of merit (a combination of Seebeck coefficient, electrical and thermal conductivities directly related to the efficiency of energy conversion) has increased by more than one order of magnitude since the end of the twentieth century (Mahan et al. 1997). This was in particular achieved in superlattices in which one dimension is smaller than the mean free path of phonons but larger than that of electrons. More generally, it is desirable to combine a macroscopic approach of transport phenomena, to design the most efficient devices, with a microscopic understanding of the basic phenomena.

In parallel with thermoelectricity, practical applications of thermodiffusion have also known a renewal of interest during the last decades with the possibility of producing much localized heating of fluids by means of high-power and extremely focused lasers. High temperature gradients will promote the separation of different constituents in mixtures.

Finally, since the problem of transport across membranes remains a subject of constant and wide interest, mainly in life sciences, we have reserved one section of this chapter to its analysis.

3.1 Electrical Conduction

The results presented in Chap. 2 are directly applicable to the problem of motion of electric charges in a conductor. First, we briefly study the flow of electrical carriers in a rigid conductor at rest and in absence of thermal effects. The basic variables are (1) z, the electrical charge of the carrier, say electrons, per unit mass (z should not be confused with the charge z_e per unit mass of the electrons to which it is related by $z = c_\mathrm{e} z_\mathrm{e}$, where c_e is the mass fraction of electrons) and (2) u the internal energy per unit mass. The evolution equations are the law of conservation of charges and the internal energy balance, i.e.

$$\rho \frac{\mathrm{d}z}{\mathrm{d}t} = -\nabla \cdot \boldsymbol{i}, \tag{3.1}$$

$$\rho \frac{\mathrm{d}u}{\mathrm{d}t} = \boldsymbol{E} \cdot \boldsymbol{i}, \tag{3.2}$$

where $\boldsymbol{E}$ is the electric field and $\boldsymbol{i}$ the density of conductive current (current intensity per unit transversal area); in whole generality, the latter is related to the density $\boldsymbol{I}$ of total current $\boldsymbol{I}$ by $\boldsymbol{I} = \boldsymbol{i} + \rho z \boldsymbol{v}$, where $\rho z \boldsymbol{v}$ is the convective current. In the present problem, one has $\boldsymbol{v} = 0$ so that $\boldsymbol{I} = \boldsymbol{i}$ while $\mathrm{d}/\mathrm{d}t$ will reduce to the partial time derivative.

The Gibbs' equation (2.4) simplifies as $T\,\mathrm{d}s = \mathrm{d}u - \mu_\mathrm{e}\,\mathrm{d}c_\mathrm{e}$, with μ_e the chemical potential of the electrons. In terms of the time derivative and the charge density z, Gibbs' relation takes the form

$$T \frac{\mathrm{d}s}{\mathrm{d}t} = \frac{\mathrm{d}u}{\mathrm{d}t} - \frac{\mu_\mathrm{e}}{z_\mathrm{e}} \frac{\mathrm{d}z}{\mathrm{d}t}. \tag{3.3}$$

Substitution of (3.1) and (3.2) in (3.3) yields

$$\rho \frac{\mathrm{d}s}{\mathrm{d}t} = \nabla \cdot \left(\frac{\mu_{\mathrm{e}}}{T z_{\mathrm{e}}} \boldsymbol{i} \right) + \frac{1}{T} \left(\boldsymbol{E} - \nabla \frac{\mu_{\mathrm{e}}}{z_{\mathrm{e}}} \right) \cdot \boldsymbol{i}, \tag{3.4}$$

from which it is deduced that the entropy flux and the entropy production are given, respectively, by

$$\boldsymbol{J}^s = \frac{\mu_{\mathrm{e}}}{T z_{\mathrm{e}}} \boldsymbol{i}, \tag{3.5}$$

$$T\sigma^s = \left(\boldsymbol{E} - \nabla \frac{\mu_{\mathrm{e}}}{z_{\mathrm{e}}} \right) \cdot \boldsymbol{i} \geq 0. \tag{3.6}$$

The last result suggests writing the following linear *flux–force relation*

$$\boldsymbol{E} - \nabla \frac{\mu_{\mathrm{e}}}{z_{\mathrm{e}}} = r\boldsymbol{i}, \tag{3.7}$$

or, for negligible values of $\nabla(\mu_{\mathrm{e}}/z_{\mathrm{e}})$,

$$\boldsymbol{E} = r\boldsymbol{i}, \tag{3.8}$$

which is Ohm's law with r the electrical resistivity. After introducing Ohm's law in (3.6), one obtains $T\sigma^s = ri^2$, which represents Joule's heating and from which is inferred that the electrical resistivity is a positive quantity.

In the presence of an external magnetic field $\boldsymbol{\mathcal{H}}$, interesting coupled effects appear and the resistivity is no longer a scalar, but a second-order tensor because of the anisotropy introduced by $\boldsymbol{\mathcal{H}}$. In this case, the resistivity depends on the magnetic field and the Onsager–Casimir's reciprocity relations state that $r_{ij}(\boldsymbol{\mathcal{H}}) = r_{ji}(-\boldsymbol{\mathcal{H}})$. If the external magnetic field is applied in the z-direction with all currents and gradients parallel to the x–y plane, the resistivity tensor $\mathbf{r}$ will take the form (Problem 3.2)

$$\mathbf{r} = \begin{pmatrix} r_{11}(\mathcal{H}) & r_{12}(\mathcal{H}) & 0 \\ -r_{12}(\mathcal{H}) & r_{22}(\mathcal{H}) & 0 \\ 0 & 0 & r_{33}(\mathcal{H}) \end{pmatrix}, \tag{3.9}$$

where $\mathcal{H}$ is the component in the z-direction. It follows that an electric field in the x-direction will produce a current not only in the parallel direction but also in the normal y-direction, a phenomenon known as the isothermal *Hall* effect. This effect is easily understandable, as the external magnetic field will deflect the charge carriers along a curved trajectory. The discovery, at the beginning of the 1980s, of subtle phenomena such as quantum Hall effect, boosted an enormous impetus on this topic. The dependence of the diagonal components on the magnetic field, known as magneto-resistance, has led to important applications in computer technology. These topics, however, are beyond the scope of purely macroscopic considerations.

3.2 Thermoelectric Effects

The coupling between temperature gradient and electrical potential gradient gives rise to the thermoelectric effects of Seebeck, Peltier, and Thomson. Here we present them in the framework of classical irreversible thermodynamics, outline their main physical features and, as a practical application, we will determine the efficiency of thermoelectric generators.

3.2.1 Phenomenological Laws

In presence of both thermal and electrical effects, the law of conservation of charge (3.1) remains unchanged while the energy balance equation (3.2) contains an additional term and reads as

$$\rho \frac{\mathrm{d}u}{\mathrm{d}t} = -\nabla \cdot \boldsymbol{q} + \boldsymbol{E} \cdot \boldsymbol{i}. \tag{3.10}$$

Combining (3.1) and (3.10) with the Gibbs' equation (3.3), it is a simple exercise (see Problem 3.1) to prove that the entropy flux and the rate of dissipation per unit volume $T\sigma^s$ are, respectively, given by

$$\boldsymbol{J}^s = \frac{\boldsymbol{q}}{T} - \frac{\mu_\mathrm{e}}{T z_\mathrm{e}} \boldsymbol{i}, \tag{3.11}$$

$$T\sigma^s = -\boldsymbol{J}^s \cdot \nabla T + \left[\boldsymbol{E} - \nabla \left(\frac{\mu_\mathrm{e}}{z_\mathrm{e}} \right) \right] \cdot \boldsymbol{i}. \tag{3.12}$$

Restricting to isotropic media, the corresponding phenomenological equations are

$$\boldsymbol{J}^s = -L_{11} \nabla T + L_{12} \left[\boldsymbol{E} - \nabla \left(\frac{\mu_\mathrm{e}}{z_\mathrm{e}} \right) \right], \tag{3.13}$$

$$\boldsymbol{i} = -L_{21} \nabla T + L_{22} \left[\boldsymbol{E} - \nabla \left(\frac{\mu_\mathrm{e}}{z_\mathrm{e}} \right) \right], \tag{3.14}$$

with $L_{12} = L_{21}$ in virtue of Onsager's reciprocal relations. For practical reasons, it is convenient to resolve (3.13) and (3.14) with respect to $\boldsymbol{J}^s$ and $\boldsymbol{E} - \nabla(\mu_\mathrm{e}/z_\mathrm{e})$ and to introduce the following phenomenological coefficients:

$$\lambda = T(L_{11} - L_{12}L_{21}/L_{22}) \quad \text{(heat conductivity)},$$

$$\pi = T(L_{12}/L_{22}) \quad \text{(Peltier coefficient)},$$

$$\varepsilon = L_{21}/L_{22} \quad \text{(Seebeck coefficient)},$$

$$r = 1/L_{22} \quad \text{(electrical resistivity)}.$$

With this notation, (3.13) and (3.14) take the more familiar forms

$$\boldsymbol{q} = -\lambda \nabla T + \left(\pi + \frac{\mu_\mathrm{e}}{z_\mathrm{e}} \right) \boldsymbol{i}, \tag{3.15}$$

$$\boldsymbol{E} - \nabla \left(\frac{\mu_\mathrm{e}}{z_\mathrm{e}} \right) \boldsymbol{i} = \varepsilon \nabla T + r \boldsymbol{i}, \tag{3.16}$$

with, as a consequence of the symmetry property $L_{12} = L_{21}$,

$$\pi = \varepsilon T. \tag{3.17}$$

This result is well known in thermoelectricity as the *second Kelvin relation* and has been confirmed experimentally and theoretically in transport theories.

In most applications, it is permitted to omit the contribution of the terms in $\mu_\mathrm{e}/z_\mathrm{e}$ in relations (3.15) and (3.16), which receive then a simple interpretation. It follows from (3.15) that, even in absence of a temperature gradient, a heat flux may be generated due to the presence of an electrical current: this is called the *Peltier's effect.* The coefficient $\pi = \pm(|\boldsymbol{q}|/|\boldsymbol{i}|)_{\Delta T=0}$ is a measure of the quantity of heat absorbed or rejected at the junction of two conductors of different materials kept at uniform temperature and crossed by an electric current of unit density (see Box 3.2).

Expression (3.16) exhibits the property that, even in absence of an electric current, an electrical potential $\boldsymbol{E} = -\nabla\phi$ can be created due solely to a temperature difference, this is known as Seebeck's effect: the coefficient $\varepsilon = \pm(\Delta\phi/\Delta T)_{i=0}$ is the Seebeck or thermoelectric power coefficient, it measures the electrical potential produced by a unit temperature difference, in absence of electric current. The Seebeck's effect finds a natural application in the construction of thermocouples (see Box 3.1).

Box 3.1 The Seebeck's Effect and Thermoelectric Power

The most relevant application of the Seebeck's effect is the thermocouple, which is frequently used for temperature measurements. The latter is constituted by two different materials (usually metals or alloys) forming a circuit with some points at different temperatures. Consider the situation presented in Fig. 3.1, where the two junctions between the materials A and B are at different temperatures T_1 and T_2, respectively. A voltmeter is inserted inside the system, at a position where temperature is T_0. In absence of electric current, the voltmeter indicates a value $\Delta\Phi$ for the electric potential difference, which is related to $\Delta T \equiv T_2 - T_1$, by means of $\Delta\Phi = \int_{T_1}^{T_2} (\varepsilon_\mathrm{A} - \varepsilon_\mathrm{B})\,\mathrm{d}T$. Assuming that $\varepsilon_\mathrm{A} - \varepsilon_\mathrm{B}$ is known and constant, one has $\Delta\Phi = (\varepsilon_\mathrm{A} - \varepsilon_\mathrm{B})\Delta T$, and the measurement of $\Delta\Phi$ will allow us to determine the value of ΔT. Usually, one of the two temperatures is a reference temperature (for instance, the melting point of ice) and the other one is the temperature to be measured.

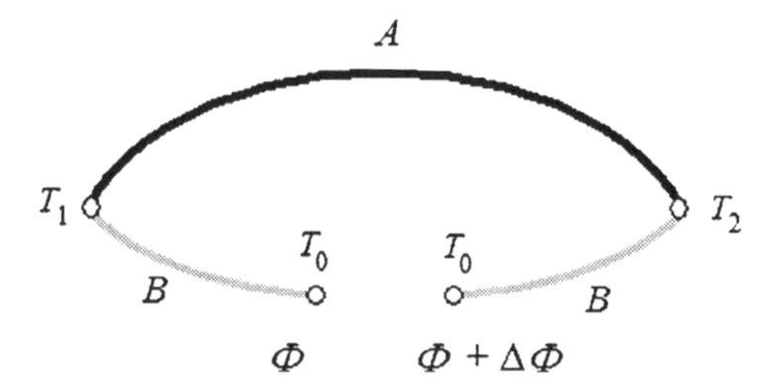

Fig. 3.1 Thermocouple

The thermoelectric power of the thermocouple is defined as the difference between the absolute thermoelectric powers of the two materials, namely $\varepsilon_{AB} = \varepsilon_B - \varepsilon_A$. Intuitively, the thermoelectric power reflects the tendency of carriers to diffuse from the hot end (characterized by a higher thermal energy whence higher average kinetic energy of the carriers) to the cold end, carrying with them their electric charge. The sign of the Seebeck coefficient ε is taken as positive when the carriers are positive charges (as for instance holes in p-doped semiconductors) and negative otherwise (as in most metals or n-doped semiconductors). This rule is not respected in some metals, like copper or silver, which are characterized by a positive thermoelectric power. The usual values of ε are of the order of $10\,\mu\mathrm{V\,K}^{-1}$ for metals and $100\,\mu\mathrm{V\,K}^{-1}$ for semiconductors. Typical examples of thermocouples are those formed by the junctions of copper/constantan (a copper–nickel alloy), nickel–chromium alloy/constantan, or platinum–rhodium alloy/platinum.

Another interesting application of the Seebeck's effect is thermoelectric generation, namely electrical power generated from a temperature difference; the discussion of this subject will be postponed later.

Box 3.2 The Peltier's Effect and Thermoelectric Cooling Systems

The Peltier's effect describes the absorption or emission of heat accompanying the flow of an electric current across the junctions of two materials A and B, under isothermal conditions (see Fig. 3.2). Assume, for instance, that the circulation of the current from A to B produces a cooling of the junction between them. The effect is reversed by reversing the direction of the electric current.

To interpret this phenomenon from a microscopic perspective is easy. If in material A the electrons are moving at an energy level lower than in material B, the transition of one electron from A to B will require the supply of some energy from the junction and therefore the passage of current is accompanied by cooling the right junction. The reverse is true by transiting from B to A: some quantity of energy is released at the left junction which

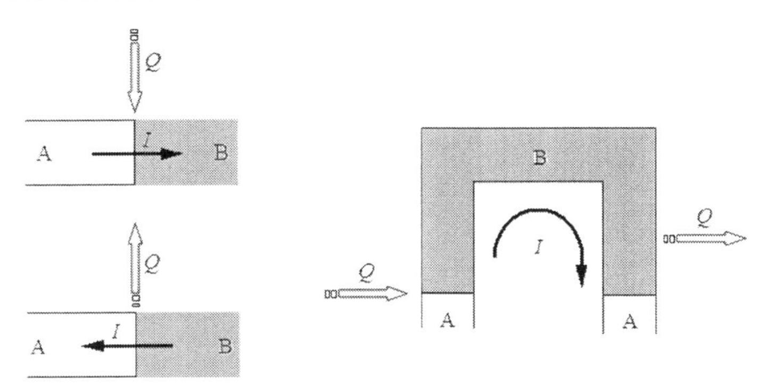

Fig. 3.2 The Peltier's effect

is heated. An example is that of electrons flowing from a low energy level in p-type semiconductors to a higher-energy level in n-type semiconductors. Materials frequently used are bismuth telluride (Bi_2Te_3) heavily doped and, more recently, the layered oxide $Na_xCo_2O_4$.

A practical application of the Peltier's effect is the thermoelectric refrigerator. If the junction A $\rightarrow$ B is cooled, when electric current flows in a given direction, it will be put inside the container to be refrigerated, whereas the junction B $\rightarrow$ A, delivering heat, is left outside. The effect of the passage of a current is thus to extract heat from the container and to deliver it to the external world. When current flows in the opposite direction, the reverse process occurs, namely extracting heat from the external world and heating the container. The analogy with usual refrigerators or heat pumps based on Carnot cycle is the following: the junction absorbing heat is analogous to the expansion of refrigerant in contact with the cold reservoir whereas the junction delivering heat is the analogous of the compression of the refrigerant in contact with the hot reservoir. Thermoelectric coolers have the advantage of not having moving parts, since there is neither compression nor expansion of the refrigerant, nor phase change. They are reliable, do not need maintenance and can be given a very compact form. Nevertheless, they present the disadvantage to consume more energy and are therefore less efficient than usual coolers.

A third thermoelectric coupling of interest is *Thomson's* heat. This effect occurs from the simultaneous presence of a temperature gradient and an electric current, and is the result of the coupling of both the Peltier's and Seebeck's effects. Substituting (3.15) and (3.16) in the energy balance equation (3.10) and taking into account of the conservation of total charge ($\nabla \cdot \boldsymbol{i} = 0$) and Kelvin's second relation, it is found that

$$\rho \frac{\mathrm{d}u}{\mathrm{d}t} = \nabla \cdot (\lambda \nabla T) + r i^2 - \boldsymbol{i} \cdot (\nabla \pi)_T + \left(\frac{\pi}{T} - \frac{\partial \pi}{\partial T} \right) \boldsymbol{i} \cdot \nabla T, \tag{3.18}$$

where subscript T means that the temperature is fixed. If the temperature remains uniform, only the Joule effect ri^2 and the Peltier's effect $\boldsymbol{i} \cdot (\nabla\pi)_T$ contribute to the dissipated heat. Now, in presence of a temperature gradient, two supplementary terms occur in (3.18): heat conduction plus a term resulting from the coupling $\boldsymbol{i} \cdot \nabla T$ of the electric current and the temperature gradient and called the Thomson's heat. The latter may be positive or negative in contrast with Joule heating, which is invariably positive.

The coefficient

$$\sigma_{\mathrm{Th}} \equiv \frac{\pi}{T} - \frac{\partial \pi}{\partial T} \tag{3.19}$$

is the so-called Thomson coefficient, which, in virtue of $\pi = \varepsilon T$ can still be cast in the form

$$\sigma_{\mathrm{Th}} = -T\,\frac{\partial \varepsilon}{\partial T}. \tag{3.20}$$

This result is referred to as the *first Kelvin relation* and exhibits the strong connection between the various thermoelectric effects.

3.2.2 Efficiency of Thermoelectric Generators

We have shortly presented in Boxes 3.1 and 3.2 some practical applications of thermoelectric materials. Motivated by the developments of research on new materials (Mahan et al. 1997), let us calculate the efficiency of thermoelectric conversion. Consider a single one-dimensional thermoelectric element of length l under steady conditions. The hot side is at a temperature T_{h} (assumed to be the upper side, at $y = l$), and the cold side (the lower side, at $y = 0$) is at temperature T_{c}. An electric current i and a quantity of heat $\dot{Q}$ enter uniformly into the hot side of the element. The efficiency η of the generator is defined as the ratio of electric power output and heat supplied per unit time.

The electric power output is the product of the electric current i and the electric field E, the latter being given by (3.16) with the term $\nabla(\mu_{\mathrm{e}}/z_{\mathrm{e}})$ neglected:

$$P_{\mathrm{el}} = iE = i\int_0^l \varepsilon(T)(\mathrm{d}T/\mathrm{d}y)\mathrm{d}y - i^2\int_0^l r(T)\,\mathrm{d}y. \tag{3.21}$$

The first term is the Seebeck one and the second one describes Joule's dissipation. The sign minus in the second term of the right-hand side arises because in the present example ∇T and $\boldsymbol{i}$ have opposite directions. The total heat flux $\dot{Q}$ supplied per unit time at the hot side is the sum of the purely conductive contribution given by Fourier's law and the Peltier term πi from (3.15) when the term $\mu_{\mathrm{e}}/z_{\mathrm{e}}$ is again neglected. Making use of the second Kelvin relation (3.17), one may write

$$\dot{Q} = \lambda_{\mathrm{h}}(T_{\mathrm{h}} - T_{\mathrm{c}})l^{-1} + iT_{\mathrm{h}}\varepsilon_{\mathrm{h}}, \tag{3.22}$$

where subscript "h" means that the mentioned quantities are evaluated at $T = T_{\mathrm{h}}$, in addition, both $\dot{Q}$ and i are supposed to be directed downwards.

The efficiency of the thermoelectric generator is given by

$$\eta \equiv \frac{P_{\mathrm{el}}}{\dot{Q}} \approx \frac{i\varepsilon(T_{\mathrm{h}} - T_{\mathrm{c}}) - i^2 rl}{\lambda(T_{\mathrm{h}} - T_{\mathrm{c}})l^{-1} + iT_{\mathrm{h}}\varepsilon}. \tag{3.23}$$

If the difference of temperatures is not very important, in such a way that the generator may be considered as almost homogeneous, the quantities λ, ε, and r may be taken as constants, as it has been done in (3.23).

It is usual to express the efficiency η as the product of Carnot's efficiency $\eta_{\mathrm{Carnot}} \equiv (T_{\mathrm{h}} - T_{\mathrm{c}})/T_{\mathrm{h}}$ and a so-called reduced efficiency η_{r}, i.e. $\eta = \eta_{\mathrm{Carnot}}\eta_{\mathrm{r}}$. We therefore write (3.23) as

$$\eta = \frac{T_{\mathrm{h}} - T_{\mathrm{c}}}{T_{\mathrm{h}}} \frac{i\varepsilon - \frac{ri^2 l}{T_{\mathrm{h}} - T_{\mathrm{c}}}}{\lambda \frac{T_{\mathrm{h}} - T_{\mathrm{c}}}{lT_{\mathrm{h}}} + i\varepsilon}. \tag{3.24}$$

The reduced efficiency may still be expressed in terms of the ratio of fluxes $x = i/[\lambda(T_{\mathrm{h}} - T_{\mathrm{c}})l^{-1}]$ as

$$\eta_{\mathrm{r}}(x) = \frac{\varepsilon x - \lambda r x^2}{T_{\mathrm{h}}^{-1} + \varepsilon x} = \frac{\varepsilon x[1 - (x\varepsilon/Z)]}{T_{\mathrm{h}}^{-1} + \varepsilon x}, \tag{3.25}$$

where Z is the so-called thermoelectric figure of merit

$$Z = \frac{\varepsilon^2}{r\lambda}, \tag{3.26}$$

which has the dimension of the reciprocal of temperature, and depends only on the transport coefficients, generally function of the temperature. Expression (3.26) of the figure of merit reflects the property that efficiency is enhanced by high values of the Seebeck coefficient ε, low values of the electric resistivity r, and heat conductivity λ. High values of ε contribute to a strong coupling between heat and electric current and low values of r and λ minimize the losses due to Joule's dissipation and to heat conduction.

In the numerator of (3.25) are competing the thermoelectric effect with Joule dissipation. Since the first one is linear in the current, or x, and the second one is non-linear in the same quantity, there exists necessarily an optimal value of the reduced efficiency. The optimum ratio of the fluxes x_{opt} corresponding to the maximum value of η_{r} is obtained from the condition $\mathrm{d}\eta_{\mathrm{r}}(x)/\mathrm{d}x = 0$; the result is

$$x_{\mathrm{opt}} = \frac{(1 + ZT)^{1/2} - 1}{\varepsilon T}. \tag{3.27}$$

The maximum efficiency is then

$$\eta_{\max} = \frac{T_{\mathrm{h}} - T_{\mathrm{c}}}{T_{\mathrm{h}}} \frac{2 + ZT - 2\sqrt{1 + ZT}}{ZT}. \tag{3.28}$$

If the ratio x is significantly different from the optimum value (3.27), the material is not efficiently converting heat energy into electric energy. Once x is optimized, the optimal current may be found, for a given value of the temperature gradient. Note that for $ZT \gg 1$, one finds back Carnot's but materials currently used in thermoelectric devices have relatively low values of ZT, between 0.4 and 1.5, so that in reality, the maximum Carnot value is far from being attained.

In general, the temperature difference between the two sides of the generator will be so high that the assumption of homogeneity used in (3.25)–(3.28) is not tenable. In this case, one may consider the thermoelectric device as formed by a series of small quasi-homogeneous elements, at different average temperatures. For instance, for two elements thermally in series, the combined efficiency is (see Problem 3.9),

$$\eta = \frac{P_{\mathrm{el},1} + P_{\mathrm{el},2}}{\dot{Q}_1} = 1 - (1 - \eta_1)(1 - \eta_2). \tag{3.29a}$$

In the continuum limit, in which the generator is constituted of many layers at different temperatures, (3.29a) must be replaced by (see Problem 3.9),

$$\eta = 1 - \exp\left[- \int_{T_{\mathrm{c}}}^{T_{\mathrm{h}}} \frac{\eta_{\mathrm{r}}(x, T)}{T} \, \mathrm{d}T \right]. \tag{3.29b}$$

Using this expression, one finds for the total efficiency of the device

$$\eta = 1 - \frac{\varepsilon_{\mathrm{c}} T_{\mathrm{c}} + x_{\mathrm{c}}^{-1}}{\varepsilon_{\mathrm{h}} T_{\mathrm{h}} + x_{\mathrm{h}}^{-1}}. \tag{3.30}$$

In this configuration, it may happen that the optimum current determined by x_{opt} in one segment (for instance, the hot side) is significantly different from the optimum value x_{opt} in another segment (for instance, the cold side); in this case, there will be no suitable current for which both parts of the generator are operating with optimal efficiency. This is a challenge in materials sciences, as x_{opt} is temperature dependent through the transport coefficients of the material (namely Z and ε) and it would be highly desirable to find materials with suitable temperature dependence of these coefficients to optimize the generation.

In terms of the coefficient Z and assuming constant transport coefficients ε, λ, and r, the maximum efficiency of the power generation is given by (see Problem 3.7)

$$\eta_{\max} = \frac{T_{\mathrm{h}} - T_{\mathrm{c}}}{T_{\mathrm{h}}} \frac{(1 + ZT_{\mathrm{av}})^{1/2} - 1}{(1 + ZT_{\mathrm{av}})^{1/2} + (T_{\mathrm{c}}/T_{\mathrm{h}})}, \tag{3.31}$$

where T_{av} is the average temperature $T_{\mathrm{av}} = \frac{1}{2}(T_{\mathrm{h}} + T_{\mathrm{c}})$. When $T_{\mathrm{c}} \approx T_{\mathrm{h}}$, the ratio $T_{\mathrm{c}}/T_{\mathrm{h}}$ in the denominator of (3.31) is close to 1 and the efficiency (3.28) is recovered.

Analogously, the coefficient of performance for refrigeration (i.e. the ratio between the heat extracted per unit time and the electric power consumed by the corresponding engine) is

$$\eta = \frac{\dot{Q}}{P_{\mathrm{el}}} = \frac{T_{\mathrm{c}}}{T_{\mathrm{h}} - T_{\mathrm{c}}} \frac{(1 + ZT_{\mathrm{av}})^{1/2} - (T_{\mathrm{h}}/T_{\mathrm{c}})}{1 + (1 + ZT_{\mathrm{av}})^{1/2}}. \tag{3.32}$$

In many practical situations, two parallel generators are used, one of n-type semiconductors (current carried by electrons, with $\varepsilon < 0$) and another with p-type semiconductors (current brought by holes, with $\varepsilon > 0$). The global efficiency may be derived from (3.23) or (3.31) by using an average for both generators, namely

$$\eta_{\mathrm{n}\|\mathrm{p}} = \frac{\eta_{\mathrm{p}} Q_{\mathrm{p}} + \eta_{\mathrm{n}} Q_{\mathrm{n}}}{Q_{\mathrm{p}} + Q_{\mathrm{n}}}, \tag{3.33}$$

where Q_{p} and Q_{n} are the amounts of heat supplied to the hot side of the p and n elements per unit time.

3.3 Thermodiffusion: Coupling of Heat and Mass Transport

By thermodiffusion is meant the coupling between heat and matter transport in binary or multi-component mixtures. In the case of a binary mixture, a natural choice of the state variables is ρ_1, ρ_2 (individual mass densities), $\boldsymbol{v}$ (barycentric velocity), and u (internal energy), but a more convenient choice is $\rho(= \rho_1 + \rho_2)$, $c_1 = (\rho_1/\rho)$, $\boldsymbol{v}$, and u. In the classical theory of irreversible processes, one is more interested by the behaviour of the barycentric velocity than by the individual velocities of the components and this is the reason why only the velocity $\boldsymbol{v}$, and not the individual velocities $\boldsymbol{v}_1$ and $\boldsymbol{v}_2$, figures among the space of variables. In absence of chemical reactions, viscosity, external body forces and energy sources, the corresponding evolution equations are given by

$$\frac{\mathrm{d}\rho}{\mathrm{d}t} = -\rho \nabla \cdot \boldsymbol{v}, \tag{3.34}$$

$$\rho \frac{\mathrm{d}c_1}{\mathrm{d}t} = -\nabla \cdot \boldsymbol{J}_1, \tag{3.35}$$

$$\rho \frac{\mathrm{d}\boldsymbol{v}}{\mathrm{d}t} = -\nabla p, \tag{3.36}$$

$$\rho \frac{\mathrm{d}u}{\mathrm{d}t} = -\nabla \cdot \boldsymbol{q} - p \nabla \cdot \boldsymbol{v}. \tag{3.37}$$

Moreover, we assume that the system is in mechanical equilibrium with zero barycentric velocity and zero acceleration, which is the case for mixtures confined in closed vessels. It follows then from (3.34) and (3.36) that the total mass density and the pressure remain uniform throughout the system and that the last term in (3.37) vanishes. Substituting the balance equations of mass fraction (3.35) and internal energy (3.37) in the Gibbs' equation

$$\frac{\mathrm{d}s}{\mathrm{d}t} = T^{-1}\frac{\mathrm{d}u}{\mathrm{d}t} - T^{-1}(\bar{\mu}_1 - \bar{\mu}_2)\frac{\mathrm{d}c_1}{\mathrm{d}t}, \tag{3.38}$$

where use has been made of $\mathrm{d}c_2 = -\mathrm{d}c_1$, results in the following balance of entropy

$$\rho\frac{\mathrm{d}s}{\mathrm{d}t} = -\nabla\cdot\{T^{-1}[\boldsymbol{q}-(\bar{\mu}_1-\bar{\mu}_2)\boldsymbol{J}_1]\}+\boldsymbol{q}\cdot\nabla T^{-1}-\boldsymbol{J}_1\cdot\nabla[T^{-1}(\bar{\mu}_1-\bar{\mu}_2)], \tag{3.39}$$

with the entropy flux given by

$$\boldsymbol{J}^s = T^{-1}[\boldsymbol{q} - (\bar{\mu}_1 - \bar{\mu}_2)\boldsymbol{J}_1], \tag{3.40}$$

and the rate of entropy production by

$$\sigma^s = \boldsymbol{q}\cdot\nabla T^{-1} - \boldsymbol{J}_1\cdot\nabla[T^{-1}(\bar{\mu}_1 - \bar{\mu}_2)]. \tag{3.41}$$

With the help of the Gibbs–Duhem's relation

$$c_1(\nabla\bar{\mu}_1)_{T,p} + c_2(\nabla\bar{\mu}_2)_{T,p} = 0, \tag{3.42}$$

where subscripts T and p indicate that differentiation is taken at constant T and p, and the classical result of equilibrium thermodynamics

$$T\nabla(T^{-1}\bar{\mu}_k) = -h_k T^{-1}(\nabla T) + (\nabla\bar{\mu}_k)_T, \tag{3.43}$$

with h_k the partial specific enthalpy of substance $k(k = 1, 2)$, we are able to eliminate the chemical potential from (3.41) of σ^s, which finally reads as

$$\sigma^s = -\boldsymbol{q}'\cdot\frac{\nabla T}{T^2} - \frac{\mu_{11}}{Tc_2}\boldsymbol{J}_1\cdot\nabla c_1. \tag{3.44}$$

The new heat flux $\boldsymbol{q}'$ is defined by $\boldsymbol{q}' = \boldsymbol{q} - (h_1 - h_2)\boldsymbol{J}_1$, and it is equal to the difference between the total flux of heat and the transfer of heat due to diffusion while the quantity μ_{11} stands for $\mu_{11} = (\partial\bar{\mu}_1/\partial c_1)_{T,p}$. The derivation of (3.44) exhibits clearly the property that the entropy production is a bilinear expression in the thermodynamic fluxes $\boldsymbol{q}'$ and $\boldsymbol{J}_1$ and forces taking the form of gradients of intensive variables, easily accessible to direct measurements. It is obvious that when the mixture reaches thermodynamic equilibrium, the heat and mass flows as well as the temperature and mass fraction gradients vanish. Expression (3.44) of σ^s suggests writing the following phenomenological relations between fluxes and forces:

$$\boldsymbol{q}' = -L_{qq}\frac{\nabla T}{T^2} - L_{q1}\frac{\mu_{11}}{Tc_2}\nabla c_1, \quad (3.45)$$

$$\boldsymbol{J}_1 = -L_{1q}\frac{\nabla T}{T^2} - L_{11}\frac{\mu_{11}}{Tc_2}\nabla c_1, \quad (3.46)$$

with the Onsager's reciprocal relation $L_{q1} = L_{1q}$ and the following inequalities resulting from the positiveness of entropy production:

$$L_{qq} > 0, \quad L_{11} > 0, \quad L_{qq}L_{11} - L_{q1}L_{1q} > 0. \quad (3.47)$$

After introducing the following identifications:

$$\frac{L_{qq}}{T^2} = \lambda \quad \text{(heat conductivity)},$$

$$\frac{L_{11}\mu_{11}}{\rho c_2 T} = D \quad \text{(diffusion coefficient)},$$

$$\frac{L_{q1}}{\rho c_1 c_2 T^2} = D_{\mathrm{F}} \quad \text{(Dufour coefficient)},$$

$$\frac{L_{1q}}{\rho c_1 c_2 T^2} = D_{\mathrm{T}} \quad \text{(thermal diffusion coefficient)},$$

the phenomenological laws take the form

$$\boldsymbol{q}' = -\lambda\nabla T - \rho T\mu_{11}c_1 D_{\mathrm{F}}\nabla c_1, \quad (3.48)$$

$$\boldsymbol{J}_1 = -\rho c_1 c_2 D_{\mathrm{T}}\nabla T - \rho D\nabla c_1. \quad (3.49)$$

Inequalities (3.47) imply in particular that $\lambda > 0$ and $D > 0$ while from Onsager's relation is inferred that

$$D_{\mathrm{F}} = D_{\mathrm{T}}. \quad (3.50)$$

This last result is a confirmation of an earlier result established by Stefan at the end of the nineteenth century. Starting from the law of conservation of momentum, Stefan was indeed able to demonstrate the above equality at least in the case of binary mixtures. By setting the gradient of the mass fraction equal to zero, (3.48) is identical to Fourier's equation so that λ can be identified with the heat conductivity coefficient. Similarly, relation (3.49) reduces to Fick's law of diffusion when temperature is uniform and therefore D represents the diffusion coefficient; in general, the phenomenological coefficients in (3.48) and (3.49) are not constant.

In multi-component mixtures, the mass flux of substance i is a linear function of not only ∇c_i but also of all the other mass fractions gradients $\nabla c_j (j \neq i)$. Such "diffusion drag" forces have been invoked to interpret some biophysical effects and play a role in the processes of separation of isotopes (see Box 3.3).

Box 3.3 The Soret's Effect and Isotope Separation

Thermal diffusion is exploited to separate materials of different molecular mass. If a fluid system is composed of two kinds of molecules of different molecular weight, and if it is submitted to a temperature gradient, the lighter molecules will accumulate near the hot wall and the heavier ones near the cold wall. This property was used in the 1940s for the separation of ^{235}U and ^{238}U isotopes in solutions of uranium hexafluoride, in the Manhattan project, leading to the first atomic bomb. Usually this process is carried out in tall and narrow vertical columns, where convection effects reinforce the separation induced by thermal diffusion: the light molecules near the hot walls have an ascending motion, whereas the heavy molecules near the cold wall sunk towards the lowest regions. This process accumulates the lightest isotope in the highest regions, from where it may be extracted. This process is simple but its consumption of energy is high, and therefore it has been substituted by other methods. However, it is still being used in heavy water enrichment, or in other processes of separation of light atoms with the purpose of, for instance, to generate carbide layers on steel, alloys or cements, thus hardening the surface and making it more resistant to wear and corrosion. Soret's effect plays also a role in the structure of flames and in polymer characterization. More recently, high temperature gradients have been produced by means of laser beams rather than by heating uniformly the walls of the container.

The Dufour's effect, the reciprocal of thermal diffusion, has not so many industrial applications. It plays nevertheless a non-negligible role in some natural processes as heat transport in the high atmosphere and in the soil under isothermal conditions but under a gradient of moisture.

The coefficient D_{T} in (3.49) is typical of thermal diffusion, i.e. the flow of matter caused by a temperature difference; such an effect is referred to as the *Soret's effect* in liquids with the quotient D_{T}/D called the Soret coefficient. The reciprocal effect, i.e. the flow of heat caused by a gradient of concentration as evidenced by (3.48) is the *Dufour's effect.* It should be observed that the cross-coefficients D_{T} and D_{F} are much smaller than the direct coefficients like the heat conductivity λ and the diffusion coefficient D. The latter turns out to be of the order of $10^{-8}\,\mathrm{m}^2\,\mathrm{s}^{-1}$ in liquids and $10^{-5}\,\mathrm{m}^2\,\mathrm{s}^{-1}$ in gases while the coefficient of thermal diffusion D_{T} varies between 10^{-12} and $10^{-14}\,\mathrm{m}^2\,\mathrm{s}^{-1}\,\mathrm{K}^{-1}$ in liquids and from 10^{-8} to $10^{-12}\,\mathrm{m}^2\,\mathrm{s}^{-1}\,\mathrm{K}^{-1}$ in gases. The Soret's effect is mainly observed in oceanography while the Dufour's effect, which is negligible in liquids, has been detected in the high atmosphere. The smallness of the coupling effects is the reason why they are hard to be observed and measured with accurateness.

Defining a stationary state by the absence of matter flow ($\boldsymbol{J}_1 = 0$), it turns out from (3.49) that

$$(\Delta T)_{\mathrm{st}} = -\frac{D}{D_{\mathrm{T}} c_1 c_2} (\Delta c_1)_{\mathrm{st}}, \tag{3.51}$$

which indicates that a difference of concentration is able to generate a temperature difference, called the *osmotic temperature*. This is typically an irreversible effect because the corresponding entropy production is non-zero as directly seen from (3.41). This effect should not be confused with the *osmotic pressure*, which expresses that a difference of concentration between two reservoirs kept at the same uniform temperature but separated by a semi-permeable membrane gives raise to a pressure drop, called the osmotic pressure. The latter is a pure equilibrium effect resulting from the property that, at equilibrium, the chemical potential $\bar{\mu}(T, p, c_1)$ takes the same value in both reservoirs so that $(\Delta\bar{\mu})_T = \Delta p/\rho_1 + \mu_{11}\Delta c_1 = 0$, and

$$\Delta p = -\rho_1 \mu_{11} \Delta c_1, \tag{3.52}$$

with ρ_1 the specific mass of the species crossing the membrane; it is directly checked that in the present situation, the entropy production (3.41) is indeed equal to zero.

The phenomena studied in this section are readily generalized to multi-component electrically charged systems, like electrolytes.

3.4 Diffusion Through a Membrane

The importance of transport of matter through membranes in the life of cells and tissues justify that we spend some time to discuss the problem. In biological membranes, one distinguishes generally between two modes of transport: purely *passive* transport due to a pressure gradient or a mass concentration gradient and *active* transport involving ionic species, electrical currents, and chemical reactions. Here we focus on some aspects of passive transport. Our objective is to present a simplified analysis by using a minimum number of notions and parameters; in that respect, thermal effects will be ignored but even so, the subject keeps an undeniable utility.

We consider the simple arrangement formed by two compartments I and II separated by a homogeneous membrane of thickness Δl, say of the order of 100 μm. Each compartment is filled with a binary solution consisting of a solvent 1 and a solute 2 (see Fig. 3.3).

The membrane is assumed to divide the system in two discontinuous subsystems that are considered as homogeneous.

3.4.1 Entropy Production

In absence of thermal gradients, it is inferred from (3.41) that the rate of dissipation, measured per unit volume of the membrane, will take the form

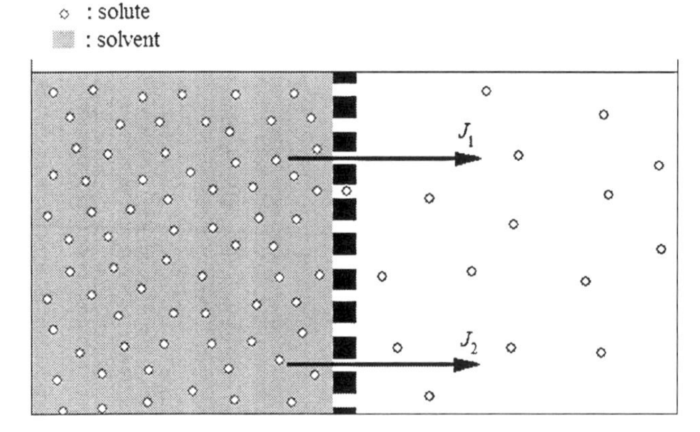

Fig. 3.3 System under study consisting in a membrane separating a binary solution

$$T\sigma^s = -\boldsymbol{J}_1 \cdot \nabla\bar{\mu}_1 - \boldsymbol{J}_2 \cdot \nabla\bar{\mu}_2. \tag{3.53}$$

After integration over the thickness Δl, the rate of dissipation per unit surface, denoted as Φ, can be written as

$$\Phi = -J_1\Delta\bar{\mu}_1 - J_2\Delta\bar{\mu}_2, \tag{3.54}$$

where $\Delta\bar{\mu}_i$ designates the difference of chemical potential of species i across the membrane, J_1 and J_2 are the flows of solvent and solute, respectively. Instead of working with J_1 and J_2, it is more convenient to introduce the total volume flow J_V across the membrane and the relative velocity J_D of the solute with respect to the solvent, defined, respectively, by

$$J_\mathrm{V} = \bar{v}_1 J_1 + \bar{v}_2 J_2, \tag{3.55}$$

$$J_\mathrm{D} = v_2 - v_1, \tag{3.56}$$

the quantities $\bar{v}_1$ and $\bar{v}_2$ stand for the partial specific volumes of the solvent and the solute, v_1 and v_2 are their respective velocities given by $v_1 = \bar{v}_1 J_1$ and $v_2 = \bar{v}_2 J_2$. With the above choice of variables, (3.54) reads as (Katchalsky and Curran 1965; Caplan and Essig 1983)

$$\Phi = -J_\mathrm{V}\Delta p - J_\mathrm{D}\Delta\pi, \tag{3.57}$$

where $\Delta p = p_\mathrm{I} - p_\mathrm{II}$ and $\Delta\pi = c_2(\Delta\bar{\mu}_2)_p$ is the osmotic pressure; the quantity $(\Delta\bar{\mu}_2)_p$ is that part of the chemical potential depending only on the concentration and defined from $\Delta\bar{\mu}_2 = V_2\Delta p + (\Delta\bar{\mu}_2)_p$, c_2 is the number of moles of the solute per unit volume. For ideal solutions, one has $(\Delta\bar{\mu}_2)_p = RT\Delta c_2$.

3.4.2 Phenomenological Relations

By assuming linear relations between thermodynamic fluxes and forces, one has

$$J_{\mathrm{V}} = L_{\mathrm{VV}}\Delta p + L_{\mathrm{VD}}\Delta\pi, \tag{3.58}$$

$$J_{\mathrm{D}} = L_{\mathrm{DV}}\Delta p + L_{\mathrm{DD}}\Delta\pi. \tag{3.59}$$

The advantage of (3.58) and (3.59) is that they are given in terms of parameters that are directly accessible to experiments. The corresponding Onsager's relation is $L_{\mathrm{VD}} = L_{\mathrm{DV}}$, whose main merit is to reduce the number of parameters from four to three.

To better apprehend the physical meaning of the phenomenological coefficients L_{VV}, L_{DV}, L_{DD}, and L_{VD}, let us examine some particular experimental situations. First consider the case wherein the concentration of the solute is the same on both sides of the membrane such that $\Delta\pi = 0$. If a pressure difference Δp is applied, one will observe according to (3.58) a volume flow proportional to Δp; the proportionality coefficient L_{VV} is called the *mechanical filtration coefficient* of the membrane: it is defined as the volume flow produced by a unit pressure difference between the two faces of the membrane. A further look on relation (3.59) indicates that even in absence of a concentration difference ($\Delta\pi = 0$), there will be a diffusion flow $J_{\mathrm{D}} = L_{\mathrm{DV}}\Delta p$ caused by the pressure difference Δp. This phenomenon is known in colloid chemistry under the name of ultrafiltration and the coefficient L_{DV} is the *ultrafiltration coefficient*. An alternative possibility is to impose $\Delta p = 0$ but different solute concentrations in compartments I and II. In virtue of (3.59), the osmotic difference $\Delta\pi$ will produce a flow of diffusion $J_{\mathrm{D}} = L_{\mathrm{DD}}\Delta\pi$ and L_{DD} is identified as the *permeability coefficient*: it is the diffusional mobility induced by a unit osmotic pressure $\Delta\pi = 1$. Another effect related to (3.58) is the occurrence of a volume flow $J_{\mathrm{V}} = L_{\mathrm{VD}}\Delta\pi$ caused by a difference of osmotic pressure at uniform hydrostatic pressure: the coupling coefficient L_{VD} is referred to as the *coefficient of osmotic flow.*

The above discussion has clearly shown the importance of the coupling coefficient $L_{\mathrm{DV}} = L_{\mathrm{VD}}$; by ignoring it one should miss significant features about motions across membranes. The importance of this coefficient is still displayed by the osmotic pressure experiment illustrated by Fig. 3.4.

The two phases I (solvent + solute) and II (solvent alone) are separated by a semi-permeable membrane only permeable to the solvent. The height of the solution in the capillary tube gives a measure of the final pressure difference obtained when the volume flow J_{V} vanishes, indeed from (3.58) one obtains

$$(\Delta p)_{J_{\mathrm{V}}=0} = -\frac{L_{\mathrm{VD}}}{L_{\mathrm{VV}}}\Delta\pi. \tag{3.60}$$

This result indicates that, contrary to what is sometimes claimed, Δp is not a measure of the osmotic pressure, this is only true if $L_{\mathrm{VD}} = -L_{\mathrm{VV}}$.

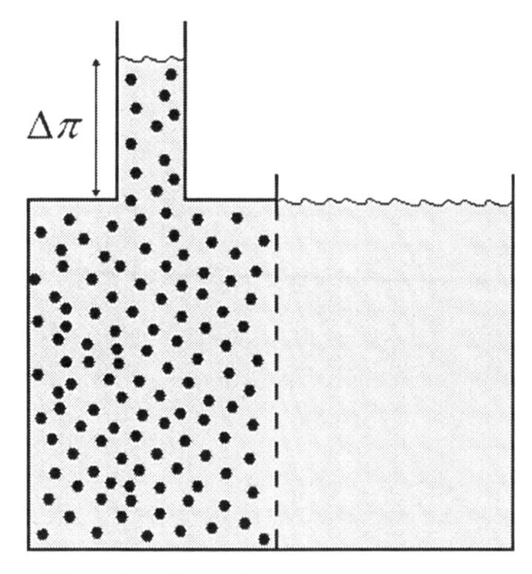

Fig. 3.4 Osmotic pressure experiment

This condition is met by so-called *ideal* semi-permeable membranes whose property is to forbid the transport of solute whatever the values of Δp and $\Delta\pi$. For membranes, which are permeable to the solute, it is experimentally found that

$$r \equiv -\frac{L_{\mathrm{VD}}}{L_{\mathrm{VV}}} < 1. \tag{3.61}$$

This ratio that is called the *reflection coefficient* tends to zero for *selective* membranes, like porous gas filters, and has been proposed to act as a measure of the *selectivity* of the membrane. For $r = 1$ (ideal membranes), all the solute is reflected by the membrane, for $r < 1$, some quantity of solute crosses the membrane, while for $r = 0$ the membrane is completely permeable to the solute. To clarify the notion of membrane selectivity, let us go back to the situation described by $\Delta\pi = 0$. In virtue of (3.58) and (3.59), one has

$$r = -\frac{L_{\mathrm{VD}}}{L_{\mathrm{VV}}} = -\left(\frac{J_{\mathrm{D}}}{J_{\mathrm{V}}}\right)_{\Delta\pi=0}, \tag{3.62}$$

or, in terms of the velocities introduced in relation (3.56),

$$\left(\frac{2v_2}{v_1 + v_2}\right)_{\Delta\pi=0} = 1 - r. \tag{3.63}$$

For an ideal semi-permeable membrane ($r = 1$), one has $v_2 = 0$ and the solute will not cross the membrane; for $r = 0$ ($v_1 = v_2$) the membrane is not selective and allows the passage of both the solute and the solvent; for negative values of $r(v_2 > v_1)$, the velocity of the solute is greater than that of the solvent and this is known as negative anomalous osmosis, which is a characteristic of the transport of electrolytes across charged membranes.

Table 3.1 Phenomenological coefficients for two biological membranes

Membrane	Solute	Solvent	ω (10^{-15} mol dyn^{-1} s^{-1})	r	L_{VV} (10^{-11} cm^3 dyn^{-1} s^{-1})
Human blood cell	Methanol	Water	122	–	–
	Urea	Water	17	0.62	0.92
Toad skin	Acetamide	Water	4×10^{-3}	0.89	0.4
	Thiourea	Water	5.7×10^{-4}	0.98	1.1

A final parameter of interest, both in synthetic and biological membranes, is the *solute permeability coefficient*

$$\omega = \frac{c_2}{L_{VV}}(L_{VV}L_{DD} - L_{VD}^2). \tag{3.64}$$

For ideal semi-permeable membranes for which $L_{VD} = -L_{VV} = -L_{DD}$, one has $\omega = 0$, and for non-selective membranes ($r = 0$), it is found that $\omega = c_2 L_{DD}$.

The interest of irreversible thermodynamics is to show clearly that three parameters are sufficient to describe transport of matter across membrane and to provide the relationships between the various coefficients characterizing a semi-permeable membrane. In Table 3.1 are reported some values of these coefficients for two different biological membranes (Katchalsky and Curran 1965).

3.5 Problems

3.1. *Entropy flux and entropy production.* Determine (3.11) and (3.12) of the entropy flux and the entropy production in the problem of thermoelectricity.

3.2. *Onsager's reciprocal relations.* In presence of a magnetic field, Onsager's relations can be written as $\mathbf{L}(\boldsymbol{\mathcal{H}}) = \mathbf{L}^{\mathrm{T}}(-\boldsymbol{\mathcal{H}})$. Decomposing $\mathbf{L}$ in a symmetric and an antisymmetric part $\mathbf{L} = \mathbf{L}^{\mathrm{s}} + \mathbf{L}^{\mathrm{a}}$, show that $\mathbf{L}^{\mathrm{s}}(\boldsymbol{\mathcal{H}}) = \mathbf{L}^{\mathrm{s}}(-\boldsymbol{\mathcal{H}})$ and $\mathbf{L}^{\mathrm{a}}(\boldsymbol{\mathcal{H}}) = -\mathbf{L}^{\mathrm{a}}(-\boldsymbol{\mathcal{H}})$.

3.3. *Thermoelectric effects.* The Peltier coefficient of a couple Cu–Ni is $\pi_{\mathrm{Cu-Ni}} \approx -5.08\,\mathrm{mV}$ at 273 K, $\pi_{\mathrm{Cu-Ni}} \approx -6.05\,\mathrm{mV}$ at 295 K, and $\pi_{\mathrm{Cu-Ni}} \approx -9.10\,\mathrm{mV}$ at 373 K. Evaluate (a) the heat exchanged in the junction by Peltier's effect when an electric current of 10^{-2} A flows from Cu to Ni at 295 K; (b) idem when the current flows from Ni to Cu. (c) The two junctions of a thermocouple made of Cu and Ni are kept at 305 and 285 K, respectively; by using the Thomson relation, determine the Seebeck coefficient and estimate the electromotive force developed by the thermocouple.

3.4. *Thermocouple.* The electromotive force $\Delta\xi$ of a Cu–Fe thermocouple is given by $\Delta\xi(\mu V) \approx -13.4\Delta T - 0.01375(\Delta T)^2$, where $\Delta T = T_R - T$, the reference temperature being $T_R = 273\,\mathrm{K}$. (a) Evaluate the Peltier heat exchanged in a Cu–Fe junction at 298 K when an electric current of 10^{-3} A flows from Cu to Fe. (b) By using the Thomson relation, evaluate the difference of Thomson coefficients of these two metals at 298 K.

3.5. *Evaluation of the maximum efficiency of a thermoelectric element.* Assume a finite temperature difference and constant transport coefficients. (a) Starting from the definition of the ratio of fluxes $x \equiv i/\lambda\nabla T$, and from the steady state form of (3.18), prove that

$$\frac{\mathrm{d}x^{-1}}{\mathrm{d}T} = T\frac{\mathrm{d}\varepsilon}{\mathrm{d}T} - x\lambda r.$$

(b) Taking into account this relation, and for ε independent of T, show that the values of x at the cold and hot boundaries of the systems, x_c and x_h, respectively, may be written in terms of $x_{\mathrm{opt,av}}$ (Snyder and Ursell 2003)

$$x_h^{-1} = x_{\mathrm{opt,av}}^{-1} - \tfrac{1}{2}\Delta T(\lambda r)x_{\mathrm{opt,av}},$$
$$x_c^{-1} = x_{\mathrm{opt,av}}^{-1} + \tfrac{1}{2}\Delta T(\lambda r)x_{\mathrm{opt,av}},$$

where $x_{\mathrm{opt,av}}$ is the value for x_{opt} found in (3.27) for the average temperature $T_{\mathrm{av}} = \frac{1}{2}(T_h + T_c)$. (c) Introducing these values in (3.30), show finally (3.31), i.e.

$$\eta = \frac{T_h - T_c}{T_h}\frac{(1 + ZT_{\mathrm{av}})^{1/2} - 1}{(1 + ZT_{\mathrm{av}})^{1/2} + (T_c/T_h)}.$$

3.6. *The figure of merit of a thermoelectric material.* The figure of merit for thermoelectric materials is defined by $Z \equiv \varepsilon^2/\lambda r$, where ε is the Seebeck coefficient, r is the electrical resistivity, and λ is the thermal conductivity. (a) Check that the dimension of this combination is the inverse of temperature. (b) The figure of merit of a junction of two thermoelectric materials in a thermocouple is defined as

$$Z_{\mathrm{tc}} \equiv \frac{\varepsilon_{\mathrm{AB}}^2}{[(\lambda/\sigma)_A^{1/2} + (\lambda/\sigma)_B^{1/2}]^2}.$$

Show that the efficiency of the thermocouple, defined as the ratio of the electric power delivered and the rate of heat supplied to the hot junction is

$$\eta_{\mathrm{tc}} = \frac{T_1 - T_2}{\frac{3T_1+T_2}{2} + \frac{4}{Z_{\mathrm{tc}}}}.$$

3.7. *Heat engines in series.* Consider two Carnot heat engines in series: the first one works between heat reservoirs at T_1 and T_2 and the second one between reservoirs at T_2 and T_3, assume in addition that the whole amount of heat Q_2 delivered by the first engine to the reservoir T_2 is transferred to the second engine. (a) Show that one recovers (3.31), namely

$$\eta = \frac{P_{\mathrm{el},1} + P_{\mathrm{el},2}}{Q_1} = 1 - (1 - \eta_1)(1 - \eta_2).$$

(b) In (3.29b) it has been considered the continuum limit of high number of Carnot engines working between reservoirs in series at temperatures $T_i + \mathrm{d}T$ and T_i, with T_i the temperature of the cold reservoir corresponding to ith Carnot engine, T_i ranging from T_c to T_h. Show that the previous expression leads to (3.29b) by writing

$$\eta = 1 - \prod_{i=1}^{\infty} (1 - \eta_i) \approx 1 - \exp\left[-\int_{T_\mathrm{c}}^{T_\mathrm{h}} \frac{\eta(T)}{T} \, \mathrm{d}T \right].$$

3.8. *Thermophoresis.* Thermophoresis in a quiescent fluid is described by the phenomenological equation

$$J = -D[\nabla c + S_\mathrm{T} c(1 - c)\nabla T],$$

where D is the diffusion coefficient and $S_\mathrm{T} \equiv D_\mathrm{T}/D$ is the Soret coefficient, D_T being the thermal diffusion coefficient. (a) Show that for concentrations much less than unity, the concentration of the solute at temperature T is given by

$$c = c_0 \exp[-S_\mathrm{T}(T - T_0)],$$

where c_0 is the concentration at temperature T_0. For a DNA sample, it was found that $S_\mathrm{T} \approx 0.14\,\mathrm{K}^{-1}$ around $T = 297\,\mathrm{K}$ (Braun and Libchaber 2002). (b) Calculate c/c_0 for two regions with a temperature difference $\Delta T = 2.5\,\mathrm{K}$.

3.9. *Cross-effects in membranes.* Assume that the flows of water J_1 and solute J_2 across a membrane are, respectively, expressed as

$$J_1 = L_{11}\Delta p + L_{12}\Delta c_2,$$
$$J_2 = L_{21}\Delta p + L_{22}\Delta c_2,$$

where Δp and Δc_2 are, respectively, the differences of pressure and solute concentration between both sides of the membrane. Is the Onsager's reciprocal relation $L_{12} = L_{21}$ applicable? Explain.

3.10. *Transport of charged ions across membranes. The Nernst equation.* Transport of charged ions, mainly H^+, Na^+, K^+, Ca^{2+}, and Cl^-, across membranes plays a crucial role in many biological processes. Consider a membrane with both sides at concentrations c_in, c_out, and voltages $\mathcal{V}_\mathrm{in}$, $\mathcal{V}_\mathrm{out}$. An argument similar to that leading to (3.58) and (3.59) yields, in the isothermal case, the following result for the diffusion flux J (number of ions which cross the membrane per unit area and unit time)

$$\boldsymbol{J} = -\tilde{D}(\nabla\mu + q\nabla\mathcal{V}),$$

q designating the electrical charge. (a) Show that the voltage difference for which both sides of the membrane are at equilibrium ($J = 0$) with respect to the transport of a given ionic species i (assumed to behave as an ideal substance, i.e. $\mu = \mu^{(0)} + k_{\mathrm{B}}T \ln c$) is given by

$$(\mathcal{V}_{\mathrm{in}} - \mathcal{V}_{\mathrm{out}})_{\mathrm{eq}} = \frac{k_{\mathrm{B}}T}{q} \ln \frac{c_{\mathrm{out}}}{c_{\mathrm{in}}}.$$

This equation is known as Nernst equation. (b) Assume that the membrane is immersed in solutions of NaCl–water with concentration 1 M (where M stands for mole per litre) and 0.1 M and that only Na^+ may cross the membrane. Calculate the difference of voltage between both sides of the membrane at equilibrium. (c) In typical biological cells at rest, the K^+ concentration are $c_{\mathrm{in}} = 150\,\mathrm{mM}$, $c_{\mathrm{out}} = 5\,\mathrm{mM}$, and the voltages are $V_{\mathrm{out}} = 0\,\mathrm{mV}$, $V_{\mathrm{in}} = -70\,\mathrm{mV}$. Evaluate the Nernst potential for K^+ at $T = 310\,\mathrm{K}$, and determine whether K^+ will move towards or outwards the cell.

3.11. *Transport of charged ions across membranes: the Goldmann equation.* In the derivation of the Nernst equation, it is assumed that the membrane is only permeable to one ionic species. In actual situations, several ionic species may permeate through the membrane. If one considers the transport of several ionic species as K^+, Na^+, and Cl^-, which are the most usual ones, the equilibrium voltage difference is given by the Goldmann equation

$$(\mathcal{V}_{\mathrm{in}} - \mathcal{V}_{\mathrm{out}})_{\mathrm{eq}} = \frac{k_{\mathrm{B}}T}{q} \ln \frac{P_{\mathrm{K}} c_{\mathrm{K,out}} + P_{\mathrm{Na}} c_{\mathrm{Na,out}} + P_{\mathrm{Cl}} c_{\mathrm{Cl,in}}}{P_{\mathrm{K}} c_{\mathrm{K,in}} + P_{\mathrm{Na}} c_{\mathrm{Na,in}} + P_{\mathrm{Cl}} c_{\mathrm{Cl,out}}},$$

where P_{K}, P_{Na}, and P_{Cl} refer to the relative permeabilities of the corresponding ions across the membrane. Note that when only one species crosses the membrane (i.e. when the permeabilities vanish for two of the ionic species), the Goldmann equation reduces to the Nernst relation. (a) Derive the Goldmann equation. *Hint*: Take into account the electroneutrality condition. (b) It is asked why the Cl concentrations appear in the equation in a different way as the concentrations of the other two ions, namely c_{in} instead of c_{out} and vice versa. (c) In the axon at rest, i.e. in the long cylindrical terminal of the neurons along which the output electrical signals may propagate, the permeabilities are $P_{\mathrm{K}} \approx 25P_{\mathrm{Na}} \approx 2P_{\mathrm{Cl}}$, whereas at the peak of the action potential (the electric nervous signal), $P_{\mathrm{K}} \approx 0.05P_{\mathrm{Na}} \approx 2P_{\mathrm{Cl}}$. The dramatic increase in the sodium permeability is due to the opening of sodium channels when the voltage difference is less than some critical value. Evaluate the Goldmann potential in both situations, by taking for the concentrations: $c_{\mathrm{K,in}} = 400\,\mathrm{mM}$, $c_{\mathrm{K,out}} = 20\,\mathrm{mM}$, $c_{\mathrm{Na,in}} = 50\,\mathrm{mM}$, $c_{\mathrm{Na,out}} = 440\,\mathrm{mM}$, and $c_{\mathrm{Cl,in}} = 50\,\mathrm{mM}$, $c_{\mathrm{Cl,out}} = 550\,\mathrm{mM}$.

Chapter 4
Chemical Reactions and Molecular Machines

Efficiency of Free-Energy Transfer and Biology

Chemical reactions are among the most widespread processes influencing life and economy. They are extremely important in biology, geology, environmental sciences, and industrial developments (energy management, production of millions of different chemical species, search for new materials). Chemical kinetics is a very rich but complex topic, which cannot be fully grasped by classical irreversible thermodynamics, whose kinetic description is restricted to the linear regime, not far from equilibrium.

Despite this limitation, this formalism is able to yield useful results, especially when dealing with coupled reactions. Indeed, the study of coupled processes is one of the most interesting features of non-equilibrium thermodynamics. In particular, the various couplings – between several chemical reactions, between chemical reactions and diffusion, and between chemical reactions and active transport processes in biological cells – receive a simple and unified description. Non-equilibrium thermodynamics is able to provide clarifying insights to some conceptually relevant problems, particularly in biology, which are rarely investigated in other formalisms.

In that respect, in the forthcoming we will focus on the problem of efficiency of free-energy transfer between coupled reactions, and the description of biological molecular motors. These engines may be modelled by some particular cycles of chemical reactions, generalizing the triangular chemical scheme proposed by Onsager in his original derivation of the reciprocal relations. Special attention will also be devoted to the effects arising from the combination of chemical reactions and diffusion. From a biological perspective, this coupling provides the basis of cell differentiation in the course of development of living organisms. The ingredients of this extremely complex phenomenon, leading to spatial self-organization far from equilibrium, may be qualitatively understood by considering a simplified version of the mechanism of coupling between autocatalytic chemical reactions and diffusion.

Biological applications of non-equilibrium thermodynamics have been the subject of several books (Katchalsky and Curran 1965; Nicolis and Prigogine 1977; Hill 1977; Caplan and Essig 1983; Westerhoff and van Dam 1987; Jou and Llebot 1990; Nelson 2004). Here, we present a synthesis of the main ideas and, for pedagogical purpose, we start the analysis with the problem of one single chemical reaction.

4.1 One Single Chemical Reaction

Equilibrium thermodynamics provides a general framework to analyse the equilibrium conditions of chemical reactions, and it leads in a natural way to the concept of equilibrium constant and its modifications under changes of temperature and pressure, as recalled in Sect. 1.7. A clear understanding of these effects is crucial to establish the range of temperature and pressure for the optimization of industrial processing. However, this optimization cannot be carried out without taking into account kinetic effects. For instance, lowering temperature has generally as a consequence to slow down the reaction velocity. On the other side, it is known from equilibrium thermodynamics that, for exothermic reactions, the efficiency of conversion of reactants into products increases when temperature is lowered. Thus, lowering temperature has two opposite effects: a lowering of the velocity of reactions and an increase of efficiency. Optimization will therefore result from a compromise between equilibrium factors and kinetic factors. We will return to this topic in Chap. 5. Here, we emphasize some kinetic aspects of chemical reactions, from the point of view of non-equilibrium thermodynamics.

To illustrate our approach consider, as in Chap. 1, the reaction of synthesis of hydrogen chloride

$$H_2 + Cl_2 \rightleftarrows 2HCl. \tag{4.1}$$

Since we will work in terms of local quantities, it is convenient to introduce the mass fractions c_k defined as $c_k \equiv m_k/m$, with m_k the mass of component k and m the total mass. In virtue of the law of definite proportions (1.70), the change of c_k during the time interval $\mathrm{d}t$ may be written as

$$\rho\frac{\mathrm{d}c_k}{\mathrm{d}t} = \nu_k M_k \frac{\mathrm{d}\xi}{\mathrm{d}t}, \quad k = 1, 2, \ldots, n, \tag{4.2}$$

with ν_k the stoichiometric coefficient of species k, M_k its molar mass, and $\mathrm{d}\xi/\mathrm{d}t$ the velocity of reaction per unit volume. This equation can be deduced from (1.70) when the mole numbers are expressed in terms of mass fractions and the degree of advancement is given per unit volume.

Besides the mass of the n species, the other relevant variable is the specific internal energy u, because of the exchange of heat with the outside. The time evolution equation for the internal energy u is given by the first law

$$\rho\frac{\mathrm{d}u}{\mathrm{d}t} = -\nabla \cdot \boldsymbol{q}, \tag{4.3}$$

where $\boldsymbol{q}$ is the heat flux vector expressing the amount of heat exchanged between the system and the outside world acting as a reservoir.

Now, we want to obtain expressions for the velocity of the reaction in terms of the corresponding thermodynamic force. This will be achieved within the general framework set out in Chap. 2, demanding that we previously determined the expression of the entropy production. To determine it, we start from the Gibbs' relation

$$\frac{\mathrm{d}s}{\mathrm{d}t} = \frac{1}{T}\frac{\mathrm{d}u}{\mathrm{d}t} - \sum_k \frac{\bar{\mu}_k}{T}\frac{\mathrm{d}c_k}{\mathrm{d}t}, \tag{4.4a}$$

where $\bar{\mu}_k$ is the chemical potential of the species k measured per unit mass. Making use of (4.2) and the relation $M_k\bar{\mu}_k = \mu_k$, with μ_k the chemical potential per unit mole, one obtains

$$\rho\frac{\mathrm{d}s}{\mathrm{d}t} = \frac{1}{T}\rho\frac{\mathrm{d}u}{\mathrm{d}t} - \sum_k \nu_k\frac{\mu_k}{T}\frac{\mathrm{d}\xi}{\mathrm{d}t}. \tag{4.4b}$$

We know from Sect. 1.7 that for *ideal systems*, i.e. mixture of ideal gases or dilute solutions, the chemical potential per unit mole μ_k is given by

$$\mu_k = \mu_k^{(0)}(T,p) + RT\ln x_k, \tag{4.5}$$

where $x_k \equiv N_k/N$ is the mole fraction while $\mu_k^{(0)}$ is independent of the composition.

At this stage, it is relevant to introduce the notion of affinity defined in (1.76) as

$$\mathcal{A} \equiv -\sum_k \nu_k\mu_k. \tag{4.6}$$

As seen below, the reaction will proceed forwards if $\mathcal{A} > 0$ or backwards if $\mathcal{A} < 0$: hence the name of affinity. At fixed values of temperature and pressure, $\mathcal{A}$ is only function of the mole fractions of species, and it was shown in Sect. 1.7 that

$$\mathcal{A} = RT\ln\frac{K(T,p)}{x_1^{\nu_1}x_2^{\nu_2}\cdots}, \tag{4.7a}$$

with $K(T,p)$ the equilibrium constant. Note that this constant may also be written as $K(T,p) = (x_1^{\nu_1})_{\mathrm{eq}}(x_2^{\nu_2})_{\mathrm{eq}}\cdots$, because at equilibrium $\mathcal{A} = 0$ and then the term inside the logarithm must be 1. It is useful, for further purposes in this chapter, to rewrite (4.7a) as

$$\mathcal{A} = RT\ln\frac{(x_1^{\nu_1})_{\mathrm{eq}}(x_2^{\nu_2})_{\mathrm{eq}}\cdots}{x_1^{\nu_1}x_2^{\nu_2}\cdots}. \tag{4.7b}$$

A third way of expressing the affinity is in terms of the molar concentrations, i.e. in the number of moles per unit volume, namely $\tilde{c}_k \equiv N_k/V$. In this case, (4.7b) may be expressed as

$$\mathcal{A} = RT \ln \frac{(\tilde{c}_1^{\nu_1})_{\text{eq}}(\tilde{c}_2^{\nu_2})_{\text{eq}} \cdots}{\tilde{c}_1^{\nu_1} \tilde{c}_2^{\nu_2} \cdots}. \tag{4.7c}$$

To derive the expression of the rate of entropy production, we introduce the balance equations of mass (4.2) and energy (4.3) in Gibbs' relation (4.4a) and (4.4b). Since in chemical reactions, the temperature is generally assumed to remain uniform, one obtains

$$\rho \frac{\mathrm{d}s}{\mathrm{d}t} = -\nabla \cdot \frac{\boldsymbol{q}}{T} + \frac{\mathcal{A}}{T} \frac{\mathrm{d}\xi}{\mathrm{d}t}. \tag{4.8}$$

By comparing this expression with the general form of the entropy balance equation $\rho \mathrm{d}s/\mathrm{d}t = -\nabla \cdot \boldsymbol{J}^s + \sigma^s$, it follows that the expression of the entropy flux is

$$\boldsymbol{J}^s = \frac{\boldsymbol{q}}{T}, \tag{4.9}$$

while the entropy production is identified as

$$\sigma^s = \frac{\mathcal{A}}{T} w > 0, \tag{4.10}$$

i.e. a bilinear expression in the thermodynamic force $\mathcal{A}/T$ and the flux $w = \mathrm{d}\xi/\mathrm{d}t$, which is the velocity of reaction. Truly, the reaction must be considered as the net effect of a forward flux w_+ going from reactants to products minus a backward flux w_- going from products to reactants, namely $w = w_+ - w_-$.

According to the second law, σ^s is always a positive quantity and zero in equilibrium. Thus when $\mathcal{A}$ is positive (respectively, negative), the velocity $w = w_+ - w_-$ is also positive (negative) and the reaction will take place from left to right (right to left). At equilibrium, the affinity $\mathcal{A}$ is zero and the net flux w will be zero, i.e. the flux w_+ towards the right is equal to the flux w_- towards the left.

For a single chemical reaction, the phenomenological relation between flux and force in the linear regime is

$$w = l \frac{\mathcal{A}}{T}, \tag{4.11}$$

where l is a phenomenological coefficient, which in (4.16) is related to the microscopic rate constants. On the condition that $l > 0$, this is the simplest way to guarantee the positiveness of the corresponding entropy production given by

$$\sigma^s = l \left(\frac{\mathcal{A}}{T} \right)^2. \tag{4.12}$$

However, it should be stressed that the linear flux–force relation (4.11) is only valid close to equilibrium. This is easily seen by comparing this result with the classical kinetic expression for the reaction rate which, in the case of the synthesis of HCl, reads as

$$w = w_+ - w_- = k_+ \tilde{c}_{\mathrm{H}_2} \tilde{c}_{\mathrm{Cl}_2} - k_- \tilde{c}_{\mathrm{HCl}}^2 = k_+ \tilde{c}_{\mathrm{H}_2} \tilde{c}_{\mathrm{Cl}_2} \left[1 - \frac{1}{K} \frac{\tilde{c}_{\mathrm{HCl}}^2}{\tilde{c}_{\mathrm{H}_2} \tilde{c}_{\mathrm{Cl}_2}} \right], \tag{4.13}$$

where k_+ and k_- are, respectively, the rate constants corresponding to the forward and backward reactions, and $K = k_+/k_- = (\tilde{c}^2_{\mathrm{HCl}})_{\mathrm{eq}}/[(\tilde{c}_{\mathrm{H_2}})_{\mathrm{eq}}(\tilde{c}_{\mathrm{Cl_2}})_{\mathrm{eq}}]$. In terms of the affinity, and in virtue of (4.7c), relation (4.13) can be expressed as

$$w = k_+\tilde{c}_{\mathrm{H_2}}\tilde{c}_{\mathrm{Cl_2}}\left[1 - \exp(-\mathcal{A}/RT)\right]. \tag{4.14}$$

For reactions near equilibrium, the affinity $\mathcal{A}$ is very small and it is justified to make a Taylor expansion around $\mathcal{A} = 0$ so that

$$w = k_+\frac{(\tilde{c}_{\mathrm{H_2}})_{\mathrm{eq}}(\tilde{c}_{\mathrm{Cl_2}})_{\mathrm{eq}}}{R}\frac{\mathcal{A}}{T}. \tag{4.15}$$

This result indicates clearly that the linear flux–force law (4.11) is only true very close to equilibrium and that the classical theory of irreversible processes does not cope with situations far from equilibrium. By comparison of the phenomenological relation (4.11) with the kinetic corresponding law (4.15), we are able to identify the phenomenological coefficient l as

$$l = k_+\frac{(\tilde{c}_{\mathrm{H_2}})_{\mathrm{eq}}(\tilde{c}_{\mathrm{Cl_2}})_{\mathrm{eq}}}{R}, \tag{4.16}$$

which confirms that l is a positive quantity as all the factors of (4.16) are positive.

The phenomenological relation (4.11) allows us to determine the behaviour of the degree of advancement ξ in the course of time. Indeed, if $\mathcal{A}$ is only depending on ξ, one can write (4.11) under the form

$$\frac{\mathrm{d}\xi}{\mathrm{d}t} = \frac{l}{T}\mathcal{A} = \frac{l}{T}\left[\mathcal{A}_{\mathrm{eq}} + \left(\frac{\partial\mathcal{A}}{\partial\xi}\right)(\xi - \xi_{\mathrm{eq}})\right] = \frac{l}{T}\left(\frac{\partial\mathcal{A}}{\partial\xi}\right)(\xi - \xi_{\mathrm{eq}}), \tag{4.17}$$

after use is made of $\mathcal{A}_{\mathrm{eq}} = 0$. This result suggests defining a relaxation time τ by

$$\tau \equiv -\frac{T}{l(\partial\mathcal{A}/\partial\xi)_{\mathrm{eq}}} > 0. \tag{4.18}$$

In this expression, there appear two different factors: one of them is the kinetic phenomenological coefficient l, whose positiveness is demanded by the second law; the other one $(\partial\mathcal{A}/\partial\xi)_T$ is related to the thermodynamic stability of the system and is negative, as shown in Sect. 1.7.

Finally, introducing (4.18) in relation (4.17) leads to

$$\tau\frac{\mathrm{d}\xi}{\mathrm{d}t} = -(\xi - \xi_{\mathrm{eq}})\ (t > 0), \tag{4.19}$$

whose solution is an exponentially decreasing function

$$\xi - \xi_{\mathrm{eq}} = (\xi_0 - \xi_{\mathrm{eq}})\exp(-t/\tau), \tag{4.20}$$

where ξ_0 is the initial value of the degree of advancement. Within the limit of an infinite time, the system tends of course to its equilibrium state.

Equation (4.20) is particularly interesting, because it provides some general information about the relaxational process without a detailed knowledge of the kinetics. If the stability condition $(\partial \mathcal{A}/\partial\xi)_{\mathrm{eq}} < 0$ is not satisfied, the relaxation time will be negative, indicating that the degree of advancement will not relax to equilibrium, but will increase exponentially.

4.2 Coupled Chemical Reactions

The study of coupled reactions is not merely a formal generalization of the above considerations, it is also the basis of discussion of various relevant concepts. For instance, if two processes are coupled, one of them may proceed in the opposite direction it would have if it were alone. We will stress this important feature in the next sections. Another interesting application of the coupling between chemical reactions is provided by the triangular scheme $A \rightleftarrows B \rightleftarrows C \rightleftarrows A$ which will serve as a basis to establish Onsager's reciprocal relations.

4.2.1 General Formalism

The main modifications with respect to the results of Sect. 4.1 are the expressions of the mass balance and the entropy production. Indeed, the variation of the mass fractions c_k of a given species is no longer due to one single reaction, but depends on the several chemical reactions in which the given species participates; it is now expressed as

$$\frac{\mathrm{d}c_k}{\mathrm{d}t} = \sum_{j=1}^{r} \nu_{kj} M_k \frac{\mathrm{d}\xi_j}{\mathrm{d}t}, \quad k = 1, 2, \ldots, n, \tag{4.21}$$

where subscript j refers to the r different chemical reactions, each of which has its own degree of advancement ξ_j, and ν_{kj} is the stoichiometric coefficient of substance k in reaction j. By following the same steps as in Sect. 4.1, it is found that, under isothermal conditions, the entropy production is

$$\sigma^s = \sum_j \frac{\mathcal{A}_j}{T} w_j. \tag{4.22}$$

It is worth noting that the second law imposes that the whole sum in (4.22), but not that each individual term, must be positive. Some terms may indeed be negative, provided that their sum remains positive. Within the assumption of linear flux–force relations, one has

$$w_i = \sum_j L_{ij} \frac{\mathcal{A}_j}{T}, \tag{4.23}$$

with, in virtue of Onsager's reciprocal relations,

$$L_{ij} = L_{ji}. \tag{4.24}$$

In Sect. 4.2.2, a particular derivation of these symmetry relations for a triangular reaction scheme is presented.

4.2.2 Cyclical Chemical Reactions and Onsager's Reciprocal Relations

A demonstration of the reciprocal relations (4.24), based on the cycle of chemical reactions A $\rightleftarrows$ B $\rightleftarrows$ C $\rightleftarrows$ A depicted in Fig. 4.1, was proposed by Onsager in his celebrated papers of 1931. In Sect. 4.4, we will use a modified form of this scheme as a model for chemically driven molecular motors.

Let k_i $(i = 1, 2, 3)$ be the kinetic constants, w_i the respective velocities of the reactions, which, according to the mass action law, are given by $w_1 = k_1\tilde{c}_A - k_{-1}\tilde{c}_B$, $w_2 = k_2\tilde{c}_B - k_{-2}\tilde{c}_C$, $w_3 = k_3\tilde{c}_C - k_{-3}\tilde{c}_A$, and $\mathcal{A}_i$ the respective affinities, i.e. $\mathcal{A}_1 = \mu_A - \mu_B$, $\mathcal{A}_2 = \mu_B - \mu_C$, $\mathcal{A}_3 = \mu_C - \mu_A$; for simplicity, all the stoichiometric coefficients are supposed to be equal to one. Since the process is cyclic, only two reactions are independent and $\mathcal{A}_1 + \mathcal{A}_2 + \mathcal{A}_3 = 0$. It will be shown that, when the flux–force relations are of the form

$$w_1 - w_3 = L_{11}\mathcal{A}_1 + L_{12}\mathcal{A}_2, \tag{4.25a}$$

$$w_2 - w_3 = L_{21}\mathcal{A}_1 + L_{22}\mathcal{A}_2, \tag{4.25b}$$

the Onsager's reciprocal relation $L_{12} = L_{21}$ is automatically satisfied near equilibrium, if the principle of detailed balance is valid. This principle states

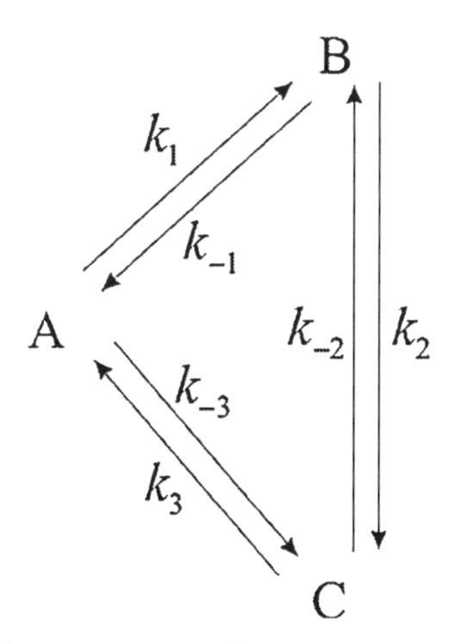

Fig. 4.1 Triangular reaction scheme

that transition A to B has the same probability as transition B to A, from which it follows that, in equilibrium, $w_1 = w_2 = w_3 = 0$. Note that this assumption does not follow directly from the phenomenological equations, since non-zero but equal values $w_1 = w_2 = w_3 \neq 0$ are compatible with $\mathcal{A}_1 = \mathcal{A}_2 = \mathcal{A}_3 = 0$. The mathematical details of the proof that $L_{12} = L_{21}$ are found in Box 4.1.

Box 4.1 Mathematical Details of Onsager's Demonstration

In virtue of (4.7c), the affinities $\mathcal{A}_1$ and $\mathcal{A}_2$ may be written as

$$\mathcal{A}_1 = RT \ln \left(\frac{\tilde{c}_A}{(\tilde{c}_A)_{eq}} \frac{(\tilde{c}_B)_{eq}}{\tilde{c}_B} \right), \quad \mathcal{A}_2 = RT \ln \left(\frac{\tilde{c}_B}{(\tilde{c}_B)_{eq}} \frac{(\tilde{c}_C)_{eq}}{\tilde{c}_C} \right), \tag{4.1.1}$$

where use has been made of the results $K_1(T,p) = (\tilde{c}_B)_{eq}/(\tilde{c}_A)_{eq}$ and $K_2(T,p) = (\tilde{c}_C)_{eq}/(\tilde{c}_B)_{eq}$. By assuming that $\tilde{c}_i = (\tilde{c}_i)_{eq} + \delta\tilde{c}_i$, with $\delta\tilde{c}_i \ll (\tilde{c}_i)_{eq}$, the affinities may be expressed up to the first order in the deviations $\delta\tilde{c}_i$ from equilibrium as

$$\mathcal{A}_1 = RT \left(\frac{\delta\tilde{c}_A}{(\tilde{c}_A)_{eq}} - \frac{\delta\tilde{c}_B}{(\tilde{c}_B)_{eq}} \right), \quad \mathcal{A}_2 = RT \left(\frac{\delta\tilde{c}_B}{(\tilde{c}_B)_{eq}} - \frac{\delta\tilde{c}_C}{(\tilde{c}_C)_{eq}} \right), \tag{4.1.2}$$

and, from $\mathcal{A}_3 = -(\mathcal{A}_1 + \mathcal{A}_2)$,

$$\mathcal{A}_3 = RT \left(\frac{\delta\tilde{c}_C}{(\tilde{c}_C)_{eq}} - \frac{\delta\tilde{c}_A}{(\tilde{c}_A)_{eq}} \right). \tag{4.1.3}$$

Furthermore, according to the mass action law, one has

$$w_1 - w_3 = (k_1 + k_{-3})\tilde{c}_A - k_{-1}\tilde{c}_B - k_3\tilde{c}_C, \tag{4.1.4}$$

$$w_2 - w_3 = k_2\tilde{c}_B + k_{-3}\tilde{c}_A - (k_{-2} + k_3)\tilde{c}_C. \tag{4.1.5}$$

We now take into consideration the restrictions coming from the detailed balance. From $w_1 = w_2 = w_3 = 0$, it is inferred that

$$k_1(\tilde{c}_A)_{eq} = k_{-1}(\tilde{c}_B)_{eq}, \quad k_2(\tilde{c}_B)_{eq} = k_{-1}(\tilde{c}_C)_{eq}, \quad k_3(\tilde{c}_C)_{eq} = k_{-3}(\tilde{c}_A)_{eq}. \tag{4.1.6}$$

Using (4.1.6) to eliminate k_{-1}, k_{-2}, and k_{-3} in (4.1.4) and (4.1.5), one obtains

$$w_1 - w_3 = k_1(\tilde{c}_A)_{eq} \left(\frac{\delta\tilde{c}_A}{(\tilde{c}_A)_{eq}} - \frac{\delta\tilde{c}_B}{(\tilde{c}_B)_{eq}} \right) + k_3(\tilde{c}_C)_{eq} \left(\frac{\delta\tilde{c}_A}{(\tilde{c}_A)_{eq}} - \frac{\delta\tilde{c}_C}{(\tilde{c}_C)_{eq}} \right), \tag{4.1.7}$$

$$w_2 - w_3 = k_2(\tilde{c}_B)_{eq} \left(\frac{\delta\tilde{c}_B}{(\tilde{c}_B)_{eq}} - \frac{\delta\tilde{c}_C}{(\tilde{c}_C)_{eq}} \right) + k_3(\tilde{c}_C)_{eq} \left(\frac{\delta\tilde{c}_A}{(\tilde{c}_A)_{eq}} - \frac{\delta\tilde{c}_C}{(\tilde{c}_C)_{eq}} \right). \tag{4.1.8}$$

In view of the results (4.1.2) and (4.1.3), the above equations may be written as

$$w_1 - w_3 = \frac{k_1}{RT}(\tilde{c}_A)_{eq}\mathcal{A}_1 - \frac{k_3}{RT}(\tilde{c}_C)_{eq}\mathcal{A}_3$$
$$= \frac{1}{RT}[k_1(\tilde{c}_A)_{eq} + k_3(\tilde{c}_C)_{eq}]\mathcal{A}_1 + \frac{k_3}{RT}(\tilde{c}_C)_{eq}\mathcal{A}_2, \quad (4.1.9)$$

$$w_2 - w_3 = \frac{k_2}{RT}(\tilde{c}_B)_{eq}\mathcal{A}_2 - \frac{k_3}{RT}(\tilde{c}_C)_{eq}\mathcal{A}_3$$
$$= \frac{k_3}{RT}(\tilde{c}_C)_{eq}\mathcal{A}_1 + \frac{1}{RT}[k_2(\tilde{c}_B)_{eq} + k_3(\tilde{c}_C)_{eq}]\mathcal{A}_2, \quad (4.1.10)$$

where use has been made of $\mathcal{A}_3 = -(\mathcal{A}_1 + \mathcal{A}_2)$. Note that (4.1.9) and (4.1.10) have been derived exclusively by reference to chemical kinetics and the principle of detailed balance.

A look on the coefficients of $\mathcal{A}_2$ in (4.1.9) and $\mathcal{A}_1$ in (4.1.10), which referring to (4.25a) and (4.25b), can be identified as L_{12} and L_{21}, respectively, leads to the conclusion that

$$L_{12} = L_{21}. \quad (4.26)$$

Let us now determine the corresponding results from classical irreversible thermodynamics. The entropy production σ^s is given by

$$T\sigma^s = w_1\mathcal{A}_1 + w_2\mathcal{A}_2 + w_3\mathcal{A}_3, \quad (4.27)$$

or, in terms of independent affinities,

$$T\sigma^s = (w_1 - w_3)\mathcal{A}_1 + (w_2 - w_3)\mathcal{A}_2, \quad (4.28)$$

which suggests the following constitutive relations:

$$w_1 - w_3 = L_{11}\mathcal{A}_1 + L_{12}\mathcal{A}_2, \quad (4.29)$$
$$w_2 - w_3 = L_{21}\mathcal{A}_1 + L_{22}\mathcal{A}_2. \quad (4.30)$$

Comparison with relations (4.1.9) and (4.1.10) yields

$$L_{12} = L_{21} = \frac{k_3}{RT}(\tilde{c}_C)_{eq}, \quad (4.31)$$

which completes the proof that the reciprocal relations are equivalent to the principle of detailed balance.

It should be realized that these results are only valid near equilibrium, where the net flux along the cycle is zero. In molecular machines, the above results are not directly transposable because the net flux along the cycle is different from zero.

4.3 Efficiency of Energy Transfer

A particularly appealing problem in non-equilibrium thermodynamics is to determine the efficiency with which free energy is exchanged between coupled chemical reactions. Indeed, in many biological situations, the transfer of free energy (or free-energy transduction, as is often referred to in textbooks on biochemistry) is of decisive importance. It is particularly true in microscopic molecular motors – one of the frontiers of current research in biophysics – which take free energy from some process (for instance, a chemical reaction) and convert it into work (for instance, to carry out another chemical reaction or a transport process). Note, however, that this subject cannot be studied from the classical perspective of thermal engines, because biological systems are often at homogeneous temperature, so that Carnot's efficiency would be zero. In contrast, the essential ideas of such isothermal free-energy transfer can be well interpreted in the framework of linear non-equilibrium thermodynamics.

Consider two coupled reactions, for which the dissipated energy is

$$T\sigma^s = w_1\mathcal{A}_1 + w_2\mathcal{A}_2. \tag{4.32}$$

The corresponding phenomenological equations describing the kinetics of the reactions are

$$w_1 = L'_{11}\mathcal{A}_1 + L'_{12}\mathcal{A}_2, \tag{4.33}$$

$$w_2 = L'_{21}\mathcal{A}_1 + L'_{22}\mathcal{A}_2. \tag{4.34}$$

The above results remain applicable when one of the processes, say 2, is a transfer of matter as considered in Sect. 4.4; such a situation is frequent in biological cells. To guarantee the positiveness of $T\sigma^s$, either both terms at the right-hand side of (4.32) are positive or one of them, say $w_2\mathcal{A}_2$, is negative but its absolute value is smaller than the first one, assumed to be positive. As a consequence, process 2 will evolve in opposite direction of its natural tendency, for instance mass transport from a region of low to high concentration. A relevant question is the efficiency of the energy transfer from the first to the second reaction. Such kind of coupling is frequently met in biology. For instance, in oxidative phosphorylation, the free energy delivered from the oxidation of a substrate will serve to the phosphorylation of ADP (adenosine diphosphate) to give ATP (adenosine triphosphate), which is a basic molecule for metabolic energy exchanges. How much energy is transferred from oxidation to phosphorylation is an interesting problem or, stated more generally, what is the *efficiency of energetic conversion* (Katchalsky and Curran 1965; Jou and Llebot 1990; Westerhoff and van Dam 1987; Criado-Sancho and Casas-Vázquez 2004).

In the above problem, reaction 1 is supplying some free energy, a part of which is taken up by reaction 2 to proceed against its usual direction, whereas the remaining part is lost under the form of heat. The objective is to obtain the maximum efficiency from such an energy transfer (or energy transduction). The problem is parallel to that of Carnot in the context of thermal engines. In the latter, heat was partially converted into work and the remaining was dissipated. Here, the energy delivered by one reaction is partially taken up by the other reaction and the remaining part is dissipated.

In the following, we suppose that the fluxes w_1 and w_2 are both positive and that the forces are of opposite signs: $\mathcal{A}_1 > 0$ and $\mathcal{A}_2 < 0$. The *degree of coupling* of the two processes is conveniently quantified by the coefficient q defined by

$$q \equiv \frac{L'_{12}}{(L'_{11}L'_{22})^{1/2}}. \tag{4.35}$$

From the requirement that the dissipated energy is positive definite, one has $L'_{11}L'_{22} \geq (L'_{12})^2$, and it follows that q will take values between -1 and $+1$: for $q = 1$, the processes are completely coupled; for $q = 0$, they are totally uncoupled. When $q < 0$, the processes are not compatible in the sense that process (1) with $w_1\mathcal{A}_1 > 0$ is unable to produce a process like (2) with a positive flux w_2 and a negative force $\mathcal{A}_2$.

In relation with coupled processes, two further quantities have been introduced: the *stoichiometric coefficient* w_2/w_1 and the *efficiency* η of the conversion defined by

$$\eta \equiv -\frac{w_2\mathcal{A}_2}{w_1\mathcal{A}_1}. \tag{4.36}$$

Introducing the quantities $x \equiv \mathcal{A}_2/\mathcal{A}_1, Z \equiv (L'_{22}/L'_{11})^{1/2}$, and using (4.33) and (4.34), it is easily checked that

$$\eta = -\frac{Zx(q + Zx)}{1 + qZx}, \tag{4.37}$$

which is maximum, namely, $\mathrm{d}\eta/\mathrm{d}x = 0$, at

$$x_{\max} = \frac{(1 - q^2)^{1/2} - 1}{qZ}, \tag{4.38}$$

thus leading to

$$\eta_{\max} = \frac{q^2}{[1 + (1 - q^2)^{1/2}]^2}. \tag{4.39}$$

This result is worth to be underlined: in some aspects, it can be compared with the classical Carnot's result for the maximum efficiency of heat engines, as it puts out an upper bound for the efficiency of energy transfer between both reactions or, in general terms, between two coupled processes. In heat engines, Carnot's maximum efficiency depends only on the values of the temperature of the heat reservoirs; here, the result (4.39) has a similar universality: it depends only on the coupling coefficient but not on the individual phenomenological coefficients. It is observed that the maximum of η

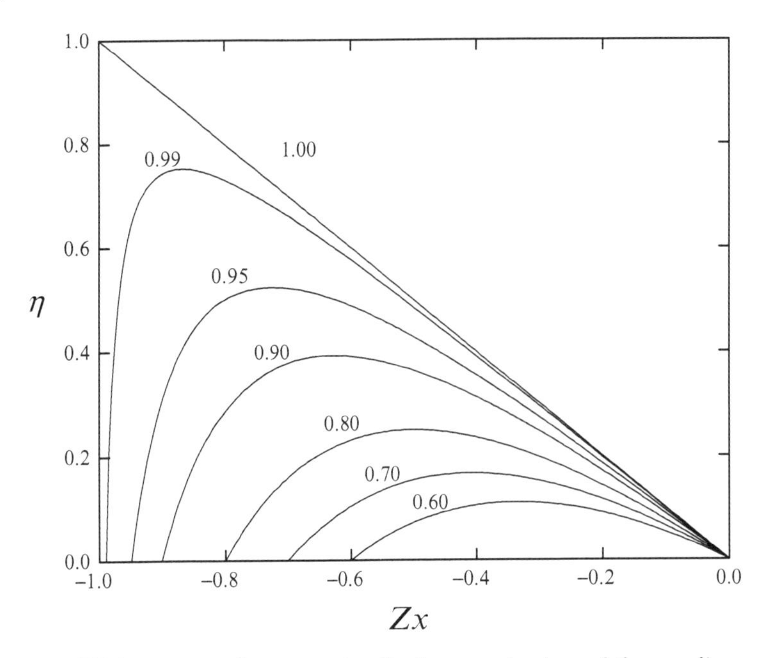

Fig. 4.2 Efficiency η vs. the parameter Zx for several values of the coupling coefficient q

diminishes with decreasing values of q and vanishes for $q = 0$. This is logical because, in absence of coupling, the efficiency of the energy transfer should be zero: this means that all the free energy delivered by reaction 1 would be dissipated into heat, so that reaction 2 cannot take place against its usual direction.

In Fig. 4.2, the values of the efficiency η as a function of the quantity Zx for several values of the coupling coefficient are reported. For $q = 1$, the maximum of η is 1; in virtue of (4.36), this situation corresponds to $w_1\mathcal{A}_1 = -w_2\mathcal{A}_2$ and consequently to a zero entropy production (reversible process).

The above considerations provide a clearer insight on the nature and importance of the various phenomenological coefficients, and they may be extended to cover a wider domain of situations, as the thermoelectric generators in Sect. 3.2 (Van den Broeck 2005).

4.4 Chemical Reactions and Mass Transport: Molecular Machines

The general developments of Sect. 4.3 will be applied to the study of molecular machines, mainly pumps across membranes and motors along filaments which are briefly introduced in Box 4.2. In classical non-equilibrium thermodynamics,

Box 4.2 More About Biological Molecular Engines
An illustration of coupling between chemical reactions and transport is the transport of macromolecules or vesicles along filaments of microtubules by kinesin or dinein, two well-known molecular motors; these processes contribute also to the formation of other organized intracellular motions, for example the separation of chromosomes during cell division (see, for instance, Nelson 2004). These motors must overcome the viscous resistance, and this is achieved by using the free energy of chemical reactions. Another example of chemically driven motion is that of the actin/myosin system in myofibrils, to which is due the muscle motility. In this system, bundles of myosin molecules are interleaved with actin filaments; upon activation, myosin crawls along the actin fibres, shortening the muscle fibre, the process being fuelled by the hydrolysis of ATP.

Most cells are characterized by membrane pumps for the active transport of Na^+ (towards the exterior) and K^+ (towards the interior); this is especially important in nerve cells, wherein the action potential is due to a transfer of Na^+ and K^+ across the excited membrane of the axon of neurons. Other kinds of pumps in biological cells are pumps of Ca^{2+} which are found in muscles, and also in the synapses, and pumps of H^+, which are important in oxidative phosphorylation in mitochondria and in photosynthesis in chloroplasts. On the other side, the flow of H^+ across some molecular engines may gear molecular motors, as for instance the flagellar motor of some bacteria, which convert the electrochemical potential jump of protons into mechanical torque for swimming.

under linear and isotropic conditions, chemical reactions (scalar processes) are not coupled with mass transport (a vector process). However, the presence of anisotropy may introduce a coupling between such processes, a frequent situation in biological problems, as in active transport of ions or molecules across cellular or tissue membranes, or along lengthy microscopic filaments.

These kinds of transport are attributed to enzymes, which act as molecular engines. The study of microscopic machines, natural or artificial, is at the frontiers of nowadays science. In the past, the calculation of the maximum efficiency of heat engines has represented a decisive contribution to equilibrium thermodynamics. Similarly, the determination of the efficiency of energy transfer between coupled processes may be considered as one of the main achievements of linear irreversible thermodynamics.

As an illustration, consider the active transport of sodium through a cellular membrane. Cells at rest state have an internal electrical potential of $-70\,\mathrm{mV}$ as compared to the external potential, taken as reference ($0\,\mathrm{mV}$). Since the internal concentration of sodium is low as compared to the external concentration, sodium has a natural tendency to flow towards the interior of the cell. Now it is observed that the cell pumps out as much sodium as the ingoing one but, of course, this transport cannot be spontaneous, because it takes place against the natural tendency of matter to flow from regions of higher to lower electrochemical potential. This transport is made at the expenses of the energy supplied by the hydrolysis of ATP, the usual fuel in biological organisms.

We are in presence of a coupling between a chemical reaction, characterized by an affinity $\mathcal{A}$ and a chemical flux w, and a transport of matter; it is shown in Sect. 2.7 on matter diffusion that the latter is described by a force, the difference of chemical potential $\Delta\mu$ between the inner and the outer regions, and a flux of matter (sodium in the present case) J. The corresponding rate of dissipation is given by

$$T\sigma^s = \mathcal{A}w + J\Delta\mu, \tag{4.40}$$

with $\mathcal{A}w > 0$, because the hydrolysis of ATP takes place along the natural tendency given by $\mathcal{A} > 0$ and $w > 0$, whereas $J\Delta\mu < 0$, because the flow of sodium ($J < 0$) is opposite to the direction of $\Delta\mu$. We are faced with an example where a reaction with positive entropy production allows for the occurrence of a simultaneous process, with a negative contribution to the entropy production.

The phenomenological laws describing the kinetics of the reaction and the rate of transport are given by

$$w = L'_{11}\mathcal{A} + L'_{12}\Delta\mu, \tag{4.41}$$

$$J = L'_{21}\mathcal{A} + L'_{22}\Delta\mu, \tag{4.42}$$

with $L'_{12} = L'_{21}$ according to Onsager's reciprocal relations. The coefficient L'_{21} in (4.42) is responsible for the coupling between chemical reaction and mass transport.

For instance, the experimental values of the phenomenological coefficients for the transport across the membrane of a frog bladder are the following: $L'_{22} = 104$, $L'_{12} = 5.40$, and $L'_{11} = 0.37$ in units of $\mathrm{mol^2\,cm^{-2}\,s^{-1}\,kcal^{-1}}$. The coupling parameter q defined by (4.35) is given by $q = 0.86$ and the maximum efficiency calculated from (4.39) is $\eta_{\max} = 33\%$.

In so-called *linear motors*, i.e. motors which transport matter along lengthy linear macromolecules, as kinesin along microtubules, the resistance is due to viscous friction rather than to a jump of the electrochemical potential. In Box 4.3, the modus operandi of such a molecular motor is presented. The expression for the dissipated energy is then

$$T\sigma^s = \mathcal{A}w + vF, \tag{4.43}$$

v being the velocity of the machine along the filament and F the viscous resistance. The phenomenological laws are, in the linear approximation,

$$w = L_{11}\mathcal{A} + L_{12}F, \tag{4.44}$$

$$v = L_{21}\mathcal{A} + L_{22}F, \tag{4.45}$$

with the coefficient L_{22} inversely proportional to the viscosity of the medium. Since molecular machines are of very small dimensions, the energy related to their several steps is comparable or smaller than the thermal noise. Their behaviour is therefore not deterministic but stochastic, in contrast with that of macroscopic engines. The understanding of the relations between thermal fluctuations and the behaviour of molecular machines is a subject of current interest, as explained in Box 4.4.

Box 4.3 Biological Molecular Motors with Three Configurations
A model for chemically driven transport may be based on a modification of the triangular reaction scheme of Sect. 4.2.2. An abstract illustration is provided by the following molecular motor running through three different configurations, denoted as M_1, M_2, and M_3 (there would be no difficulty to add more configurations), and working in a cyclic way. We assume that the motor is fuelled with ATP and hydrolyses it, changing the configuration at each chemical step. Thus, instead of a closed reaction triangle, one has the scheme

$$(M_1) + \mathrm{ATP} \rightarrow (M_2) \cdot \mathrm{ATP}, \tag{4.3.1}$$

$$(M_2) \cdot \mathrm{ATP} \rightarrow (M_3) \cdot \mathrm{ADP} + P_i, \tag{4.3.2}$$

$$(M_3) \cdot \mathrm{ADP} \rightarrow (M_1) + \mathrm{ADP}. \tag{4.3.3}$$

To be explicit, configuration 1 bounds one molecule of ATP, which produces a change to configuration M_2. In a second step, ATP is hydrolysed: the inorganic phosphate P_i is liberated and the macromolecule changes to configuration M_3, with ADP still bound to it. Finally ADP is liberated from configuration M_3 and the motor returns to the initial configuration M_1, where the cycle may start again. At each cycle, the motor does some work: either it advances a step along a macromolecule (as for instance kinesin along tubulin or myosin along actin) or it translocates some molecule (as for instance in membrane ionic pumps). For more biological details, the reader is referred, for instance, to Nelson (2004).

Finally, the triangular scheme would be

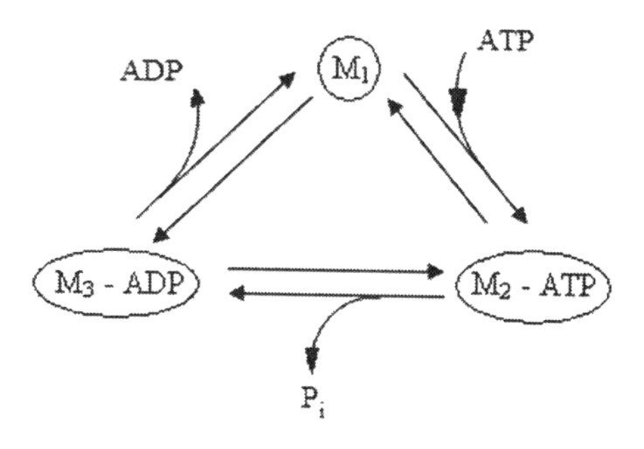

It is assumed that the kinetic rate constants of reactions (4.3.1)–(4.3.3) and k_{-1}, k_2, and k_3 in the triangular scheme do not depend on the ATP nor ADP concentration, whereas $k_1 = k_1^{(0)}[\mathrm{ATP}], k_{-3} = k_{-3}^{(0)}[\mathrm{ADP}]$, and $k_{-2} = k_{-2}^{(0)}[\mathrm{P}]$ (where $k_1^{(0)}, k_{-3}^{(0)}$, and $k_{-2}^{(0)}$ are constant values) are pseudo-first-order rate coefficients into which the concentrations of ATP, ADP, and $\mathrm{P_i}$ are incorporated. This makes an essential difference with the triangular reaction scheme presented in Sect. 4.2.2. Indeed when the hydrolysis reaction

$$\mathrm{ATP} \rightarrow \mathrm{ADP} + \mathrm{P_i} \tag{4.3.4}$$

is in equilibrium, the cycle has not net motion; however, when the affinity of this reaction is positive, the cycle will turn in the direction 123, whereas if the affinity is negative, it will move in the opposite direction 321. Such kind of model machine may both hydrolyse ATP and produce some work pumping ions from lower to higher electrochemical potential (active transport) or, in the reverse way, it may draw some energy from the ions crossing it from higher to lower electrochemical potential and produce ATP.

Box 4.4 Brownian Motors

Molecular motors are sensitive to thermal noise which might be used, to some extent, to produce useful work, provided some structural and thermodynamic conditions are satisfied. Instead of performing a simple random Brownian motion, these motors may rectify, to some extent, the thermal noise. The modelling of such molecular motors has been much developed since 1990 (Magnasco 1994; Astumian and Bier 1994; Jülicher et al. 1997; Astumian 1997) (Fig. 4.3).

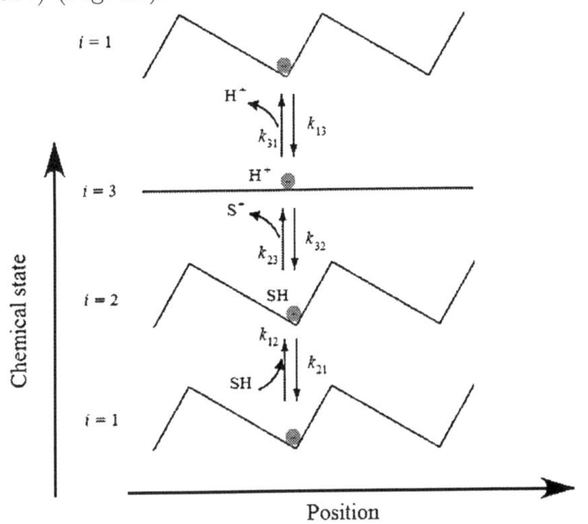

Fig. 4.3 Transport driven by a chemical reaction (Astumian 1997)

As an example, we analyse a simple model proposed by Astumian (1997) for chemically driven transport. The author studies the one-dimensional motion of a charged particle along a periodic lattice of dipoles arranged head to tail (as found in many biological macromolecules, like tubulin or actin). The interaction between the particle and the linear arrangement of dipoles is assumed to be merely electrostatic (in biology, the couplings may also be conformational). The potential energy profile has essentially the form of a sawtoothed function as that shown in Fig. 4.3 (if the particle is negatively charged, the minima of the sawtooth will correspond to the positive ends of the dipoles and the maxima to the negative ends). Because of thermal noise, the particle undergoes Brownian motion along the line, and occasionally has enough energy to pass one of the neighbouring barriers. However, despite the anisotropy of the potential profile, the probabilities of moving to the right or to the left are equal.

Assume, instead, that the particle catalyses a chemical reaction $SH \rightleftarrows S^- + H^+$, with the product particles being charged; S designates some substract species. Now, the amplitude of the potential will depend on whether H^+ or S^- are bound to the particle, resulting in a coupling between the chemical reaction and the diffusion of the particle along the line. However, when the particle is bound only to H^+ (assume that the ionized substract S^- has been expelled to the bulk), the particle becomes uncharged and there is no longer a barrier opposing its motion. Therefore, the particle may move along the line. Once H^+ is liberated, the charge of the particle constitutes again a barrier. However, it may be shown that an asymmetric potential whose barriers are fluctuating in height gives raise to a directional motion of particles. The corresponding flux is given by (Astumian and Bier 1994)

$$J \approx \frac{\gamma k_{\mathrm{B}} T}{4}\left[\operatorname{erf}\left(\frac{\alpha}{2}\zeta\right) - \operatorname{erf}\left(\frac{1-\alpha}{2}\zeta\right)\right],$$

where α is the asymmetry parameter of the potential, and $\zeta \equiv (\gamma L^2/D)^{1/2}$ while γ, L, and D are typical parameters of the model. If the sawtoothed potential is symmetric ($\alpha = 0.5$), $J = 0$ and there is no net flow.

To summarize, it turns out that molecular machines can convert free energy into motion under the conditions that this mechanism is out of equilibrium and structurally asymmetric. Moreover, molecular engines cannot increase their speed without limit as the energy supply is increased, because the speed saturates at some limiting value. This means that linear transport relations are not indefinitely valid, but break down for high values of fluxes and forces.

4.5 Autocatalytic Reactions and Diffusion: Morphogenesis

Another consequence of the coupling between chemical reactions and diffusion is the occurrence of patterns, i.e. spatial inhomogeneities, in systems far from equilibrium. Although this topic is studied in detail in Chap. 6, we find it interesting to present here an introductory version based on a model proposed by Turing (1952), who settled the foundations of morphogenesis, i.e. the transition of an originally homogeneous system into an inhomogeneous one. This is a characteristic feature of non-equilibrium thermodynamics, because near equilibrium, transport phenomena (matter diffusion, thermal conduction) tend to make the system spatially homogeneous in the long run.

Consider two fluid subsystems at the same pressure and temperature, which are separated by a permeable membrane. In each of them, the following autocatalytic reaction will take place between substances A and X

$$\mathrm{A} + 2\mathrm{X} \rightarrow 3\mathrm{X}. \tag{4.46}$$

In this reaction, a substance A is transformed into the substance X, at a rate which depends on the concentration of X. The molar concentrations of X in both subsystems, $\tilde{c}_{\mathrm{X}_1}$ and $\tilde{c}_{\mathrm{X}_2}$, respectively, tend to become equal due to diffusion of X across the membrane. The diffusion flux of X from subsystem 1 to subsystem 2 is given by

$$J_{1\rightarrow 2} = \alpha(\tilde{c}_{\mathrm{X}_1} - \tilde{c}_{\mathrm{X}_2}), \tag{4.47}$$

where α is a constant proportional to the permeability and the area of the membrane.

We will show that, under some conditions, the difference of concentrations $\tilde{c}_{\mathrm{X}}$ between both subsystems will increase because of the autocatalytic reaction (4.46). First, we note that the rate of production of substance X is higher where the molar concentration $\tilde{c}_{\mathrm{X}}$ is higher, as it follows from the mass action law

$$\frac{\mathrm{d}\tilde{c}_{\mathrm{X}}}{\mathrm{d}t} = k\tilde{c}_{\mathrm{A}}(\tilde{c}_{\mathrm{X}})^2, \tag{4.48}$$

where k is a positive kinetic constant.

Taking diffusion into account, the evolution of $\tilde{c}_{\mathrm{X}_1}$ as a function of time in subsystem 1 is given by

$$\frac{\mathrm{d}\tilde{c}_{\mathrm{X}_1}}{\mathrm{d}t} = -\alpha(\tilde{c}_{\mathrm{X}_1} - \tilde{c}_{\mathrm{X}_2}) + k\tilde{c}_{\mathrm{A}}(\tilde{c}_{\mathrm{X}_1})^2, \tag{4.49}$$

whereas the evolution of $\tilde{c}_{\mathrm{X}_2}$ in subsystem 2 is

$$\frac{\mathrm{d}\tilde{c}_{\mathrm{X}_2}}{\mathrm{d}t} = -\alpha(\tilde{c}_{\mathrm{X}_1} - \tilde{c}_{\mathrm{X}_2}) + k\tilde{c}_{\mathrm{A}}(\tilde{c}_{\mathrm{X}_2})^2. \tag{4.50}$$

By subtracting both equations, one obtains for the evolution of the difference of concentrations $\tilde{c}_{\mathrm{X}_1} - \tilde{c}_{\mathrm{X}_2}$,

$$\frac{\mathrm{d}(\tilde{c}_{\mathrm{X}_1} - \tilde{c}_{\mathrm{X}_2})}{\mathrm{d}t} = -2\alpha(\tilde{c}_{\mathrm{X}_1} - \tilde{c}_{\mathrm{X}_2}) + \beta\tilde{c}_{\mathrm{A}}\left[(\tilde{c}_{\mathrm{X}_1})^2 - (\tilde{c}_{\mathrm{X}_2})^2\right]. \tag{4.51}$$

This may still be written as

$$\frac{\mathrm{d}(\tilde{c}_{\mathrm{X}_1} - \tilde{c}_{\mathrm{X}_2})}{\mathrm{d}t} = -2\alpha(\tilde{c}_{\mathrm{X}_1} - \tilde{c}_{\mathrm{X}_2}) + k\tilde{c}_{\mathrm{A}}(\tilde{c}_{\mathrm{X}_1} + \tilde{c}_{\mathrm{X}_2})(\tilde{c}_{\mathrm{X}_1} - \tilde{c}_{\mathrm{X}_2}). \tag{4.52}$$

This result indicates that the right-hand side of (4.52) becomes positive, thus implying an increase of the difference of concentrations, when the molar concentration of A becomes higher than the critical value

$$(\tilde{c}_{\mathrm{A}})_c = \frac{2\alpha}{k}(\tilde{c}_{\mathrm{X}_1} + \tilde{c}_{\mathrm{X}_2}). \tag{4.53}$$

Note that, since reaction (4.46) is irreversible, i.e. it only goes to the right, the concentration of A corresponding to equilibrium is zero. In the problem under study, $\tilde{c}_{\mathrm{A}}$ remains constant because it is supplied from the outside at the same rate in both subsystems. If the supply rate is sufficiently high in such a way that the concentration of A exceeds the critical value, the concentration $\tilde{c}_{\mathrm{X}}$ will tend to be higher in one subsystem than in the other and will contribute to the reinforcement of the non-homogeneity of the system.

This mechanism has gained much interest in biology. Indeed, an actual problem in biology is the development of an embryo from the first fertilized cell of the individual and this process implies cell differentiation at the various stages of development. Though all the cells of the individual contain the same information in their DNA, the parts of this information which are really read are not the same in all cells, but they depend on control proteins that determine which genes are transduced into the corresponding proteins. In the above model, one could interpret X as the control molecule determining the reading of some gene. When the system is far enough from equilibrium, the inhomogeneity in the value of $\tilde{c}_{\mathrm{X}}$ makes that, in one part of the organism where $\tilde{c}_{\mathrm{X}}$ is higher than some critical value, a given gene is expressed, whereas in the other regions, where $\tilde{c}_{\mathrm{X}}$ is lower, the gene is not expressed.

4.6 Problems

4.1. *Michaelis–Menten's relation for enzymatic reactions.* An important type of biochemical reactions are reactions changing a substrate S into a product P, catalysed by an enzyme E, through the intermediate formation of an enzyme–substrate complex ES. The reaction may be expressed as

$$\mathrm{E} + \mathrm{S} \rightleftarrows \mathrm{ES} \to \mathrm{E} + \mathrm{P}.$$

The first step is reversible and the second one is irreversible. The corresponding kinetic constants are k_1 and k_{-1} for the forward and backward steps of the first reaction, and k_2 for the (forward) step of the second reaction. The velocity of reaction $w \equiv \mathrm{d}\tilde{c}_P/\mathrm{d}t = -\mathrm{d}\tilde{c}_S/\mathrm{d}t$ obeys the Michaelis–Menten expression

$$w = w_{\max}\frac{\tilde{c}_{\mathrm{S}}}{K_{\mathrm{M}} + \tilde{c}_{\mathrm{S}}},$$

where $w_{\max}$ is the saturation velocity and K_{M} a characteristic constant indicating the value of the substrate molar concentration $\tilde{c}_{\mathrm{S}}$ for which the velocity is half the maximum value. (a) Show that, according to the mass action law, the value of the concentration of the complex ES in the steady state (for which the enzyme concentration $\tilde{c}_{\mathrm{E}}$ is constant) is given by

$$\tilde{c}_{\mathrm{ES}} = \frac{k_1\tilde{c}_{\mathrm{S}}}{k_{-1} + k_2 + k_1\tilde{c}_{\mathrm{S}}}.$$

(b) It follows from the mass action law that the velocity of the reaction is $w = \mathrm{d}\tilde{c}_{\mathrm{P}}/\mathrm{d}t = k_2\tilde{c}_{\mathrm{ES}}$. Combine this result with the previous one and show that Michaelis–Menten expression is recovered. Identify $w_{\max}$ and K_{M} in terms of the kinetic constants k_1, k_{-1}, k_2, and of the enzyme concentration $\tilde{c}_{\mathrm{E}}$. Discuss the dependence of $w_{\max}$ with $\tilde{c}_{\mathrm{E}}$ and the relevance of the fact that the last step is totally irreversible (i.e. the backward reaction is lacking). (c) Assume that for a given enzymatic reaction $w_{\max} = 0.085\,\mathrm{mM\,s^{-1}}$ and $K_{\mathrm{M}} = 6.5\,\mathrm{mM}$. Obtain the velocity w for $\tilde{c}_{\mathrm{S}} = 2.5$, 10, and 25 mM.

4.2. *Application of Michaelis–Menten expression to simple molecular motors.* The kinetics of simple, tightly coupled chemical motors with at least one irreversible step is expected to be described by a Michaelis–Menten expression, discussed in the previous problem. Assume, for instance, the following scheme

$$\mathrm{motor}(x) + \mathrm{ATP} \rightleftarrows \mathrm{ATP}\cdot\mathrm{motor} \rightarrow \mathrm{motor}(x + \Delta x) + \mathrm{ADP}.$$

In this reaction, the hydrolysis of ATP gives the energy necessary for moving the motor from a position x to $x + \Delta x$ along some filament, the motor could be kinesin and the corresponding filament could be a microtubule, in which case $\Delta x = 8\,\mathrm{nm}$. The speed with which the motor moves along the filament can be cast in the form

$$v = v_{\max}(F)\frac{\tilde{c}_{\mathrm{ATP}}}{K_{\mathrm{M}}(F) + \tilde{c}_{\mathrm{ATP}}},$$

wherein the maximum velocity $v_{\max}$ and the characteristic constant K_{M} depend on the load force acting on the motor. Indeed, it has been found that, for kinesin along microtubules, the numerical values are $v_{\max} \approx 815\,\mathrm{nm\,s^{-1}}$ and $K_{\mathrm{M}} \approx 90\,\mathrm{\mu M}$ at $F = 1\,\mathrm{pN}$, and $v_{\max} \approx 400\,\mathrm{nm\,s^{-1}}$ and $K_{\mathrm{M}} \approx 300\,\mathrm{\mu M}$ at $F = 5.5\,\mathrm{pN}$. (a) Compare the velocity at $\tilde{c}_{\mathrm{ATP}} = 1$, 10, and 100 mM, under both load charges. (b) Compare the useful power delivered by the motor, given by Fv, in the several situations mentioned above.

4.3. *Conditions of maximum efficiency.* Consider a transport process across a membrane, when the affinity of the hydrolysis of ATP is $\mathcal{A}_1$ and assume that the transported substance may be considered as ideal and uncharged. Show that the ratio of internal and external concentrations of the substance, under the conditions of optimum free-energy exchange between ATP energy and transfer, is given by

$$\ln \frac{c_{\text{in}}}{c_{\text{out}}} = \frac{\mathcal{A}_1}{RT} \frac{(1-q^2)^{1/2} - 1}{qZ},$$

the coupling coefficients q and Z are defined in Sect. 4.3, namely $q \equiv L'_{12}/(L'_{11}L'_{22})^{1/2}$ and $Z \equiv (L'_{11}/L'_{22})^{1/2}$, L'_{ij} being the phenomenological coefficients of the linear constitutive equations relating the velocity of the chemical reaction (taken as flux 1) and the rate of active transport (taken as flux 2) with the corresponding conjugated thermodynamic forces.

4.4. *Kinetics of triangular reaction.* An example of a triangular reaction scheme is provided by the triangular catalytic isomerization of the three butenes C_3H_8 (namely, forms (A), (B), and (C) shown below), was studied by W. O. Haag and H. Pines (J. Am. Chem. Soc. **82** (1960) 387 and 2488). They obtained for the corresponding kinetic constants in the presence of Al_2O_3 catalyst: $k_1/k_3 = 2.4$, $k_{-2}/k_{-1} = 1.0$, and $k_{-3}/k_2 = 0.4$. Analyse whether these results are compatible with the statement that the global equilibrium in the triangular reactions corresponds to equilibrium in each of the three individual reactions (see Chartier et al. 1975).

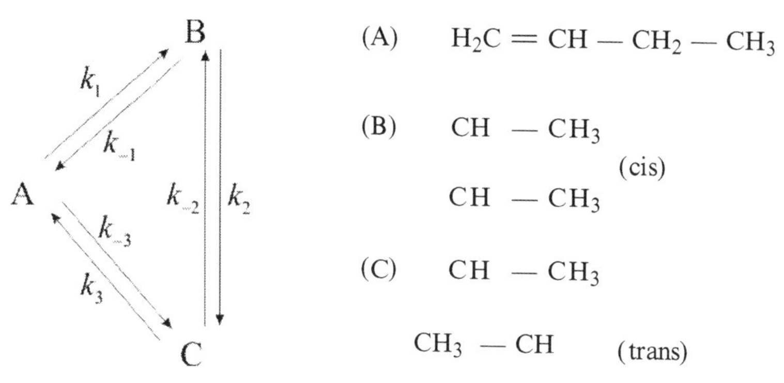

4.5. *Autocatalytic reactions and diffusion.* Discuss the instability problem analysed in Sect. 4.5 when one assumes that $A+X \rightarrow 2X$ instead of $A+2X \rightarrow 3X$, as in (4.46).

4.6. *Entropy production and degree of coupling in kinesin motor.* Kinesin, a typical molecular machine which moves along microtubules, takes on average one 8-nm step, corresponding to one monomer of tubulin, every 10 ms. A single ATP molecule is hydrolyzed per step. The chemical energy relased by ATP hydrolysis is typically 20 k_BT, and the motor uses 12 k_BT with each step (Bustamante et al. 2005). a) Compute the power and the force done by the kinesin, and the hydrolysis velocity reaction. b) What is the efficiency of the machine? c) Evaluate the entropy production of the machine in terms of k_B per second. d) If this efficiency is the maximum one, evaluate from equation (4.39) the degree of coupling between ATP hydrolysis and kinesin motion, according to the scheme provided in (4.44) and (4.45) for linear motors.

4.7. *Bacterial flagellar motor.* Swimming bacteria are usually propelled by a rotary motor embedded in the cell envelope which drives rotating flagellar filaments. These motors are typically powered by the inflow of protons across the cell membrane, and have very interesting properties which confer bacteria with the possibility of chemiotaxis, i.e., swimming towards higher nutrient concentrations. We simplify the complexity of this motor and assume that the power delivered by the protons, namely $Power = J\Delta\mu$, is applied with unit efficiency to propel the cell. Here J is the proton flux across the motor (number of protons per second) and $\Delta\mu$ is the socalled protonmotive force, defined as

$$\Delta\mu = \Delta V + 2.303(k_B T/e)\Delta\text{pH}$$

ΔV being the membrane potential (difference of potential across the membrane) and ΔpH being $\Delta\text{pH} = \log(c_{H+,out}/c_{H+,in})$, with c_{H+} being the proton concentration. Assume that a bacterium is a sphere of 1 μm of radius, and its friction coefficient is given by the Stokes law, i.e. $\zeta = 6\pi\eta r$, with r the radius of the cell and η the viscosity of the surrounding fluid. Assume that when the membrane potential is -80 mV (inside is negative) and $\Delta\text{pH} = 0$, the bacterium moves at 25 μm/s. a) Find the proton flux necessary to propel the cell in water ($\eta = 10^{-3}\text{N}\cdot\text{s}\cdot\text{m}^{-2}$). b) Find the electrical resistance of the motor. c) Evaluate the velocity of the cell when it swims in a solution with pH $= 6$ (assume that inside the cell pH $= 7$). d) Compare the total entropy production due to viscous dissipation in both cases.

4.8. *Degree of coupling and maximum power.* In expression (4.39) was given the maximum efficiency in terms of the degree of coupling q of two coupled chemical reactions. In some circumstances, however, as for instance in hunting or fleeing, so important for survival, the organism is interested in obtaining the maximum power instead of the maximum efficiency. a) Write the power output $-w_2A_2$ in the normalized form $-w_2A_2/L_{11}A_1^2$ and show that it may be expressed as $-Zx(Zx+q)$; b) Show that the corresponding efficiency at the value of q for which the power output is maximum is

$$\eta_{\text{max power}} = (q^2/4)[1(q^2/2)]^{-1}.$$

c) Compare this value with the maximum efficiency.

4.9. *Effects of the breaking of detailed balance on the Onsager relations.* Assume that in the triangular cyclical set of chemical reactions presented in Section 4.2.2, detailed balance was broken, i.e., instead of having $w_1 = w_2 = w_3 = 0$ one had $w_1 = w_2 = w_3 = w$. How would the phenomenological coefficients L_{12} and L_{21} appearing in the constitutive equations (4.25) be related? (i.e. express $L_{12} - L_{21}$ in terms of w).

Chapter 5
Finite-Time Thermodynamics

Economy, Ecology, and Heat Engines

One of the motivations of the early developments of thermodynamics was the optimization of heat engines, as steam engines, which transform partially heat into work, and which led to the first industrial revolution between the end of the eighteenth century and the beginning of the nineteenth century. This provides an interesting illustration of how an engineering problem coupled with economical needs stimulated the occurrence of a new fundamental science. Nowadays, we live in a similar situation: we are experiencing an industrial revolution based on information processing and miniaturized engines and, on the other side, we are conscious of the need to incorporate ecological restrictions into the economical progress. To convey some flavour of these problems, we present in this chapter applications of non-equilibrium thermodynamics to heat engines working with a finite non-vanishing rate. This approach is referred to as *finite-time thermodynamics* and received a strong impetus during the last quarter of the twentieth century.

In 1824, Carnot arrived to the conclusion that the efficiency of a heat engine operating between a hot and a cold heat reservoir is maximum in reversible processes, and that it depends only on the temperature of the two reservoirs, but not on the working substance. Here, Carnot's ideas are extended by including considerations not only on the efficiency but also on the power developed by the engine when the cycle is performed in a finite time. This issue is of importance from the economical perspective, because what the society needs is an amount of work in a finite time, rather than work supplied at an infinitesimal small rate, as in reversible engines.

Though the mutually conflicting demands in the maximization of efficiency and power are especially illustrative of the conceptual problems dealt with in this chapter, we also compare other optimization criteria, as for instance, those based on the minimization of entropy production, and on maximization of the power minus the product of temperature and entropy production. This diversity of criteria illustrates the conceptual subtleties arising in finite-time thermodynamics. Furthermore, the time constraints must be included in realistic models of engines to optimize their design and performance.

Finite-time thermodynamics deals with a very wide collection of situations. In the present chapter, we only present some simple and pedagogical illustrations, based on Carnot's engine, but many other problems have been investigated in finite-time thermodynamics as, for instance, how to optimize a thermodynamic cycle when its total duration is fixed. In this case, the several branches of the cycle may be performed at different rates, and the problem is to find the rates of each process in such a way that some optimization criterion (power output maximization, for instance) is satisfied. Finite-time thermodynamics is not limited to Carnot cycle, but it encompasses many other cycles (Otto, Brayton, Stirling, Diesel, etc.), and it is concerned not only with ideal gases but also with other kinds of fluids, as polytropic fluids. Particular studies have been devoted to refrigerators, heat pumps, engines with heat regeneration, several coupled engines, and have been applied to solar engines, photovoltaic cells, or to the evaluation of the maximum efficiency of the conversion from solar energy to wind energy in the planet. In short, finite-time thermodynamics allows us to understand in a simple way the essential role of time constraints on the thermodynamic cycles (Salamon et al. 1980; Ma 1985; Salamon and Sieniutycz 1991; Bejan 1996; Hoffmann et al. 1997). This is a clear illustration of the importance of including non-equilibrium thermodynamics both in the assessment and design of heat engines as well as in the understanding of some natural phenomena.

5.1 The Finite-Time Carnot Cycle

Let us briefly recall the essentials of the classical analysis of Carnot's engine. The latter is operating between two heat reservoirs at temperatures T_1 and T_2 ($T_1 > T_2$), it takes an amount of heat Q_1 from the hot one, transforms a part of it into work W, and delivers the remaining heat Q_2 to the cold reservoir. Since the engine operates cyclically, Q_1, Q_2, and W refer, respectively, to the amounts of heat, absorbed and delivered, and the work carried out by the engine during one cycle.

The internal energy of the engine does not change during one cycle, because it is a state function, so that the first law takes the form

$$Q_1 = W + Q_2. \tag{5.1}$$

If we ask about the efficiency η of the cycle – namely the work produced W divided by the heat supplied Q_1 to the engine – one has

$$\eta \equiv \frac{W}{Q_1} = 1 - \frac{Q_2}{Q_1}. \tag{5.2}$$

Kelvin–Planck formulation of the second law states the impossibility of building a cyclical heat engine, which transforms heat completely into work. Indeed, the maximum value of η is limited by the minimum value of the heat

Q_2 delivered to the cold reservoir. Therefore, to evaluate the maximum efficiency, the second law must be taken into account. Accordingly, the global balance of entropy is never negative, i.e.

$$\Delta S_{\text{tot}} = \Delta S_1 + \Delta S_2 + \Delta S_{\text{engine}} \geq 0, \tag{5.3}$$

where ΔS_1, ΔS_2, and ΔS_{engine} are the variations of entropy, respectively, in the reservoirs 1 and 2 and in the engine during one cycle. Since the engine operates cyclically, and recalling that the entropy is a function of state, one has $\Delta S_{\text{engine}} = 0$. Moreover, the temperature of each reservoir remains constant, so that, according to Clausius' definition of entropy, the variations of their respective entropies are $\Delta S_1 = -Q_1/T_1$ and $\Delta S_2 = Q_2/T_2$. Therefore, (5.3) becomes

$$\Delta S_{\text{tot}} = -\frac{Q_1}{T_1} + \frac{Q_2}{T_2} = \Delta_{\text{irr}} S \geq 0. \tag{5.4}$$

The equality sign corresponds to reversible processes and $\Delta_{\text{irr}} S$ stands for the entropy produced during one cycle by irreversible processes (as for instance thermal resistance or mechanical and hydrodynamical friction). After combining (5.2) and (5.4), it follows that the efficiency obeys the inequality

$$\eta \leq 1 - \frac{T_2}{T_1}. \tag{5.5}$$

This expression shows that the efficiency is maximum when the engine operates reversibly: indeed, in this case the relation (5.5) becomes an equality, providing the so-called *Carnot's efficiency*, depending only on the temperature of the sources but not on any detail of the working substance or the engine itself.

However, in order that the engine operates reversibly, one must proceed infinitely slowly, with the result that the power delivered by the reversible engine is in fact zero. This is of no practical interest because, in actual situations, a non-vanishing power is desirable, which implies to obtain a finite amount of work in a finite time. Anyway, one should avoid spending unrealistic efforts to improve engines beyond their actual physical limits; in that respect, the result (5.5) indicates that a way to enhance the efficiency is to raise as much as possible the temperature of the hot reservoir and to reduce the temperature of the cold reservoir.

A cycle carried out in a finite time means a reduction of the efficiency, and therefore one is faced with an alternative: maximization of efficiency or maximization of power. To study this problem, we will follow in this introductory section the didactical example proposed by Curzon and Ahlborn (1975).

5.1.1 Curzon–Ahlborn's Model: Heat Losses

Curzon and Ahlborn assume that the rate of heat exchanged between the working substance at a temperature T_1' and the heat reservoir at a temperature T_1 is described by Newton's law of heat transport, namely

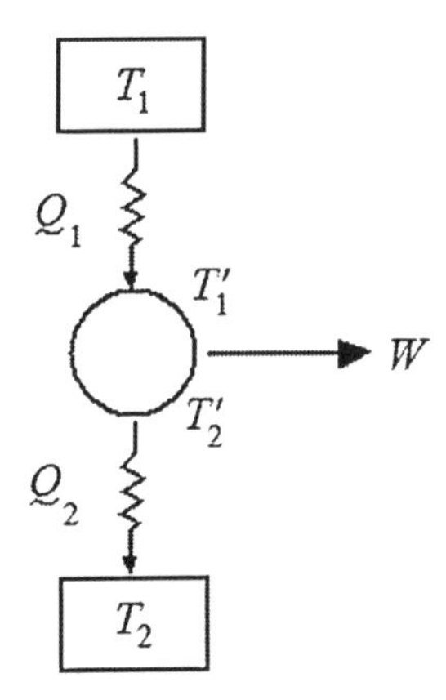

Fig. 5.1 Sketch of the Curzon–Ahlborn's model of heat engine with finite thermal resistance between the working fluid and the heat reservoirs

$$\frac{dQ_1}{dt} = \alpha_1(T_1 - T_1'), \tag{5.6}$$

where the coefficient α_1 depends on the thermal conductivity of the wall separating the working fluid from the source and on geometrical factors, as the area and the width of the wall. Analogously, the rate of heat delivered to the cold source is

$$\frac{dQ_2}{dt} = \alpha_2(T_2 - T_2'), \tag{5.7}$$

where α_2 plays a role analogous to α_1 in the heat exchange between the working fluid and the cold reservoir. A sketch of the Curzon–Ahlborn's model is provided in Fig. 5.1.

It follows from (5.6) and (5.7) that the times required for the engine to absorb an amount of heat Q_1 and to deliver the heat Q_2 are, respectively, given by

$$t_1 = \frac{Q_1}{\alpha_1(T_1 - T_1')}, \quad t_2 = \frac{Q_2}{\alpha_2(T_2' - T_2)}. \tag{5.8}$$

By writing these expressions, we have assumed that the temperatures T_1' and T_2' of the working substance remain constant during the heat exchange process. Furthermore, for the sake of simplicity, it is admitted that the time spent in the adiabatic parts of the cycle is much shorter than the time required for absorbing or delivering the heat. Physically, this implies the absence of frictional losses and inertial effects (e.g. one works with a frictionless piston of zero mass). In this case, the total duration of the cycle is

$$t_{\text{cycle}} \approx t_1 + t_2. \tag{5.9}$$

This theoretical framework has the advantage of allowing explicit calculations of all the quantities related to the way the heat engine is operating, and to introduce explicitly the notion of time. In comparison with reversible thermodynamics, one makes a step forward by capturing more realistic features of the heat engine.

The power P delivered by the engine during one cycle is the ratio of the work W that is performed and the total duration of the cycle, i.e.

$$P = \frac{W}{t_{\text{cycle}}}. \tag{5.10}$$

Now, we want to obtain both efficiency and power as a function of the duration of the cycle. To do this, we need the expression of the entropy produced during a cycle, which is given by (5.4). On the other hand, we know from Chap. 2 that, when an amount of heat Q flows from a system at temperature T to another at temperature T', the amount of entropy produced is

$$\Delta_{\text{irr}} S = -Q \left(\frac{1}{T} - \frac{1}{T'} \right). \tag{5.11}$$

The total entropy generated during one cycle is assumed to result exclusively from irreversible heat exchanges between the reservoirs and the working substance, all other possible sources of irreversibility, as for instance viscosity, are neglected for the moment. In the above problem, one has

$$\Delta_{\text{irr}} S = -Q_1 \left(\frac{1}{T_1} - \frac{1}{T_1'} \right) - Q_2 \left(\frac{1}{T_2'} - \frac{1}{T_2} \right). \tag{5.12}$$

Making use of (5.4), one obtains

$$-\frac{Q_1}{T_1'} + \frac{Q_2}{T_2'} = 0. \tag{5.13}$$

Compared to the equivalent result (5.4), wherein the temperatures T_1 and T_2 of the sources appear, in (5.13) the temperatures of the working fluid during the isothermal expansion and compression are coming out. The analogy between (5.13) and (5.4) for the reversible situation (i.e. with $\Delta_{\text{irr}} S = 0$) underlines that the working fluid itself is behaving in a reversible way, and that the only source of irreversibility is the heat transfer between systems at different temperatures (i.e. between the working fluid and the corresponding heat reservoirs). This is called an *endoreversible heat engine* (de Vos 1992; Andresen 1996; Hoffmann et al. 1997), because irreversibilities find their origin in the contact with the sources but not in internal causes. In particular, for a given compression ratio $V_{\max}/V_{\min}$ of an ideal gas and using the first law of thermodynamics, the values of Q_1 and Q_2 are

$$Q_1 = RT_1' \ln(V_{\max}/V_{\min}), \quad Q_2 = RT_2' \ln(V_{\max}/V_{\min}). \tag{5.14}$$

We now study the conditions under which the power becomes a maximum. The work per cycle W is equal to the difference $Q_1 - Q_2$, and will be the larger

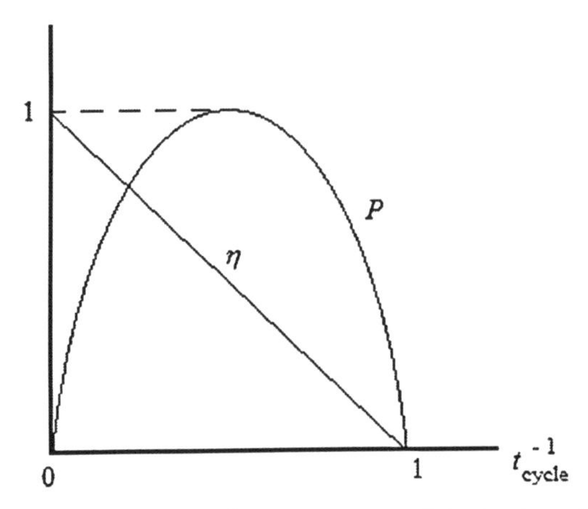

Fig. 5.2 Normalized efficiency and power output of the heat engine as a function of the inverse of the duration of the cycle

when the higher is Q_1 and the lower is Q_2. According to (5.13), this requires to increase T_1' and to reduce T_2', because $Q_1/Q_2 = T_1'/T_2'$. But if T_1' increases and T_2' decreases, the rate of heat exchanged with the sources, expressed by (5.6) and (5.7), diminishes. Under these circumstances, one has slow cycles with high efficiency but small power: each cycle produces much work but one cycle takes a long time. If the rate of heat is increased by lowering T_1' and rising T_2', efficiency is reduced: each cycle produces less work but it takes a shorter time. When the rate is too high, the work per cycle becomes very low in such a way that the power decreases and becomes eventually zero. Between these extreme situations with a vanishing power, there is another one with a maximum power output, as sketched in Fig. 5.2. In this figure, the zero values of power output correspond either to very low speed per cycle (high efficiency but small power) or to very high speed per cycle (low efficiency and low power).

To determine the values of T_1' and T_2' corresponding to the maximum power output, at fixed values of the temperatures T_1 and T_2 of the heat reservoirs, let us write explicitly the expression for the power output, namely

$$P(T_1', T_2') = \frac{Q_1 - Q_2}{t_{\text{cycle}}} = \frac{Q_1 - Q_2}{\frac{Q_1}{\alpha_1(T_1 - T_1')} + \frac{Q_2}{\alpha_2(T_2' - T_2)}}. \tag{5.15}$$

Eliminating Q_1 and Q_2 in terms of T_1' and T_2' by using (5.13), one obtains

$$P(T_1', T_2') = \frac{T_1' - T_2'}{\frac{T_1'}{\alpha_1(T_1 - T_1')} + \frac{T_2'}{\alpha_2(T_2' - T_2)}}. \tag{5.16}$$

After introducing the variables $x_1 \equiv T_1 - T_1'$ and $x_2 \equiv T_2' - T_2$, (5.16) may be rewritten in the more compact form

$$P(x_1, x_2) = \frac{\alpha_1 \alpha_2 x_1 x_2 (T_1 - T_2 - x_1 - x_2)}{\alpha_2 T_1 x_2 + \alpha_1 T_2 x_1 + (\alpha_1 - \alpha_2) x_1 x_2}. \tag{5.17}$$

The conditions for P being extremum are

$$\frac{\partial P(x_1, x_2)}{\partial x_1} = 0, \quad \frac{\partial P(x_1, x_2)}{\partial x_2} = 0. \tag{5.18}$$

This leads to two equations for the values of x_1 and x_2 corresponding to the maximum power,

$$\alpha_2 T_1 x_2 (T_1 - T_2 - x_1 - x_2) = x_1(\alpha_2 T_1 x_2 + \alpha_1 T_2 x_1 + (\alpha_1 - \alpha_2) x_1 x_2), \tag{5.19}$$

$$\alpha_1 T_2 x_1 (T_1 - T_2 - x_1 - x_2) = x_2(\alpha_2 T_1 x_2 + \alpha_1 T_2 x_1 + (\alpha_1 - \alpha_2) x_1 x_2). \tag{5.20}$$

It is verified that

$$x_{2,\max} = \left(\frac{\alpha_1 T_2}{\alpha_2 T_1}\right)^{1/2} x_{1,\max} \tag{5.21}$$

with

$$x_{1,\max} = \frac{T_1 - (T_1 T_2)^{1/2}}{1 + (\alpha_1/\alpha_2)^{1/2}}, \quad x_{2,\max} = \frac{(T_1 T_2)^{1/2} - T_2}{1 + (\alpha_2/\alpha_1)^{1/2}}. \tag{5.22}$$

In particular, it is found from (5.21) and (5.22) that

$$T_1' = C T_1^{1/2}, \quad T_2' = C T_2^{1/2}, \quad C \equiv \frac{(\alpha_1 T_1)^{1/2} + (\alpha_2 T_2)^{1/2}}{\alpha_1^{1/2} + \alpha_2^{1/2}}. \tag{5.23}$$

Introducing (5.22) in (5.17), we obtain for the maximum power output

$$P_{\max} = \frac{\alpha_1 \alpha_2}{(\alpha_1^{1/2} + \alpha_2^{1/2})^2} (T_1^{1/2} - T_2^{1/2})^2. \tag{5.24}$$

The efficiency as a function of x_1 and x_2 is

$$\eta(x, x) = 1 - \frac{T_2'}{T_2'} = 1 - \frac{T_2 + x_2}{T_1 - x_1}. \tag{5.25}$$

Bringing it in the expressions (5.22) yields the important relation

$$\eta_{\text{max power}} = 1 - \left(\frac{T_2}{T_1}\right)^{1/2}. \tag{5.26}$$

It is worth to stress that, in contrast with Carnot's result (5.5), the efficiency at maximum power output relates to the square root of the ratio of temperatures. Both relations (5.24) and (5.26) are of high interest, especially the second one, which is seen to depend only on the values of the temperatures of the reservoirs, in contrast to the first one, which is also function of the heat transfer coefficient introduced in (5.6) and (5.7).

As an illustration, consider a power station working, for instance, between heat reservoirs at 565 and 25°C and having an efficiency of 36%. We want to evaluate this power station from the thermodynamic point of view. It follows from (5.5) that the Carnot's maximum efficiency is 64.1%; our first opinion on the quality of the power station would be rather negative if Carnot's efficiency is our standard for evaluation. However, according to (5.26), its efficiency at maximum power is 40%. Thus, if the objective of the power station is to work at maximum power output, it is seen that the efficiency is not bad compared to the 40% efficiency corresponding to this situation. We can therefore conclude that, to have realistic standards for evaluation of an actual heat engine, one needs to go beyond Carnot's efficiency and to incorporate finite-time considerations.

We have stressed that (5.26) is independent of the heat transfer coefficients, and therefore it is rather general. However, its generality is not so universal as Carnot's result (5.5), because it has been derived by assuming that thermal conduction is the only source of irreversibility, thus neglecting other possible dissipative processes, and also because it depends on the specific form adopted for the heat transfer law. For instance, instead of Newton's law taken in (5.6) and (5.7), one could assume a transfer law of the form

$$\frac{\mathrm{d}Q_1}{\mathrm{d}t} = \beta_1 \left(\frac{1}{T_1'} - \frac{1}{T_1} \right), \quad \frac{\mathrm{d}Q_2}{\mathrm{d}t} = \beta_2 \left(\frac{1}{T_2} - \frac{1}{T_2'} \right). \tag{5.27}$$

In this case, it may be shown that the efficiency at the maximum power is given by

$$\eta(\text{at } P_{\max}) = \frac{1 + (\beta_1/\beta_2)^{1/2}}{2 + \left(1 + \frac{T_2}{T_1}\right)(\beta_1/\beta_2)^{1/2}} \left(1 - \frac{T_2}{T_1}\right). \tag{5.28}$$

Compared to (5.26), this expression depends now explicitly on the heat transfer coefficients. Several other heat transfer laws have been proposed in the literature (e.g. Hoffmann et al. 1997).

5.1.2 Friction Losses

In the previous examples, the presence of internal dissipation has been ignored. Dissipation could be, for instance, produced by friction of the piston along the walls, heat leaks due to imperfect insulation of the reservoirs, internal losses in the working fluid, such as turbulence or chemical reactions. A simple way to introduce friction is to assume that the total force acting on the piston is

$$F = pA - fv, \tag{5.29}$$

where pA is the force exerted by the gas on the piston of area A and fv represents a frictional viscous force, v being the velocity of displacement of the piston and f the friction coefficient. This model was investigated by Rebhan (2002), who assumed that heat transfer effects are arbitrarily fast and that the only source of dissipation is friction. This is the situation analysed below.

During a displacement $\mathrm{d}x$, the work performed by the piston will be

$$\mathrm{d}W = p\,\mathrm{d}V - fv^2(t)\mathrm{d}t, \tag{5.30}$$

and the work corresponding to one cycle is

$$W = Q_1 - Q_2 - f\int_c v^2(t)\mathrm{d}t, \tag{5.31}$$

the integration being performed over the duration of one cycle. The computation of the last term depends on the time dependence of the velocity. Let us qualitatively write

$$W = Q_1 - Q_2 - \frac{4f\delta D^2}{\tau}, \tag{5.32}$$

where $2D$ is the total distance covered by the piston during one cycle, τ is the duration of one cycle, and δ is a fixed numerical parameter characterizing the dependence of the velocity with respect to time (Rebhan 2002). As a result, the power of the engine will be expressed by

$$P = \frac{Q_1 - Q_2}{\tau} - \frac{4f\delta D^2}{\tau^2}. \tag{5.33}$$

For given values of Q_1 and Q_2, P is a function of τ only. The maximum power is then obtained from the condition that $\mathrm{d}P(\tau)/\mathrm{d}\tau = 0$, which yields for the optimum time

$$\tau_{\mathrm{opt}} = \frac{8f\delta D^2}{Q_1 - Q_2}. \tag{5.34}$$

Introducing this result into (5.33), it is found that the maximum power is

$$P_{\mathrm{max}} = \frac{(Q_1 - Q_2)^2}{16f\delta D^2}. \tag{5.35}$$

The efficiency is given by

$$\eta = \frac{W}{Q_1} = \frac{Q_1 - Q_2}{Q_1} - \frac{4f\delta D^2}{\tau Q_1}. \tag{5.36}$$

Substituting the value for the optimum time, one obtains

$$\eta(\max P) = \frac{1}{2}\left(1 - \frac{Q_2}{Q_1}\right) = \frac{\eta_{\mathrm{Carnot}}}{2}, \tag{5.37}$$

which is of course lower than Carnot's efficiency, but also lower than the Curzon–Ahlborn's result (5.26) derived for heat losses only.

Just like (5.26), expression (5.37) does not depend explicitly on the friction coefficient. Combined effects of heat losses and friction lead to rather cumbersome mathematical relations, which are certainly interesting from a practical perspective, but which do not add essential new ideas to the above considerations.

5.2 Economical and Ecological Constraints

In practical situations, there will be a compromise between maximum efficiency and maximum power, and the priorities will be settled by economical and ecological considerations. Note that, from an economic point of view, the efficiency of the engine should be written as

$$\eta_{\text{economic}} \equiv \frac{W \times \Pr(W)}{Q_1 \times \Pr(Q_1)}, \tag{5.38}$$

where $\Pr(W)$ is the price at which the unit of work is sold and $\Pr(Q_1)$ is the price at which the unit of heat which must be supplied to the engine is bought. The economical efficiency will incorporate non-physical information, as the prices of fuel and work (electrical work, for instance). As a matter of fact, the actual economical efficiency is much more complicated than (5.38), because it should also include, for instance, the investment for building the plant and operating costs (as, for instance, the salaries of the workers, other indirect costs as taxes, etc.). Anyway, the price of fuel plays generally a crucial role: if fuel is expensive and the work produced by the engine may only be sold at low price, then the situation with the highest efficiency, i.e. taking most advantage of the fuel, will be the most suitable one. On the contrary, if fuel is cheap and if the work may be sold at high price, the situation of maximum power will be preferred from an economic point of view. Furthermore, the higher the power output is, the faster the investments will be recovered, whereas the higher the efficiency, the faster the fuel expenses will be recovered.

Now, if ecological constraints are incorporated, the criteria may change. For instance, the lowest is the efficiency, the least ecological will be the station, because it will consume more fuel to produce the same amount of work and therefore will produce more pollution. In short, it can be said that the situation with the highest efficiency is the most ecological one, whereas, if fuel is sufficiently cheap, the operation at maximum power output will be the most interesting from the economical standpoint. In general, the situation is rather complex, and will depend on particular circumstances, as for instances, on the rate of demand in energy.

Another example where a similar conflict between efficiency and power arises is in exothermic chemical reactions. Indeed, it is known from equilibrium thermodynamics that the efficiency of the reaction (i.e. the number of moles which are produced for a given initial quantity of reactants) is increased when temperature is lowered, according to the Le Chatelier's principle discussed in Chap. 1. However, from a kinetic perspective, the lower the temperature is, the lower will be the reaction rate. Thus, lowering temperature has two opposite effects: it increases the efficiency but it diminishes the reaction rate. Which temperature is optimal cannot be decided simply from thermodynamic arguments: economical and ecological aspects must be included in the analysis. From an economic standpoint, if the price of the reactant is low, the most satisfactory approach is the fastest one; however, from the ecological point of view, the faster the procedure is, the higher is the consumption of reactant and the higher is the pollution produced by the factory.

Both power and efficiency must therefore be considered in the assessment of actual heat engines. Up to now, we have discussed only two criteria, namely power optimization and efficiency optimization. Other elements may be taken into account to convey the subtleties of the interplay between economical and ecological needs.

One proposal is to minimize the entropy production, which is related to the loss of available work by $W_{\text{lost}} = T_2 \Delta_{\text{irr}} S$; the rate of entropy production over one cycle is, after use is made of expressions (5.12) for $\Delta_{\text{irr}} S$ and (5.8) and (5.9) for t_{cycle},

$$\sigma^s \equiv \frac{\Delta_{\text{irr}} S}{t_{\text{cycle}}} = \frac{Q_1 \left(\frac{1}{T_1'} - \frac{1}{T_1} \right) + Q_2 \left(\frac{1}{T_2} - \frac{1}{T_2'} \right)}{\frac{Q_1}{\alpha(T_1 - T_1')} + \frac{Q_2}{\alpha(T_2' - T_2)}}. \tag{5.39}$$

In terms of x_1 and x_2 defined under (5.16), this may be written as

$$\sigma^s(x_1, x_2) = \frac{\alpha_1 \alpha_2}{T_1 T_2} \frac{x_1^2 x_2 T_2 + x_1 x_2^2 T_1}{\alpha_1 T_2 x_1 + \alpha_2 T_1 x_2 + (\alpha_1 - \alpha_2) x_1 x_2}. \tag{5.40}$$

Another ecological criterion, which is often used, is the maximization of the quantity

$$\Psi = \text{power} - T_2 \times \text{entropy production} = P - T_2 \Delta_{\text{irr}} S. \tag{5.41}$$

By doing so, a high power (a factor of economical interest) is promoted, together with low dissipation (a factor of ecological interest), in such a way that one combines both economical and ecological aims. Physically, (5.41) is related to the useful work available in a process, which is given by $W = W_{\text{R}} - T_2 \Delta_{\text{irr}} S$, W_{R} being the work performed by the reversible process and $T \Delta_{\text{irr}} S$ the work dissipated as heat by irreversibilities. Dividing both terms of this expression by the cycle duration t_{cycle} gives back (5.41). From the expressions (5.17) for the power output and (5.39) for the rate of entropy production, one obtains

$$\Psi(x_1, x_2) = \frac{\alpha_1 \alpha_2}{T_1} \frac{T_1(T_1 - T_2) - (T_1 + T_2) x_1 - 2 T_1 x_2}{\alpha_1 T_2 x_1 + \alpha_2 T_1 x_2 + (\alpha_1 - \alpha_2) x_1 x_2} x_1 x_2. \tag{5.42}$$

The values of the variables for which this quantity becomes a maximum, in the particular case $\alpha_1 = \alpha_2$, are

$$x_{1,\max} = \frac{T_1}{2} \frac{T_1 - T_2}{T_1 + T_2 + 2T_1 A}, \quad x_{2,\max} = A x_{1,\max}, \tag{5.43}$$

where

$$A \equiv \left[\frac{T_2(T_1 + T_2)}{2T_1^2} \right]^{1/2}. \tag{5.44}$$

Introducing (5.43) into (5.25), it is found that the corresponding efficiency is

$$\eta(\text{at}\, \Psi_{\max}) = \eta_{\text{Carnot}} \frac{1 + 2\frac{T_2}{T_1} + \frac{3}{2}\sqrt{2}\left(\frac{T_2}{T_1} + \frac{T_2^2}{T_1^2}\right)^{1/2}}{1 + 3\frac{T_2}{T_1} + 2\sqrt{2}\left(\frac{T_2}{T_1} + \frac{T_2^2}{T_1^2}\right)^{1/2}}. \tag{5.45}$$

As an illustration, Table 5.1 gives the values of efficiency, power, entropy production, and dissipation (product of lowest temperature times entropy production) for a Carnot's heat engine with $T_1 = 400\,\text{K}$, $T_2 = 300\,\text{K}$, $V_{\max}/V_{\min} = 2$, $\alpha = 209\,\text{W K}^{-1}$, under different operational conditions, as considered by Angulo-Brown (1991). Here, it has been taken into account that $Q_1 = RT_1' \ln(V_{\max}/V_{\min})$ and $Q_2 = RT_2' \ln(V_{\max}/V_{\min})$.

Note the practical advantage of the Ψ optimization criterion: though it reduces somewhat the power availability with respect to the maximum power (from 375 to 294 W, i.e. a 25% reduction approximately), it lowers appreciably the entropy production (from 1.08 to 0.3 W K^{-1}) and the dissipation (from 325 to 93 W). Therefore, it yields a satisfactory value for the power output, which is economically interesting, and it considerably brings down the dissipation with respect to the maximum power situation, an appreciated result from the ecological point of view.

Table 5.1 Values of efficiency, power, entropy production, and dissipation for a Carnot's heat engine with $T_1 = 400\,\text{K}$, $T_2 = 300\,\text{K}$, $V_{\max}/V_{\min} = 2$, $\alpha = 209\,\text{W K}^{-1}$, under operation conditions defined by (a) maximum efficiency (Carnot), (b) maximum power output, (c) minimum entropy production, and (d) maximum Ψ (from Angulo-Brown, J. Appl. Phys. **69** (1991) 7465)

	Maximum efficiency	Maximum power	Minimum entropy production	Maximum Ψ
Efficiency (%)	25	13	23	19.2
Power (W)	0	375	114	294
Entropy production (W K^{-1})	0	1.08	0.30	0.31
Dissipation (W)	0	325	90	93

5.3 Earth's Atmosphere as a Non-Equilibrium System and a Heat Engine

Planetary atmospheres are fascinating non-equilibrium thermodynamic systems, whose study has benefited by the impressive observational data provided by space missions. It is well recognized that Earth's atmosphere is of utmost importance for our survival: atmosphere warming and the climate change are among the main topics in the current agenda of Earth's sciences and a topic of much social, economical, and political concern since the last decade. The atmosphere is continuously subject to energy flow, because it absorbs, reflects, transforms, and stores shortwave solar radiation energy, and releases back to the outer space longwave, infrared radiation energy. It represents therefore a paradigmatic example of a non-equilibrium system (Peixoto and Oort 1984; Kleidon and Lorenz 2005). In this section, we present a simplified version of the Earth energy balance, with special attention to the understanding of global warming, and we apply finite-time thermodynamics to evaluate the efficiency of the atmosphere as a heat engine, converting thermal radiation energy into wind kinetic energy.

5.3.1 Earth's Energy Balance

We present here an outline of the main factors influencing the average Earth temperature, based on a rather simplified modelling of the global energy balance. On the one hand, the Earth absorbs an incoming radiation power

$$P_{\mathrm{abs}} = (1 - r)\pi R_{\mathrm{E}}^2 I_{\mathrm{S}}, \tag{5.46}$$

where R_{E} is the radius of the Earth and r is the reflectivity (also called *albedo*), whose mean current value is 0.35. The quantity I_{S} ($\approx 1{,}370\,\mathrm{W\,m^{-2}}$) is the so-called *solar constant*, namely the maximum radiation flux arriving at the upper part of the atmosphere, in a region which is directly facing the Sun, and averaged over all latitudes and seasons and over the day/night alternation. The combination πR_{E}^2 is thus the transversal area of the Earth, intercepting the solar radiation. In virtue of Wien's law of radiation, stating that $\lambda_{\mathrm{max}} = 2{,}897 T^{-1}\,\mu\mathrm{m}$, with T the temperature of the body emitting radiation, the radiation arriving from the Sun, whose surface temperature is $\approx 6{,}000\,\mathrm{K}$, is shortwave. Except for reflection and absorption in the clouds, the atmosphere is essentially transparent to this kind of radiation, which is directly striking the surface of the Earth.

The power emitted by the Earth is

$$P_{\mathrm{emit}} = 4\pi R_{\mathrm{E}}^2 e\sigma T_{\mathrm{E}}^4, \tag{5.47}$$

where e is the emissivity, whose approximate value is 0.58, σ the Stefan–Boltzmann constant, and T_{E} the average temperature of the Earth, a quantity on which we will centre our attention. The numerical factor 4 in (5.47) assumes that the Earth is radiating over the whole surface, and not only the surface facing to the Sun. According to Wien's law, the radiation emitted by the Earth is longwave, and behaves as infrared radiation. As commented below, several gases in the atmosphere absorb and re-emit this type of radiation, in contrast to the incoming radiation.

To find the average temperature of the Earth, the emitted and absorbed powers are made equal, i.e.

$$(1-r)\pi R_{\mathrm{E}}^2 I_{\mathrm{S}} = 4\pi R_{\mathrm{E}}^2 e\sigma T_{\mathrm{E}}^4, \tag{5.48}$$

from which follows that

$$(1-r)\frac{I_{\mathrm{S}}}{4} = e\sigma T_{\mathrm{E}}^4, \tag{5.49}$$

and the resulting average temperature of Earth is given by

$$T_{\mathrm{E}} = \left[\frac{(1-r)I_{\mathrm{S}}}{4e\sigma}\right]^{1/4}, \tag{5.50}$$

and found to be of the order of 290 K. In the balance equation (5.48), some internal sources of heat – volcanoes and radioactive sources – have been neglected, but should be incorporated in a more detailed analysis. According to (5.50), the Earth reflectivity r and emissivity e have a deep influence on the Earth temperature; for instance, a reduction of these parameters implies an increase of the average temperature. If there was no greenhouse effect, whose effect is to reduce e, the temperature of the Earth would be some 30°C colder than it is today.

The reflectivity, or albedo, r and the emissivity e are fractional quantities with values comprised between 0 and 1; for a perfect black body, one would have $r = 0$ and $e = 1$. The albedo r and emissivity e depend on several variables, mainly temperature and composition. The actual value of r results from a balance between the extension of ices, snows, deserts, and clouds, which have high values of albedo (high reflectivity) and the surface occupied by forests and seas, with low values.

5.3.2 Global Warming

Global warming is nowadays the object of an enormous amount of research at the theoretical, observational, and numerical levels. The role of human activities and their implications on climatic changes actually deserves an increasing international interest. We will focus here on some qualitative ideas which allow seizing the key features of these phenomena. Truly, the first studies about global warming were carried out by Arrhenius in 1896, but they did not receive a widespread attention until the end of the 1960s, when the first data on the increase of CO_2 in the atmosphere became available.

The average value of Earth's temperature is given by (5.50). Concerning the emissivity e, its value depends especially on the concentration of those gases sensible to infrared radiation. The major natural infrared absorbers in the atmosphere are water vapour, clouds, and carbon dioxide. The current concerns are related with CO_2 and some other gases, mainly CH_4, N_2O, and CFC, whose concentration has considerably increased during the last century as a consequence of human activities. For instance, the abundance of CO_2 in the atmosphere has increased from 315 ppmv (parts per million in volume) in 1950 to 350 ppmv in 1990, as a consequence of fossil fuel burning and deforestation (for the sake of comparison, let us mention that in 1750 the CO_2 atmospheric concentration was some 280 ppmv). CO_2 remains in the atmosphere for about 500 years. The increase in CH_4 has been much less (1% per year, at the current rate). This gas is produced by bacteria that decompose organic matter in oxygen-poor environments, as marshes or rice paddles, burning of forests and grasslands, or the guts of termites in the digestive tracts of cattle, sheep, pigs, and other livestock. CH_4 remains in the atmosphere for 7–10 years and each molecule is some 25 times more efficient than a molecule of carbon dioxide in warming the atmosphere. Other gases with much effect are CFC (chlorofluorocarbons), which also deplete ozone in the stratosphere. The main sources are leaking air conditioners and refrigerators, or evaporation of some industrial solvents. CFC remain in the atmosphere for 65–110 years, and a molecule of CFC has 10,000–20,000 times the impact on global warming compared to a molecule of carbon dioxide. Nitrous oxide (N_2O) is released mainly from the breakdown of nitrogen fertilizers in soil, and nitrate-contaminated groundwater. Its average stay in the atmosphere is 150 years and the effect on global warming from each molecule is 230 times more important than a CO_2 molecule. The main contributions to the greenhouse effect are 54% CO_2, 18% CH_4, 14% CFC, and 6% N_2O.

The net planetary radiative forcing is dominated by the presence of the water vapour whose contribution to the total value of e is of the order of 60%, and whose concentration is strongly related to that of CO_2. Note that a relatively small variation of temperature may produce an important feedback in the water vapour concentration. Another possible multiplicative feedback arises from the relationship between ice extension and the albedo r: a warmer planet will have less ice meaning a reduction in the reflectivity, which in turn will produce an increase of the temperature, thus leading to a further reduction of ice extension and so on.

The main uncertain factor influencing global warming and its impact on climate concerns the role of clouds on the radiative balance because clouds contribute both to warming, by absorbing infrared radiation, and to cooling, by reflecting a part of the incoming radiation. The relative importance of these two antagonistic contributions depends on the altitude of the clouds: for low ones, warming plays the leading part, whereas for high ones, cooling is the dominating effect. Another source of query is the response of the ocean to variations of the concentration of the greenhouse gases and to perturbations in the oceanic currents distributing the heat from the Equator to the Poles. Furthermore, since the oceans have an enormous thermal inertia, the warming process is expected to be slow; this means that it may take many years before warming effects become perceptible.

The estimations about the CO_2 atmospheric concentration are deeply bound to the evolution of CO_2 emissions. It is estimated that, in 2100, CO_2 concentration could be close to 550 ppmv. Solutions are not easy to find because of the clash of interests between the necessity to reduce the use of fossil fuels and the increasingly widespread and growing need for energy, thus leading to serious policy confrontations.

Useful information about climate change may also be drawn from a study of the global entropy balance of Earth. The convective motions of upward hot air and down falling cold air and several phenomena observed in our everyday life imply the appearance of some structure and order, which must be compatible with a global increase of entropy of the universe. It turns out that, globally, the Earth is a very dissipative system. Indeed, it receives photons at $T_S \approx 6{,}000\,\mathrm{K}$ and emits them at $T_E \approx 290\,\mathrm{K}$. For each incoming photon, some 20 photons are re-emitted to the outer space. Other sources of entropy production, as for instance heat transfer between hot and cold regions, or turbulence of winds and oceanic currents are unavoidable and should be incorporated in more detailed analyses. Several models taking into account the role of entropy production on the climate have been proposed. Based on simplified atmospheric models, Nicolis and Nicolis (1980) and Paltridge (1981) explored whether or not the climate is governed by an extremal principle related with the entropy production. They found reasonable values for the most significant climatic variables by maximizing the transport part of the entropy production. We will not enter into this rather specialized topic, which throws anyway a light on the contribution of the second law to a better understanding of the global climate system. For more recent models, see for instance Pujol and Llebot (1999) and Ozawa et al. (2003).

5.3.3 Transformation of Solar Heat into Wind Motion

In this section, we will apply finite-time thermodynamics to the Earth atmosphere, considered as a heat engine driven by solar radiation and producing useful work in the form of wind motion. On the yearly average, the quantity of heat transferred per unit time by solar radiation to the surface of the Earth is, per unit area,

$$\dot{Q}_S = \frac{1-r}{4} I_S \approx 223\,\mathrm{W\,m^{-2}}, \tag{5.51}$$

where, as in Sect. 5.3.2, I_S is the solar constant and r is the albedo. Since the average energy of wind motion is approximately $7\,\mathrm{W\,m^{-2}}$, the efficiency of conversion of solar radiation energy into mechanical energy of wind is about $7/223 \approx 3\%$.

A rough upper bound estimation for the conversion of solar heat into work is obtained by taking for the temperature of the hot reservoir $T_1 = 313\,\mathrm{K}$, corresponding to the average Equator temperature of 40°C, and for the temperature of the cold reservoir $T_2 = 233\,\mathrm{K}$, the average temperature at the Poles. After introducing these values in the expression of the Carnot's efficiency, one finds 25.5%, which is much higher than the 3% observed value. Applying the maximum power efficiency of the Curzon–Ahlborn's engine, the result falls down to 14%. However, instead of Curzon–Ahlborn's approach, we prefer to follow an analysis by Gordon and Zarmi (1989), which is more illustrative and pedagogical, by keeping nevertheless simplicity.

These authors investigate that the global-scale motion of wind in convective Hadley cells, namely those cells in which hot air rises towards colder regions along the upper atmosphere, falls down on the surface of the Earth, and moves back to hotter regions along the Earth surface. Gordon and Zarmi model these various motions by means of a Carnot's engine, with two isothermal and two adiabatic branches. The first isothermal corresponds to the atmosphere absorbing solar energy at low altitudes, at temperature T_1', not exactly equal to the Earth surface temperature T_1. The second isothermal branch, at temperature T_2', is related to the atmosphere at high altitudes, where air rejects heat to the universe. The adiabatic branches correspond to the rising and falling air currents. Furthermore, it is assumed that the time spent on the isothermal branches is equal, and that the time spent on the adiabatic branches is negligible. The values of the hotter T_1' and colder temperatures T_2' of the working fluid will be determined to maximize the power conversion, in analogy with Sect. 5.1.

In the present case, the heat exchange mechanisms are of radiative type, and therefore the analysis of Sect. 5.1, based on Newton's law of transfer, is not directly applicable. This justifies that we re-examine the problem. The power (per unit area) available for work will be

$$P = \dot{Q}_1 - \dot{Q}_2 = \dot{Q}_{\mathrm{S}} - \sigma\left(\frac{T_1'^4}{2} + \frac{T_2'^4}{2}\right), \tag{5.52}$$

with the identifications $\dot{Q}_1 = \dot{Q}_{\mathrm{S}} - \frac{1}{2}\sigma T_1'^4$, $\dot{Q}_2 = \frac{1}{2}\sigma T_2'^4$, and $\sigma = 5.67 \times 10^{-8}\,\mathrm{W\,m^{-2}\,K^{-4}}$ the Stefan–Boltzmann constant, an upper dot means time derivation. The factor $\frac{1}{2}$ in (5.52) is a consequence of the assumption that air spends half the time on one isotherm and half the time on the other. We know from Sect. 5.1 that this does not exactly reflect the reality, but since we are only interested in obtaining some orders of magnitude, it is not extremely important.

On the other side, the entropy balance may be expressed as

$$-\frac{\dot{Q}_1}{T_1'} + \frac{\dot{Q}_2}{T_2'} = -\frac{\dot{Q}_{\mathrm{S}}}{T_1'} + \frac{\sigma T_1'^3}{2} + \frac{\sigma T_2'^3}{2} = 0, \tag{5.53}$$

where it is assumed that the cycle is endoreversible and the entropy flux is simply $\dot{Q}_i/T$, with $\dot{Q}_i$ given by the same relations as in (5.52).

The temperatures T_1' and T_2' are derived by maximizing the mechanical power (5.52), subject to the constraint (5.53). Since $\dot{Q}_S$ is fixed, we may maximize

$$\Phi(T_1', T_2') \equiv \frac{\sigma}{2}\left(T_1'^4 + T_2'^4\right) + \lambda\left(\frac{\dot{Q}_S}{T_1'} - \frac{\sigma T_1'^3}{2} - \frac{\sigma T_2'^3}{2}\right), \tag{5.54}$$

where λ is a Lagrange multiplier taking into account condition (5.53). The final results are

$$T'_{1,\max} = 229\,\mathrm{K}, \quad T'_{2,\max} = 192\,\mathrm{K}, \quad \text{and} \quad P_{\max} = 17.1\,\mathrm{W\,m^{-2}}. \tag{5.55}$$

The resulting efficiency is 7%, much closer to the actual one (3%) than the Carnot's and Curzon–Ahlborn's naïve estimations.

In more sophisticated formulations, the hot isotherm is considered to be on the dayside of the Earth and the cold one on the nightside. Dissipation of the wind energy is also introduced in some analyses. Furthermore, instead of assuming a homogeneous model for the Earth, one may consider several convective cells (the so-called *Hadley cells*) from the Equator to the Pole. Including all these refinements yields results that are very close to the actual observed values.

5.4 Problems

5.1. *Heat engine with heat leak.* Assume that, parallel to the Curzon–Ahlborn's engine studied in Sect. 5.1, there is a direct heat leak from the hot source at temperature T_1 and the cold source at temperature T_2. Then the efficiency must be defined as

$$\eta(\tau) \equiv \frac{W(\tau)}{Q_{\text{leak}}(\tau) + Q_1(\tau)} = \frac{W(\tau)}{Q_{\text{leak}}(\tau) + \frac{W(\tau)}{1-\tau}},$$

where $\tau \equiv T_2'/T_1'$ and $Q_{\text{leak}}(\tau) = K_{\text{leak}} t_{\text{total}}(\tau)(T_1 - T_2)$, with K_{leak} the heat transfer coefficient of the leak and $t_{\text{tot}}(\tau)$ the total duration of the cycle for the Curzon–Ahlborn's engine. Note that, in contrast with the latter, Carnot's efficiency is never reached, because due to the heat leak, Q_{leak} diverges, and the efficiency of this very slow cycle is very low due to the high value of the heat leak. Find the maximum efficiency and the maximum power for this engine.

5.2. *The Curzon–Ahlborn's engine.* Show that, in the Curzon–Ahlborn's engine, the ratio t_1/t_2 associated to the two isotherm branches is given by $t_1/t_2 = (\alpha_2/\alpha_1)^{1/2}$.

5.3. *Maximizing Curzon–Ahlborn's engine.* Assume that the total duration of the cycle given by (5.8) and (5.9) is fixed. Then, maximizing the power is equivalent to maximize the difference $Q_1 - Q_2$. Performing this maximization, by introducing the constraint on the duration of the cycle by means of a Lagrange multiplier λ, i.e. by imposing

$$\delta\left[Q_1 - Q_2 + \lambda\left(\frac{Q_1}{\alpha_1(T_1 - T_1')} + \frac{Q_2}{\alpha_2(T_2' - T_2)}\right)\right] = 0$$

and recalling (5.13), show that

$$\frac{x_{1,\max}}{x_{2,\max}} = \left(\frac{\alpha_2 T_1}{\alpha_1 T_2}\right)^{1/2},$$

which is the result (5.21).

5.4. *Minimum entropy production criterion.* Show that the minimization of the total entropy production $\Delta_{\text{irr}}S$, under the condition of fixed total duration of the cycle, given by (5.8) and (5.9), yields

$$\frac{x_{1,\max}}{x_{2,\max}} = \left(\frac{\alpha_2}{\alpha_1}\right)^{1/2} \frac{T_1}{T_2}.$$

Note that this result is different from the previous problem.

5.5. *Maximization process in Curzon–Ahlborn's engine.* Assume that the environment temperature is T_0, in such a way that the loss power is $T_0\sigma$ instead of $T_2\sigma$ as taken in (5.41). By analogy with (5.41), study the maximization of $\Psi_0 \equiv P - T_0\sigma$, with T_0 constant, and show that the efficiency at maximum Ψ_0 is (Z. Yan, J. Appl. Phys. **73** (1993) 3583)

$$\eta(\text{at}\max\Psi_0) = 1 - \frac{T_2}{T_1}\left(\frac{T_1 + T_0}{T_2 + T_0}\right)^{1/2}.$$

5.6. *Finite-time analysis of refrigerators.* One of the basic quantities in refrigerators is the so-called *coefficient of performance* (COP), defined as

$$\text{COP} \equiv \frac{\text{heat extracted from the cold source}}{\text{work supplied to the engine}}.$$

Performing an analysis similar to that of the Curzon–Ahlborn's heat engine, find the coefficient of performance at maximum cooling power. *Hint*: see Agrawal and Menon 1990.

5.7. *Maximum power of solar thermal heat engines.* A simple endoreversible modelling of solar thermal heat engines is to assume that the heat transfer between the hot source and the engine is given by a Stefan–Boltzmann law

$\dot{Q}_1 = K(T_1^4 - T_1'^4)$, and that the heat transfer to the cold source is reversible. Prove that the power–efficiency relation is

$$P(\eta) = \frac{\eta K\left[(1-\eta)^4 T_1^4 - T_2^4\right]}{(1-\eta)^4},$$

with $\eta \equiv 1 - (T_2/T_1')$. Show that $4T_1'^5 - 3T_2T_1'^4 - T_1^4T_2 = 0$ corresponds to the maximum power.

5.8. *Energy conversion efficiency of a simple photovoltaic cell.* Assume that the device is characterized by only two energy levels with a distance apart equal to E_G. Photons, with energy less than E_G, simply move through the device without leaving to it any energy. Photons with energy higher than E_G excite particles from the lower to the higher levels, and perform a useful work equal to E_G whereas the excess energy is dissipated as heat. The energy efficiency may be defined as

$$\eta \equiv \frac{\text{useful power}}{\text{incident power}} = \frac{E_G \int_{E_G}^{\infty} \frac{E^2 \mathrm{d}E}{\exp(\beta E)-1}}{\int_{E_G}^{\infty} \frac{E^3 \mathrm{d}E}{\exp(\beta E)-1}},$$

where $\beta = (k_B T)^{-1}$. In the second equality, it is assumed that the energy distribution of the incoming radiation is Planck's black body distribution at temperature T. (a) Show that the maximum efficiency is achieved for materials whose energy gap E_G is $E_{G,opt} = 2.17 k_B T$. (b) Estimate E_G if $T = 5{,}760\,\mathrm{K}$, the Sun surface temperature. (c) Evaluate the maximum efficiency. Truly the actual efficiency will be less than this value (see, for instance, Landsberg and Tonge 1980; de Vos 1992).

5.9. *Average temperature of the atmosphere.* Determine the value of the average Earth temperature from (5.50) in the following cases. (a) First, assume that the reflectivity r is zero and the emissivity e is unit, corresponding to the case that the Earth behaves as a perfectly black body. (b) The reflectivity r has the actual value, 35%, and $e = 1$ (no greenhouse effect), and compare the result with the current average temperature, namely 288 K. (c) Determine the value of the emissivity e in order that Earth's average temperature has its current actual value.

5.10. *Naïve approach to global warming.* A main feature of global warming due to increase of atmospheric CO_2 concentration is that emissivity e is lowered, because CO_2 absorbs the infrared radiation emitted by the Earth. Assume, for simplicity, that $e = 0.58 - 0.53 \ln(c_{CO_2}/c_{CO_2,0})$, where c_{CO_2} is the CO_2 concentration in the atmosphere and $c_{CO_2,0} \approx 350\,\mathrm{ppmv}$ is a reference value (in parts per million in volume). Determine the temperature increase if the CO_2 concentration is raised, respectively, to 700 and 1,400 ppmv. *Comment*: This model is pedagogically illustrative but excessively simplistic, because it ignores important feedbacks of the temperature increase on the water vapour content, the cloud formation, and the albedo.

5.11. *Sun's and Earth's temperatures.* The current value of the solar constant, namely the maximum radiation flux arriving at the upper part of the atmosphere in a region, which is directly facing the Sun, is equal to 1,367 $\mathrm{W\,m^{-2}}$. (a) Calculate the temperature at the surface of the Sun, by assuming that it radiates as a black body. (b) Determine the variation of the solar constant if the temperature of the Sun is increased by 1% with respect to its current value, obtained before. (c) To which extent will the average temperature of the Earth change if the albedo and the emissivity were kept constant? (Stefan–Boltzmann constant, $\sigma = 5.67 \times 10^{-8}\,\mathrm{W\,m^{-2}}$; distance Sun–Earth $= 1.50 \times 10^{11}$ m, radius of the Sun $= 6.96 \times 10^{8}$ m).

5.12. *Carnot engine at finite time.* A Carnot engine is working between two heat reservoirs at 400 K and 300 K. a) Compare the maximum efficiency with the efficiency at maximum power. b) If the engine receives 1000 J from the hot reservoir, evaluate the entropy produced per cycle in the situation at maximum power. c) If the duration of the cycle at maximum power is 20 s, find the value of the maximum power delivered by the engine. d) If the transfer coefficients between the engine and the reservoirs do not depend on temperature, find the efficiency and power of this engine when the cycle duration is 60 seconds.

5.13. *Ice comet.* A spherical ice comet of 10^5 Kg describes an eccentric elliptical orbit around the Sun, with major radius $2 \cdot 10^8$ km and minor radius $1 \cdot 10^8$ km. Evaluate which proportion of the mass of the comet will evaporate during one orbit. Assume that the albedo coefficient is 0.8 for ice and 0.2 for liquid water.

5.14. *Naïve scaling relation for the life-span of stars.* Bigger stars have shorter lifespans than smaller stars. The physical idea is that they need higher internal pressures, and therefore higher temperatures, to resist the higher gravitational pressures. Since the energy emitted per unit time, which is proportional to the loss of mass of the star per unit time in nuclear reactions, is proportional to T^4, according to Stephan-Boltzmann's law, they will consume faster their mass. a) Make a simple estimation of the scaling exponent α relating the lifespan of the star t_{star} to the mass M of the star through $t_{\text{star}} = AM^{-\alpha}$, with A a constant. (*Hint*: equate the pressure of an ideal gas to the average gravitational pressure determined on dimensional grounds from G, M and R, the gravitational constant and the mass and radius of the star. Note that if one makes the simplistic assumption that different stars have the same density, it is found that $t_{\text{star}} = AM^{-7/3}$). b) According to this estimate, how long will last a star of twice the mass of the Sun, if the total lifespan of the Sun will be of some 10^{10} years? Will such a star be able to sustain intelligent life in its planetary system, if any?

5.15. *Minimum entropy production.* a) Find the hot and cold temperature ratios x_1 and x_2 defined in (5.16) which extremize the entropy production (5.40) (assume, for simplicity, that $\alpha_1 = \alpha_2$). b) Determine the corresponding value for the efficiency as obtained from (5.25). c) Evaluate the corresponding efficiency and compare it with Carnot's maximum efficiency and with efficiency at maximum power.

Chapter 6
Instabilities and Pattern Formation

Dissipative Structures Far from Equilibrium

The notion of irreversibility is frequently associated with a tendency of the systems towards spatial homogeneity and rest, i.e. towards equilibrium. Although this observation is true in isolated systems, it is no longer verified in closed and open systems, which may exhibit organized structures in space and time. This organization arises as a consequence of occurrence of instabilities in systems driven far from equilibrium by external forces. Instabilities may be considered as phase transition phenomena, because they manifest under the form of a discontinuous change of the physical behaviour at a critical value of an external control parameter.

Such non-equilibrium processes are, however, rather different from those occurring in equilibrium phase transitions. In equilibrium thermodynamics, there is an evident link between the second law and stability. In Chap. 1, we have derived an explicit criterion of stability by imposing that the second variation of the entropy is negative definite. In non-equilibrium, such a universal criterion of stability does not exist because it is generally not possible to construct thermodynamic potentials depending on the whole set of variables. The general problem of stability is then conventionally carried out by submitting the reference state, for instance, a stationary non-equilibrium state, to an arbitrary disturbance. These disturbances may be caused deliberately from the outside by external agents, like a temperature or pressure gradient, or induced by irregularities or imperfections of the apparatus or still by internal fluctuations of the system (see Chap. 11).

A proof of stability is based on the time variation of the perturbation: if the perturbation decays in a finite time, the reference state is *stable*, and it is *asymptotically stable* if the disturbance tends to zero in the long time limit. If the disturbance grows with time, the reference state is *unstable*. In this case, the mathematical solution of the problem cannot survive in the physical world and the behaviour of the system will be governed by another mathematical solution: it is said that the solution suffers a bifurcation. When the disturbance does neither decrease nor increase with time, one speaks about *neutral* or *marginal stability*, and the corresponding values taken by the relevant parameters are said to be *critical.*

When the perturbations remain infinitesimally small, the approach is linear. Several illustrations – like Bénard–Marangoni's convective instability, Taylor's instability, and Turing's reaction–diffusion instability – are examined in the present chapter. One of the most interesting aspects of the stability theory is the prediction of organized structures, taking the form of spatial and temporal patterns, which appear beyond the linear critical instability threshold. These patterns are also referred to as *dissipative structures*, a term introduced by Prigogine to characterize ordered systems maintained far from equilibrium by external constraints. The mechanisms underlying the development of patterns are essentially non-linear and modelled by coupled non-linear partial differential equations whose treatment is rather intricate. Various examples are treated in the second part of this chapter.

In Sects. 6.1 and 6.2, we briefly discuss the mathematical aspects of the linear and non-linear theories of stability. When the perturbations are infinitesimally small, the problem is linear and the solution may be decomposed into normal modes. This technique has been made popular by the celebrated textbook of Chandrasekhar (1961). As soon as the amplitude of the perturbations becomes finite, the linear approach is no longer appropriate and must be replaced by more sophisticated non-linear techniques or direct numerical solutions. A detailed analysis of these methods can be found in specialized monographs (Joseph 1976; Drazin and Reid 1981; Cross and Hohenberg 1993; Bodenshatz et al. 2000); here we shall essentially focus on the so-called *amplitude method* which, in our opinion, is the simplest and amongst the most powerful.

One of the most attractive aspects of the (non-linear) stability theory is the prediction of self-organization phenomena, in the form of spatial or temporal patterns, beyond the critical threshold. Pattern formations are important and appear in a multitude of natural phenomena being a central theme of research in modern physics. They provide a major feature of the dynamics of several processes observed not only in hydrodynamics but also in electricity, optics, material science, oceanography, geophysics, astrophysics, chemistry, and living systems. Thermal convection in fluids is one of the most representative problems and is discussed in Sect. 6.3. Indeed, most of the recent developments about non-equilibrium processes in the context of instabilities and chaos have been widely inspired from the study of this example.

In this chapter, the approach is more dynamical than thermodynamical, i.e. it is based on the study of evolution equations rather than on state and constitutive thermodynamic relations. However, from a conceptual point of view, it is very important to understand that thermodynamics does not preclude the formation of patterns provided the systems are driven far enough from equilibrium. Such a topic represents an important contribution towards a better knowledge of non-equilibrium systems and therefore it finds naturally a place in this book.

6.1 The Linear Theory of Stability

This theory is concerned with stability of systems subject to infinitesimally small perturbations about a reference state. Most physical and chemical systems obey evolution equations of the form

$$\frac{\mathrm{d}\boldsymbol{a}}{\mathrm{d}t} = f(\boldsymbol{a}, \nabla \boldsymbol{a}; \boldsymbol{r}, t), \tag{6.1}$$

where f is a continuous and twice differentiable function depending on the field variables $\boldsymbol{a}$, for instance the velocity $\boldsymbol{v}$ and the temperature T fields, and their gradients, $\boldsymbol{r}$ designates the position vector, and t the time. Assume that there exists a "reference" solution corresponding to some specific constraints

$$\boldsymbol{a} = \boldsymbol{a}_{\mathrm{r}}. \tag{6.2}$$

Such a solution is also called unperturbed or basic, and may be time dependent (for instance periodic) or time independent (i.e. stationary). The question now arises about the stability of this particular solution. Therefore, one examines the response of the system with respect to a disturbance $\boldsymbol{a}'(\boldsymbol{r}, t)$ so that the generalized coordinates $\boldsymbol{a}(\boldsymbol{r}, t)$ become

$$\boldsymbol{a}(\boldsymbol{r}, t) = \boldsymbol{a}_{\mathrm{r}} + \boldsymbol{a}'(\boldsymbol{r}, t), \tag{6.3}$$

where the quantity $\boldsymbol{a}'(\boldsymbol{r}, t)$ is the "perturbation" which is supposed to be small enough so that all non-linear terms can be neglected in (6.1). Substituting (6.3) in the evolution equations (6.1) and developing them around the reference solution yields a linear set of differential equations of the form

$$\frac{\mathrm{d}\boldsymbol{a}'}{\mathrm{d}t} = \mathcal{L}\boldsymbol{a}', \tag{6.4}$$

where the linear operator $\mathcal{L}$ is to be evaluated at the reference unperturbed state. Since linear stability implies stability with respect to all possible infinitesimal disturbances, we have to examine the reaction of the system to all such disturbances. Practically, this will be accomplished by expressing $\boldsymbol{a}'$ as a superposition of a complete ensemble of n normal modes, generally selected as Fourier modes in a two-dimensional wave vector space ($\boldsymbol{k} = k_x, k_y$),

$$\boldsymbol{a}'(\boldsymbol{x}, z, t) = \sum_n \boldsymbol{W}_n(z) \exp(\sigma_n t) \exp(\mathrm{i}\boldsymbol{k}_n \cdot \boldsymbol{x}), \tag{6.5}$$

and by examining the stability with respect to each individual mode n; $\boldsymbol{x}(x, y)$ is the position vector in the plane normal to the z-axis. The quantity $\boldsymbol{W}_n(z)$ is called the *amplitude* and σ_n is the growth rate of the disturbance,

$$\sigma_n = \operatorname{Re}\sigma_n + \mathrm{i}\operatorname{Im}\sigma_n, \tag{6.6}$$

generally a complex and $\boldsymbol{k}$-dependent quantity. The $\boldsymbol{k}$ dependence of the growth rate describes the spatial symmetry of the system; in rotationally invariant systems, the σ_n's will only depend on the modulus $|\boldsymbol{k}|$, whereas in anisotropic systems like nematic liquid crystals there is an angle between the direction of anisotropy and $\boldsymbol{k}$.

The requirement that the field equations have non-trivial solutions leads to an eigenvalue problem for the σ_n's. The stability problem is completely determined by the sign of the real part of the σ_n's:

- If one single $\operatorname{Re}\sigma_n > 0$, the system is unstable.
- If $\operatorname{Re}\sigma_n < 0$ for all the values of n, the system is stable.
- If $\operatorname{Re}\sigma_n = 0$, the stability is *marginal or neutral.*

In the case of marginal stability, there corresponds to each value of $\boldsymbol{k}$ a critical value of the control parameter, say the temperature difference ΔT in Bénard's problem, a characteristic velocity in flows through a pipe or the angular velocity in Taylor's problem, for which $\operatorname{Re}\sigma_n = 0$. All these critical values define a curve of marginal stability, say ΔT vs. $\boldsymbol{k}$, whose minimum $(\Delta T_{\mathrm{c}}, \boldsymbol{k}_{\mathrm{c}})$ determines the critical threshold of instability. In stability problems, it is convenient to work with non-dimensional control parameters like the Rayleigh number Ra in Bénard's instability, the Reynolds number Re for the transition from laminar to turbulent flows, or the Taylor number Ta in presence of rotation; therefore, the marginal curves and the corresponding critical values are generally expressed in terms of these non-dimensional quantities.

In several problems, it is postulated that $\operatorname{Re}\sigma = 0$ implies $\operatorname{Im}\sigma = 0$, this conjecture is called the principle of *exchange of stability*, which has been demonstrated to be satisfied in the case of self-adjoint problems; in this case a stationary state is attained after the onset of the instability. If $\operatorname{Re}\sigma = 0$ but $\operatorname{Im}\sigma \neq 0$, the onset of instability is initiated by oscillatory perturbations and one speaks of *overstability* or *Hopf bifurcation.* This kind of instability is observed, for instance, in rotating fluids or fluid layers with a deformable interface. The condition $\operatorname{Re}\sigma_n > 0$ for at least one value of n is a sufficient condition of instability; on the contrary, even when *all* the eigenvalues are such that $\operatorname{Re}\sigma_n < 0$, one cannot conclude in favour of stability as one cannot exclude the possibility that the system is unstable with respect to finite amplitude disturbances. It is therefore worth to stress that a linear stability analysis predicts only *sufficient conditions of instability.*

6.2 Non-Linear Approaches

As soon as the amplitude of the disturbance is finite, the linear approach is not appropriate and must be replaced by non-linear theories. Among them one may distinguish the "local" and the "global" ones. In the latter, the details of the motion and the geometry of the flow are omitted, instead attention is focused on the behaviour of global quantities, generally chosen as a positive definite functional. A typical example is Lyapounov's function; according to Lyapounov's theory, the system is stable if there exists a functional Z satisfying $Z > 0$ and $\mathrm{d}Z/\mathrm{d}t \leq 0$. In classical mechanics, an example of Lyapounov's function is the Hamiltonian of conservative systems. Glansdorff and Prigogine (1971) showed that the second variation of entropy $\delta^2 S$ provides an example of Lyapounov's functional in non-equilibrium thermodynamics. The main problems with Lyapounov's theory are:

1. The difficulty to assign a physical meaning to the Lyapounov's functional
2. The fact that a given situation can be described by different functionals
3. That in practice, it yields only sufficient conditions of stability

We do no longer discuss this approach and invite the interested reader to consult specialized works (e.g. Movchan 1959; Pritchard 1968; Glansdorff and Prigogine 1971). Here we prefer to concentrate on the more standard "local" methods where it is assumed that the perturbation acts at any point in space and at each instant of time. We have seen that the solution of the linearized problem takes the form $\exp(\sigma_n t)$ and that instability occurs when the growth rate becomes positive, or equivalently stated, when the dimensionless control parameter R exceeds its critical value R_{c}. For values of $R > R_{\mathrm{c}}$, the hypothesis of small amplitudes is no longer valid as non-linear terms become important and will modify the exponential growth of the disturbances. Another reason for taking non-linear terms into account is that the linear approach predicts that a whole spectrum of horizontal wave numbers become unstable. This is in contradiction with experimental observations, which show a tendency towards simple cellular patterns indicating that only one single wave number, or a small band of wave numbers, is unstable.

Non-linear methods are therefore justified to interpret the mechanisms occurring above the critical threshold. The problem that is set up is a non-linear eigenvalue problem. Unfortunately, no general method for solving non-linear differential equations in closed form has been presented and this has motivated the development of perturbation techniques. A widely used approach is the so-called *amplitude method* initiated by Landau (1965) and developed by Segel (1966), Stuart (1958), Swift and Hohenberg (1977), and many others. It is essentially assumed that the non-linear disturbances have the same form as the solution of the linear problem with an unknown time-dependent amplitude. Explicitly, the solutions will be expressed in terms of the eigenvectors $\boldsymbol{W}(z)$ of the linear problem in the form

$$\boldsymbol{a}'(\boldsymbol{x}, z, t) = A(t) \exp(\mathrm{i}\boldsymbol{k} \cdot \boldsymbol{x})\, \boldsymbol{W}(z), \tag{6.7}$$

where $A(t)$ denotes an unknown amplitude, generally a complex quantity. In the linear approximation, $A(t)$ is proportional to $\exp(\sigma t)$ and obeys the linear differential equation

$$\frac{\mathrm{d}A(t)}{\mathrm{d}t} = \sigma A(t), \tag{6.8}$$

whereas in the non-linear regime, projection of (6.8) on the space of the $\boldsymbol{W}$'s leads to a coupled system of non-linear ordinary differential equations for the amplitudes

$$\frac{\mathrm{d}A(t)}{\mathrm{d}t} = \sigma A(t) + N(A, A), \tag{6.9}$$

where $N(A, A)$ designates the non-linear contributions. Practically, the equations are truncated at the second or third order. A simple example is provided by the following Landau relation (Drazin and Reid 1981)

$$\frac{\mathrm{d}A}{\mathrm{d}t} = \sigma A - l_{\mathrm{c}} |A|^2 A, \tag{6.10}$$

where l_{c} is a complex constant depending on the system to be studied and $|A|^2 = AA^*$ with A^* the complex conjugate of A. In (6.10), one has imposed the constraint $A = -A$ reflecting the inversion symmetry of the field variables like the velocity and temperature fields. This invariance property is destroyed and additional quadratic terms in A^2 will be present when some material parameters like viscosity or surface tension are temperature dependent. To take into account some spatial effects like the presence of lateral boundaries, it may be necessary to complete the above relation (6.9) by spatial terms in A or independent terms.

By multiplying (6.10) by A^* and adding the complex conjugate equation, one arrives at

$$\frac{\mathrm{d}|A|^2}{\mathrm{d}t} = 2(\mathrm{Re}\,\sigma)|A|^2 - 2l|A|^4, \tag{6.11}$$

where l is the real part of l_{c}. If A_0 designates the initial value, the solution of (6.11) is

$$|A|^2 = \frac{A_0^2}{\frac{l}{\mathrm{Re}\,\sigma}A_0^2 + \left(1 - \frac{l}{\mathrm{Re}\,\sigma}A_0^2\right)\exp[-2(\mathrm{Re}\,\sigma)t]}. \tag{6.12}$$

(1) Let us first examine what happens for $l > 0$. When $(\mathrm{Re}\,\sigma) < 0$, the system relaxes towards the reference state $A = 0$ which is therefore stable; in contrast, for $(\mathrm{Re}\,\sigma) > 0$, the solution (6.12) tends, for $t \to \infty$, to a stationary solution $|A_{\mathrm{s}}|$ given by

$$|A_{\mathrm{s}}| = (\mathrm{Re}\,\sigma/l)^{1/2}, \tag{6.13}$$

which is independent of the initial value A_0. This is a *supercritical* stability, the reference flow becomes linearly unstable at the critical point $\mathrm{Re}\,\sigma = 0$, or equivalently at $R = R_{\mathrm{c}}$, and bifurcates on a new steady stable branch with an amplitude tending to A_{s}. When the bifurcation is supercritical, the transition between the successive solutions is continuous and is called a *pitchfork bifurcation* as exhibited by Fig. 6.1.
It is instructive to develop $\mathrm{Re}\,\sigma$ around the critical point in terms of the wave number $\boldsymbol{k}$ and the dimensionless characteristic number R so that

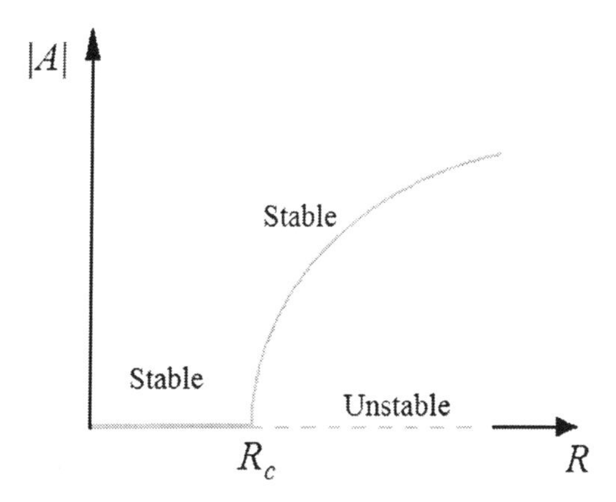

Fig. 6.1 Supercritical pitchfork bifurcation: the solution $A = 0$ is linearly stable for $R < R_c$ but linearly unstable for $R > R_c$, the branching of the curve at the critical point $R = R_c$ is called a *bifurcation*. Unstable states are represented by *dashed lines* and stable states are represented by *solid lines*

$$\mathrm{Re}\,\sigma = \alpha(R - R_c) + \beta(\boldsymbol{k} - \boldsymbol{k}_c) \cdot (\boldsymbol{k} - \boldsymbol{k}_c) + \cdots, \tag{6.14}$$

where α is some positive constant. When $R < R_c$, all perturbations are stable with $\mathrm{Re}\,\sigma < 0$; at $R = R_c$, the system is marginally stable and when R increases above R_c, the system becomes linearly unstable. Combining (6.14) with (6.13) results in

$$A_s \sim (R - R_c)^{1/2} \text{ as } R \to R_c, \tag{6.15}$$

indicating that the amplitude A_s of the steady solution is proportional to the square root of the distance from the critical point. There is a strong analogy with a phase transition of second order where the amplitude A_c of the critical mode plays the role of the order parameter and the exponent 1/2 in (6.15) is the critical exponent.

(2) Let us now examine the case $l < 0$. If $\mathrm{Re}\,\sigma > 0$, both terms of Landau's equation (6.11) are positive and $|A|$ grows exponentially; it follows from (6.12) that $|A|$ is infinite after a finite time $t = (2\mathrm{Re}\,\sigma)^{-1} \ln[1 - (\mathrm{Re}\,\sigma)/(lA_0^2)]$, however this situation never occurs in practice because in this case it is necessary to include higher-order terms in $|A|^6, |A|^8, \ldots$ in Landau's equation and generally no truncation is allowed. A more realistic situation corresponds to $\mathrm{Re}\,\sigma < 0$; now, the two terms in the right-hand side of (6.12) are of opposite sign. Depending on whether A_0 is smaller or larger than $|A_s|$ given by (6.12), we distinguish two different behaviours; for $A_0 < |A_s|$, the solution given in (6.12) shows that $|A| \approx \exp[(2\mathrm{Re}\,\sigma)t]$ and tends to zero as $t \to \infty$; in contrast for $A_0 > |A_s|$, the denominator of (6.12) becomes infinite after a time $t = (2\mathrm{Re}\,\sigma)^{-1} \ln[1 - \mathrm{Re}\,\sigma/lA_0^2]$ and $|A| \to \infty$ (see Fig. 6.2). In this case,

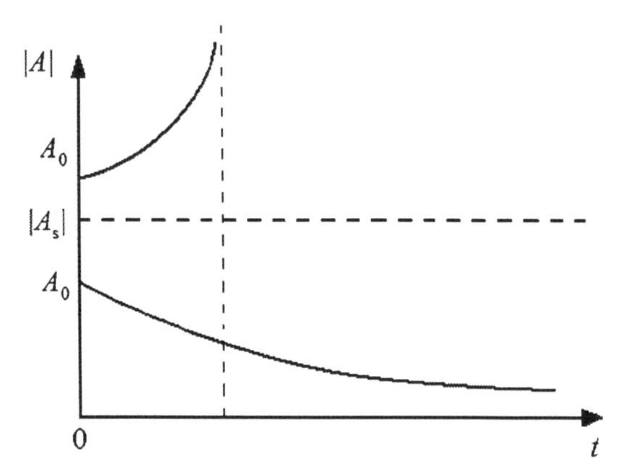

Fig. 6.2 Time dependence of the amplitude for two different initial values A_0 in the case of a subcritical instability

the reference state is stable with respect to infinitesimally small disturbances but unstable for perturbations with amplitude greater than the critical value A_s, which appears as a threshold value. This situation is referred to as *subcritical* or *metastable*, by using the vocabulary of the physicists.

In some systems, as for instance non-Boussinesq fluids, where the transport coefficients like the viscosity or the surface tension are temperature dependent, the symmetry $A = -A$ is destroyed and the amplitude equation takes the form

$$\frac{\mathrm{d}A}{\mathrm{d}t} = \sigma A + gA^2 - lA^3, \tag{6.16}$$

when A is assumed to be real, g and l are positive constants characterizing the system. This form admits three steady solutions $A_s = 0$ and $A_{1,2}$ given by

$$A_{1,2} = \frac{g \pm \sqrt{g^2 + 4\sigma l}}{2l}, \tag{6.17}$$

and they are represented in Fig. 6.3 wherein the amplitude A_s is sketched as a function of the dimensionless number R.

For $R < R_G$, the basic flow is globally stable which means that all perturbations, even large, decay ultimately; for $R_G < R < R_c$, the system admits two stable steady solutions $A_s = 0$ and the branch GD whereas CG is unstable. At $R = R_c$, the system becomes unstable for small perturbations and we are faced with two possibilities: either there is a continuous transition towards the branch CF which is called a *transcritical bifurcation*, characterized by the intersection of two bifurcation curves, or there is an abrupt jump to the stable curve DE, the basic solution "snaps" through the bifurcation to some flow with a larger amplitude. By still increasing R, the amplitude will continue to grow until a new bifurcation point is met. If, instead, the

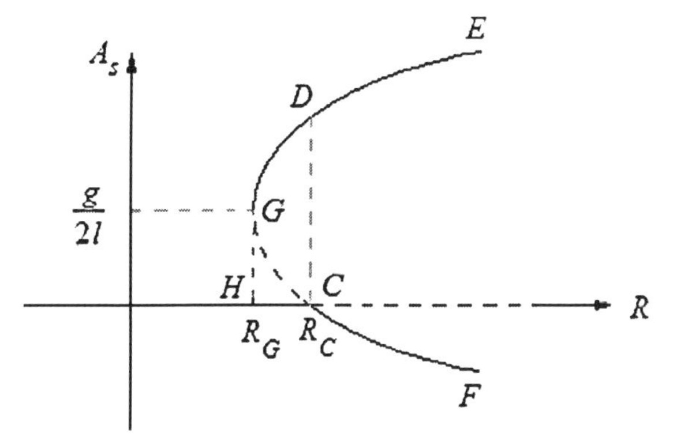

Fig. 6.3 Subcritical instability: the system is stable for infinitesimally small perturbations but unstable for perturbations with amplitude larger than some critical value. *Solid and dot lines* refer to stable and unstable solutions, respectively

amplitude is gradually decreased, one moves back along the branch EDG up to the point G where the system falls down on the basic state $A = 0$ identified by the point H. The cycle $CDGH$ is called a *hysteresis* process and is reminiscent of phase transitions of the first order.

In the present survey, it was assumed that the amplitude equation was truncated at order 3. In presence of strong non-linearity, i.e. far from the linear threshold, such an approximation is no longer justified and the introduction of higher-order terms is necessary, however this would result in rather intricate and lengthy calculations. This is the reason why model equations, like the Swift–Hohenberg equations (1977) or generalizations of them (Cross and Hohenberg 1993; Bodenshatz et al. 2000), have been recently proposed. Although such model equations cannot be derived directly from the usual balance equations of mass, momentum, and energy, they capture most of the essential of the physical behaviour and have become the subject of very intense investigations. Recent improvements in the performances of numerical analysis have fostered the resolution of stability problems by direct integration of the governing equations. Although such approaches are rather heavy, costly, and mask some interesting physical features, they are useful as they may be regarded as careful numerical control of the semi-analytical methods and associated models.

6.3 Thermal Convection

Fluid motion driven by thermal gradients, also called *thermal convection*, is a familiar and important process in nature. It is far from being an academic subject. Beyond its numerous technological applications, it is the basis for the interpretation of several phenomena as the drift of the continental plates, the Sun activity, the large-scale circulations observed in the oceans, the atmosphere, etc. As a prototype of thermal convection, we shall examine the behaviour of a thin fluid layer enclosed between two horizontal surfaces whose lateral dimensions are much larger than the width of the layer. The two horizontal bounding planes are either rigid plates or stress-free surfaces, the lower surface is uniformly heated so that the fluid is subject to a vertical temperature gradient. If the temperature gradient is sufficiently small, heat is transferred by conduction alone and no motion is observed. When the temperature difference between the two plates exceeds some critical value, the conduction state becomes unstable and motion sets in. The most influential experimental investigation on thermal convection dates back to Bénard (1900). The fluid used by Bénard was molten spermaceti, a whale's non-volatile viscous oil, and the motion was made visible by graphite or aluminium powder. In Bénard's original experiment, the lower surface was a rigid plate but the upper one was open to air, which introduces an asymmetry in the boundary conditions besides surface tension effects. The essential result of Bénard's experiment was the occurrence of a stable, regular pattern of hexagonal convection cells. Further investigations showed that the flow was ascending in the centres of the cells and descending along the vertical walls of the hexagons. Moreover, optical investigations revealed that the fluid surface was slightly depressed at the centre of the cells.

A first theoretical interpretation of thermal convection was provided by Rayleigh (1920), whose analysis was inspired by the experimental observations of Bénard. Rayleigh assumed that the fluid was confined between two free perfectly heat conducting surfaces, and that the fluid properties were constant except for the mass density. In Rayleigh's view, buoyancy is the single responsible for the onset of instability. By assuming small infinitesimal disturbances, he was able to derive the critical temperature gradient for the onset of convection together with the wave number for the marginal mode. However, it is presently recognized that Rayleigh's theory is not adequate to explain the convective mechanism investigated by Bénard. Indeed in Bénard's set up, the upper surface is in contact with air, and surface tractions originating from surface tension gradients may have a determinant influence on the onset of the flow. By using stress-free boundary conditions, Rayleigh completely disregarded this effect. It should also be realized that surface tension is not a constant but that it may depend on the temperature or (and) the presence of surface contaminants. This dependence is called the capillary or the Marangoni effect after the name of the nineteenth-century Italian investigator. The importance of this effect was only established more than 40 years later after Rayleigh's paper by Block (1956) from the experimental point of view. Pearson (1958) made the first theoretical study about the influence of the variation of surface tension with temperature on thermal convection. The predominance of the Marangoni effect in Bénard's original experiment is now admitted beyond doubt and confirmed by experiments conducted recently in space-flight missions where gravity is negligible. When only buoyancy effects are accounted for, the problem is generally referred to as Rayleigh–Bénard's instability while Bénard–Marangoni is the name used to designate surface tension-driven instability. When both buoyancy and surface tension effects are present, one speaks about the Rayleigh–Bénard–Marangoni's instability.

6.3.1 The Rayleigh–Bénard's Instability: A Linear Theory

We are going to study the instabilities occurring in a viscous fluid layer of thickness d (between a few millimetres and a few centimetres) and infinite horizontal extent limited by two horizontal non-deformable *free surfaces*, the z-axis is pointing in the opposite direction of the gravity acceleration $\boldsymbol{g}$. The fluid is heated from below with T_h and T_c, the temperatures of the lower and upper surfaces, respectively (see Fig. 6.4). The mass density ρ is assumed to decrease linearly with the temperature according to the law

$$\rho = \rho_0[1 - \alpha(T - T_0)], \tag{6.18}$$

where T_0 is an arbitrary reference temperature, say the temperature of the laboratory, and α the coefficient of thermal expansion, generally a positive quantity except for water around 4°C. For ordinary liquids, α is of the order of 10^{-3}–$10^{-4}\,\mathrm{K}^{-1}$.

When the temperature difference $\Delta T = T_\mathrm{h} - T_\mathrm{c}$ (typically not more than a few °C) between the two bounding surfaces is lower than some critical value, no motion is observed and heat propagates only by conduction inside the fluid. However by further increasing ΔT, the basic heat conductive state becomes unstable at a critical value $(\Delta T)_\mathrm{crit}$ and matter begins to perform bulk motions which, in rectangular containers, take the form of regular rolls aligned parallel to the short side as visualized in Fig. 6.5, this structure is referred to as a *roll pattern*. Note that the direction of rotation of the cells is unpredictable and uncontrollable, and that two adjacent rolls are rotating in opposite directions.

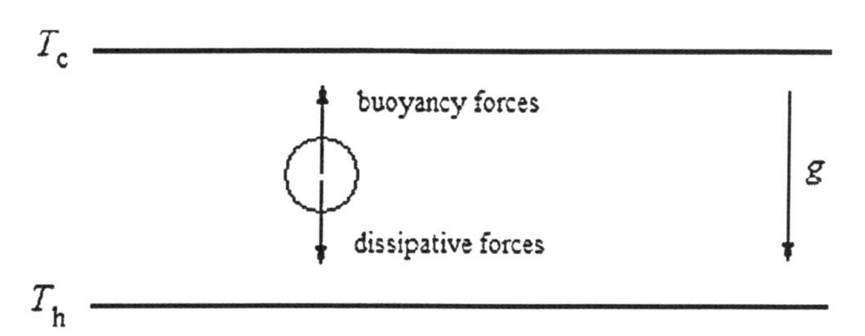

Fig. 6.4 Horizontal fluid layer submitted to a temperature gradient opposed to the acceleration of gravity $\boldsymbol{g}$

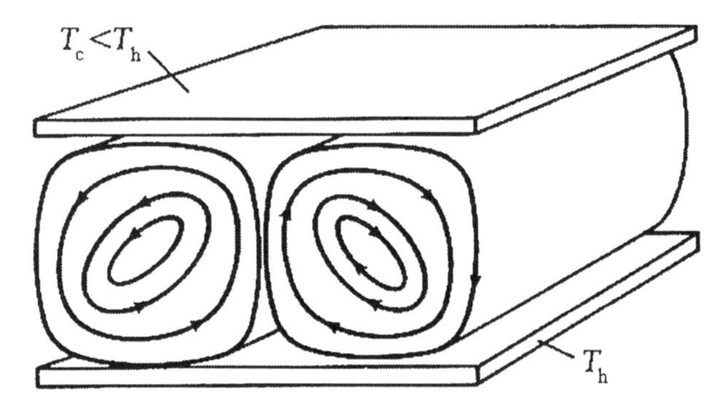

Fig. 6.5 Convective rolls in Rayleigh–Bénard's instability

A qualitative interpretation of the onset of motion is the following. By submitting the fluid layer to a temperature difference, one generates a temperature and a density gradient. A fluid droplet close to the hot lower plate has a lower density than everywhere in the layer, as density is generally a decreasing function of temperature. As long as it remains in place, the fluid parcel is surrounded by particles of the same density, and all the forces acting on it are balanced. Assume now that, due to a local fluctuation, the droplet is slightly displaced upward. Being surrounded by cooler and denser fluid, it will experience a net upward Archimede's buoyant force proportional to its volume and the temperature difference whose effect is to amplify the ascending motion. Similarly a small droplet initially close to the upper cold plate and moving downward will enter a region of lower density and becomes heavier than the surrounding particles. It will therefore continue to sink, amplifying the initial descent. What is observed in the experiments is thus the result of these upward and downward motions.

However, experience tells us that convection does not appear whatever the temperature gradient as could be inferred from the above argument. The reason is that stabilizing effects oppose the destabilizing role of the buoyancy force; one of them is viscosity, which generates a friction force directed opposite to the motion, the second one is heat diffusion, which tends to spread out the heat contained in the droplet towards its environment reducing the temperature difference between the droplet and its surroundings. This explains why a critical temperature difference is necessary to generate a convective flow: motion will start as soon as buoyancy overcomes the dissipative effects of viscous friction and heat diffusion. These effects are best quantified by the introduction of the thermal diffusion time and the viscous relaxation time

$$\tau_\chi = d^2/\chi, \quad \tau_\nu = d^2/\nu, \tag{6.19}$$

where χ is the thermal diffusivity, ν the kinematic viscosity, and d a scaling length, τ_χ is the time required by the fluid to reach thermal equilibrium with its environment, τ_ν is related to the time needed to obtain mechanical equilibrium. Another relevant timescale is the buoyant time, i.e. the time that a droplet, differing from its environment by a density defect $\delta\rho = \rho_0\alpha\Delta T$, needs to travel across a layer of thickness d,

$$\tau_{\mathrm{B}}^2 = d/(\alpha g \Delta T). \tag{6.20}$$

This result is readily derived from Newton's law of motion $\rho_0 \mathrm{d}^2 z/\mathrm{d}t^2 = g\delta\rho$ for a small volume element; a large value of ΔT means that the buoyant time is short. To give an order of magnitude of these various timescales, let us consider a shallow layer of silicone oil characterized by $d = 10^{-3}\,\mathrm{m}$, $\nu = 10^{-4}\,\mathrm{m}^2\,\mathrm{s}^{-1}$, $\chi = 10^{-7}\,\mathrm{m}^2\,\mathrm{s}^{-1}$, it is then found that $\tau_\nu = 10^{-2}\,\mathrm{s}$ and $\tau_\chi = 10\,\mathrm{s}$.

The relative importance of the buoyant and dissipative forces is obtained by considering the ratios $\tau_\nu/\tau_{\mathrm{B}}$ and $\tau_\chi/\tau_{\mathrm{B}}$ or, since they occur simultaneously, through the so-called *dimensionless Rayleigh number*,

$$Ra = \frac{\tau_\nu \tau_\chi}{\tau_{\mathrm{B}}^2} = \frac{\alpha g \Delta T d^3}{\nu\chi}. \tag{6.21}$$

The Rayleigh number can therefore be viewed as the ratio between the destabilizing buoyancy force and the stabilizing effects expressed by the viscous drag and the thermal diffusion; convection will start when Rayleigh number exceeds some critical value $(Ra)_{\mathrm{c}}$. For $Ra < (Ra)_{\mathrm{c}}$, the fluid remains at rest and heat is only transferred by conduction, for $Ra > (Ra)_{\mathrm{c}}$, there is a sudden transition to a complex behaviour characterized by the emergence of order in the system. The ratio between the dissipative processes is measured by the dimensionless Prandtl number defined as

$$Pr = \tau_\chi/\tau_\nu = \nu/\chi, \tag{6.22}$$

for gases $Pr \sim 1$, for water $Pr = 7$, for silicone oils Pr is of the order of 10^3, and for the Earth's mantle $Pr \sim 10^{23}$.

In a linear stability approach, the main problem is the determination of the marginal stability curve, i.e. the curve of Ra vs. the wave number k at $\sigma = 0$. The one corresponding to Rayleigh–Bénard's instability is derived in the Box 6.1.

Box 6.1 Marginal Stability Curve

The mathematical analysis is based on the equations of fluid mechanics written within the Boussinesq approximation. This means first that the density is considered to be constant except in the buoyancy term; second that all the material properties as viscosity, thermal diffusivity, and thermal expansion coefficient are temperature independent; and third that mechanical dissipated energy is negligible. The governing equations of mass, momentum, and energy balance are then given by

$$\nabla \cdot \boldsymbol{v} = 0,$$

$$\left(\frac{\partial \boldsymbol{v}}{\partial t} + \boldsymbol{v} \cdot \nabla \boldsymbol{v}\right) = -\frac{1}{\rho_0}\nabla p + [1 - \alpha(T - T_0)]\boldsymbol{g} + \nu\nabla^2\boldsymbol{v}, \quad (6.1.1)$$
$$\frac{\partial T}{\partial t} + \boldsymbol{v} \cdot \nabla T = \chi\nabla^2 T,$$

where use has been made of the equation of state (6.18) and where $\boldsymbol{v}(v_x, v_y, v_z)$ and p designate the velocity and pressure fields, respectively. The set (6.1.1) represents five scalar partial differential equations for the five unknowns p, T and v_x, v_y, v_z.

In the basic *unperturbed state*, the fluid is at rest and temperature is conveyed by conduction, so that the solutions of (6.1.1) are simply

$$\boldsymbol{v}_{\mathrm{r}} = 0, \quad T_{\mathrm{r}} = T_{\mathrm{h}} - \frac{\Delta T}{d}z, \quad \frac{\partial p_{\mathrm{r}}}{\partial z} = -\rho_0 g[1 - \alpha(T_{\mathrm{r}} - T_0)], \quad (6.1.2)$$

where subscript r refers to the unperturbed reference solution, z denotes the vertical coordinate measured positive upwards with $z = 0$ corresponding to the lower boundary and x, y the horizontal coordinates, and $\Delta T = T_{\mathrm{h}} - T_{\mathrm{c}}$ is the positive temperature difference between the lower and upper boundaries. Designating by $\boldsymbol{v}' = \boldsymbol{v} - \boldsymbol{0}$, $T' = T - T_{\mathrm{r}}$, $p' = p - p_{\mathrm{r}}$ the infinitesimally small perturbations of the basic state, we can linearize (6.1.1) and obtain the following set for the perturbed fields

$$\nabla \cdot \boldsymbol{v}' = 0, \quad (6.1.3)$$
$$\frac{\partial \boldsymbol{v}'}{\partial t} = -\frac{1}{\rho_0}\nabla p' + \alpha T' g \boldsymbol{e}_z + \nu\nabla^2\boldsymbol{v}', \quad (6.1.4)$$
$$\frac{\partial T'}{\partial t} + \boldsymbol{v}' \cdot \nabla T_{\mathrm{r}} = \chi\nabla^2 T', \quad (6.1.5)$$

where $\boldsymbol{e}_z$ is the unit vector pointing opposite to $\boldsymbol{g}$. We now determine the corresponding boundary conditions. If we assume that the thermal conductivity at the limiting surfaces is much higher than in the fluid itself, any thermal disturbance advected by the fluid will be instantaneously smoothed out so that T' will vanish at the bounding surfaces. Since the horizontal boundaries are assumed to be free surfaces, the shearing stress is zero at the surface; when use is made of the equation of continuity (6.1.3), this condition is identical to setting $\partial^2 v_z'/\partial z^2 = 0$ together with $v_z' = 0$, as the surfaces are non-deformable. Summarizing, the boundary conditions are

$$v_z' = \partial^2 v_z'/\partial z^2 = 0, \quad T' = 0 \text{ at perfectly heat conducting free surfaces.} \quad (6.1.6)$$

In the case of *rigid boundaries*, the no-slip condition imposes that all the components of the velocity are zero $v_x' = v_y' = v_z' = 0$, which combined with the continuity condition yields

$$v_z' = \partial v_z'/\partial z = 0, \quad T' = 0 \text{ at perfectly heat conducting rigid walls.} \quad (6.1.7)$$

At *adiabatically isolated walls*, the condition $T' = 0$ will be replaced by $\partial T'/\partial z = 0$, expressing that the surface is impermeable to heat flow. When the boundary surface is neither perfectly heat conducting nor adiabatically isolated, heat transfer is governed by Newton's cooling law $-\lambda\partial T/\partial z = h(T - T_\infty)$ where λ is the heat conductivity of the fluid, h the so-called *heat transfer coefficient*, and T_∞ the temperature of the outside world.

A further simplification of the set (6.1.3)–(6.1.5) is obtained by applying twice the operator *rot* ($\equiv \nabla\times$) on the momentum equation and using the continuity equation, we are then left with the two following relations in the two unknowns v'_z and T'

$$\frac{\partial}{\partial t}(\nabla^2 v'_z) = \alpha g \nabla_1^2 T' + \nu \nabla^4 v'_z, \tag{6.1.8}$$

$$\frac{\partial T'}{\partial t} = \beta v'_z + \chi \nabla^2 T', \tag{6.1.9}$$

where $\nabla_1^2 \equiv \partial^2/\partial x^2 + \partial^2/\partial y^2$ denotes the horizontal Laplacian whereas $\beta = \Delta T/d$. We now make the variables dimensionless by introducing the following scaling

$$X = \frac{x}{d}, \quad Y = \frac{y}{d}, \quad Z = \frac{z}{d}, \quad \hat{t} = \frac{\nu}{d^2}t, \quad w = \frac{d}{\nu}v'_z, \quad \theta = \frac{\chi}{\nu}\frac{T'}{\Delta T}, \tag{6.1.10}$$

and solve (6.1.8) and (6.1.9) with a normal mode solution of the form

$$w = W(Z)\exp[\mathrm{i}(k_x X + k_y Y)]\exp(\sigma\hat{t}), \tag{6.1.11}$$

$$\theta = \Theta(Z)\exp[\mathrm{i}(k_x X + k_y Y)]\exp(\sigma\hat{t}), \tag{6.1.12}$$

where $W(Z)$ and $\Theta(Z)$ are the amplitudes of the perturbations, k_x and k_y are the dimensionless wave numbers in the directions x and y, respectively, and σ is the dimensionless growth rate. Since the present problem is self-adjoint, it can be proved (Chandrasekhar 1961) that σ is a real quantity. There exist several ways to make the variables dimensionless, the choice made here seems to be one of the most preferred. Sometimes, it is also preferable to describe the horizontal periodicity of the flow by means of the wavelength $\lambda = 2\pi/k$ rather than the wave number as it provides a direct measure of the dimensions of the cells.

Substitution of solutions (6.1.11) and (6.1.12) in (6.1.8) and (6.1.9) leads to the following amplitude differential equations

$$(D^2 - k^2)(D^2 - k^2 - \sigma)W = Ra\, k^2\Theta, \tag{6.1.13}$$

$$(D^2 - k^2 - \sigma\, Pr)\Theta = -W, \tag{6.1.14}$$

where D stands for $\mathrm{d}/\mathrm{d}Z$ and $k^2 = k_x^2 + k_y^2$, Ra and Pr denote the dimensionless Rayleigh and Prandtl numbers defined by (6.21) and (6.22). The boundary conditions corresponding to two perfectly heat conducting free surfaces are

$$W = D^2W = 0, \quad \Theta = 0 \text{ at } Z = 0 \text{ and } Z = 1. \tag{6.1.15}$$

The latter are satisfied for solutions of the form $W = A \sin \pi Z, \Theta = B \sin \pi Z$, which substituted in (6.1.13) and (6.1.14) lead to the following algebraic equations for the two unknowns A and B:

$$\begin{aligned} A + (\pi^2 + k^2 + \sigma)B &= 0, \\ (\pi^2 + k^2)(\pi^2 + k^2 + \sigma Pr^{-1})A - k^2 Ra\, B &= 0. \end{aligned} \tag{6.1.16}$$

Non-trivial solutions demand that the determinant of the coefficients vanishes, which results in the following dispersion relation between k, σ, Ra, and Pr:

$$(\pi^2 + k^2)\sigma^2 + (\pi^2 + k^2)^2(1 + Pr)\sigma + Pr[(\pi^2 + k^2)^3 - k^2 Ra] = 0. \tag{6.1.17}$$

By setting $\sigma = 0$ in (6.1.17), one obtains the marginal curve $(Ra)_0$ vs. k determining the Rayleigh number at the onset of convection (Fig. 6.6); it is independent of the Prandtl number and given by

$$(Ra)_0 = \frac{(\pi^2 + k^2)^3}{k^2}. \tag{6.23}$$

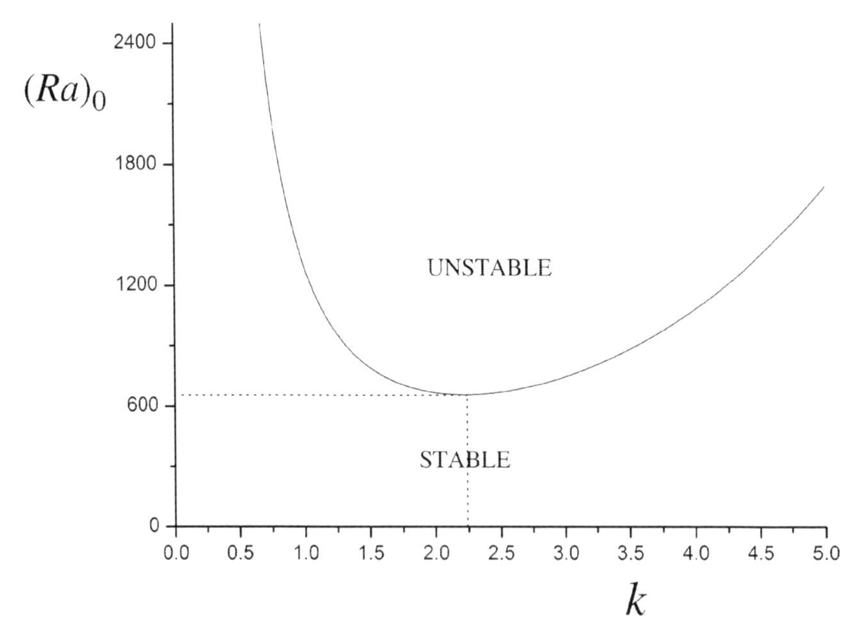

Fig. 6.6 Marginal stability curve for Rayleigh–Bénard's instability in a horizontal fluid layer limited by two stress-free perfectly heat conducting surfaces

The solutions of (6.1.17) can be written as

$$\sigma = \frac{(\pi^2 + k^2)(1 + Pr)}{2} \left\{ -1 \pm \left[1 + \frac{4(Ra - (Ra)_0)}{(Ra)_0(1 + Pr)(1 + Pr^{-1})} \right]^{1/2} \right\}, \tag{6.24}$$

which shows explicitly that the growth rate is a function of k, Ra, and Pr. For $Ra < (Ra)_0$, one has $\sigma < 0$ (stability) while for $Ra > (Ra)_0$, $\sigma > 0$ (instability). It follows also from (6.24) that σ is real when $Ra > 0$ and that $\sigma < 0$ when $Ra < 0$; this latter result indicates that the system is unconditionally stable by heating from above.

For slightly supercritical conditions, where Ra is close to $(Ra)_0$, (6.24) reads as

$$\sigma = (\pi^2 + k^2) \frac{Ra - (Ra)_0}{(Ra)_0(1 + Pr^{-1})}, \tag{6.25}$$

which shows that, for Ra larger than $(Ra)_0$, the amplitude of the disturbances amplifies exponentially and the basic state is unstable. However, for such values, the non-linear terms become important and the linear analysis ceases to be valid.

The minimum value of the marginal curve is obtained from relation (6.23) by differentiation with respect to k; setting this result equal to zero gives the critical wave number k_c at which the curve $(Ra)_0(k)$ is minimum, and the corresponding critical Rayleigh number $(Ra)_c$. For the present problem, it is found that

$$k_c = \pi/\sqrt{2} = 2.21, \ (Ra)_c = 27\pi^4/4 = 657.5 \text{ (free–free boundary conditions)}. \tag{6.26}$$

The critical Rayleigh number allows us to determine the critical temperature difference at which the system changes from the state of rest to the state of cellular motion, the critical wave number provides information about the horizontal periodicity of the patterns at the onset of convection, k_c represents the most dangerous mode picked up by the fluid. For other boundary conditions, the calculations are more complicated but the procedure remains valid, the marginal curves will have approximately the same form as in Fig. 6.6 with critical values given by

$$\begin{aligned} k_c &= 2.68, \quad (Ra)_c = 1,100.6 \quad \text{(rigid–free boundaries)}, \\ k_c &= 3.117, \quad (Ra)_c = 1,707.7 \quad \text{(rigid–rigid boundaries)}. \end{aligned}$$

As expected, stability is reinforced (larger $(Ra)_c$ value) in presence of rigid surfaces as the fluid motion is more strongly inhibited by the viscous forces. At the same time the dimensions of the cells (larger k_c value) are diminished: more energy must now be dissipated to compensate for the larger release of energy by buoyancy, clearly narrower cells are associated with greater dissipation and energy release. It is also worth to note that, for $Ra > (Ra)_c$

and according to the linear theory, a continuous spectrum of modes becomes unstable from which follows that the observed pattern should be very intricate. This is in contradiction with the observation that the flow prefers rather simple cellular forms; the reason for this discrepancy must once more be attributed to the omission of non-linear terms. Moreover, the linear theory is unable to predict the particular pattern (either rolls, squares, or hexagons) selected by the fluid; this is so because the eigenvalue problem is degenerate, which means that to one eigenvalue Ra there corresponds an infinite number of possible patterns with the same wave number k. The reason why a particular pattern is selected can only be understood from a non-linear approach, which shows that, for the present problem, two-dimensional parallel rolls are the preferred patterns as confirmed by experimental observations.

6.3.2 The Rayleigh–Bénard's Instability: A Non-Linear Theory

We now examine the behaviour of the amplitude of the disturbance beyond the critical point. Returning to the linear theory where the amplitude is supposed to behave as $A(t) \sim \exp \sigma t$, we may write that, at threshold, the relevant differential equation is

$$\frac{\mathrm{d}A}{\mathrm{d}t} = \sigma A, \tag{6.27}$$

where σ is given by (6.25). For supercritical Rayleigh numbers, the amplitude will then increase exponentially but non-linear self-interaction between modes becomes important giving raise to higher-order terms in A^n. We are then led to an amplitude equation of the form suggested by Landau, i.e.

$$\frac{\mathrm{d}A}{\mathrm{d}t} = \sigma A - lA^3, \tag{6.28}$$

where l is a positive constant to be determined from the boundary conditions whereas σ is the growth rate corresponding to the most dangerous mode k_c ($= \pi/\sqrt{2}$ for free–free boundaries). In virtue of (6.25) and for large values of Pr as in silicone oils, σ is given by

$$\sigma = 3k_\mathrm{c}^2 \varepsilon, \tag{6.29}$$

in which

$$\varepsilon = [Ra - (Ra)_\mathrm{c}]/(Ra)_\mathrm{c} \tag{6.30}$$

measures the relative distance from the critical point. There is no term in A^2 in (6.28) because it is assumed that the convective pattern is such that, by reversing the fluid velocities, the same pattern is observed; this implies that (6.28) must be invariant with respect to the symmetry $A = -A$. Of course this is only true for two-dimensional rolls and it is not so for three-dimensional patterns as hexagonal cells. Solving (6.28) for steady conditions, one obtains

$$A_{\mathrm{s}} = (\sigma/l)^{1/2} = k_{\mathrm{c}}(3\varepsilon/l)^{1/2}, \tag{6.31}$$

from which it follows that the steady amplitude A_{s} is proportional to $\varepsilon^{1/2}$; this result is typical of a supercritical bifurcation as represented on Fig. 6.1. To examine the stability of the steady solution, let us superpose an infinitesimal disturbance A' such that $A = A_{\mathrm{s}} + A'$. Substituting this expression in (6.28) and using $\mathrm{d}A_{\mathrm{s}}/\mathrm{d}t = 0$ leads to

$$\frac{\mathrm{d}A'}{\mathrm{d}t} = -2\sigma A' = -6k_{\mathrm{c}}^2\varepsilon A', \tag{6.32}$$

and, after integration,

$$A' \sim \exp[-6k_{\mathrm{c}}^2\varepsilon t]. \tag{6.33}$$

The perturbation decays for $\varepsilon > 0$, i.e. for $Ra > (Ra)_{\mathrm{c}}$, and it is concluded that two-dimensional rolls are stable for perturbations with a wave number equal to the critical one. When several modes are taken into account as in the Bénard–Marangoni's problem which is treated in Sect. 6.3.3, the problem is much more complicated because various modes will interact and instead of (6.28), we are led to a coupled set of non-linear equations of the form

$$\frac{\mathrm{d}A_n}{\mathrm{d}t} = \sigma A_n - A_n \sum_m l_{mn} A_m^2 - l A_n^3 \quad (m, n = 1, 2, \ldots). \tag{6.34}$$

An example of such kind of equations is the famous Lorenz model (1963), which is widely used in meteorology and which is at the essence of the theory of chaos. The explicit form of Lorenz set of equations is

$$\frac{\mathrm{d}A_1}{\mathrm{d}t} = Pr(A_2 - A_1), \quad \frac{\mathrm{d}A_2}{\mathrm{d}t} = rA_1 - A_2 - A_1A_3, \quad \frac{\mathrm{d}A_3}{\mathrm{d}t} = -bA_3 + A_1A_2, \tag{6.35}$$

where $r = Ra/(Ra)_{\mathrm{c}}$ and $b = 4/(1 + 4k^2)$. The coefficient A_1 describes the velocity of the fluid particles, A_2 the temperature fluctuations, and A_3 a horizontally averaged temperature mode. A detailed discussion of the Lorenz model can, for example, be found in Sparrow (1982).

When the presence of lateral walls is taken into account, the amplitude equations must include extra terms expressing the variation of the amplitudes with respect to the horizontal coordinates. An abbreviated version is the following

$$\frac{\partial A}{\partial t} = \sigma A - lA^3 + \beta \frac{\partial^2 A}{\partial x^2}, \tag{6.36}$$

which is called the Ginzburg–Landau's equation in reference to a paper by these authors on superconductivity, wherein an equation of the above form was given.

The main important result of the above analysis is that the infinite horizontal plane should be covered with straight parallel rolls of infinite length and oriented randomly. The randomness of the pattern finds its origin in the randomness of the initial disturbances in an infinite plane theory. This degeneracy may be overcome either by enclosing the fluid in boxes of finite aspect ratio or by using a grid with an imposed spacing. When the grid is slightly heated by a lamp and placed above the upper cooled surface, it induces rolls whose wavelength corresponds to the grid spacing. By increasing the Rayleigh number, one can maintain these rolls and follow their evolution in the course of time.

6.3.3 Bénard–Marangoni's Surface Tension-Driven Instability

In Sect. 6.3.2, we have shown that, in sufficiently deep fluid layers, convection sets in when buoyancy overcomes heat dissipation and viscous forces, and that only rolls are stable for small and moderate Rayleigh numbers. Alternatively, in rather shallow layers with a upper surface open to air, as in Bénard's original experience, a regular pattern of hexagons is observed as shown in Fig. 6.7.

It is now admitted that it is the temperature dependence of the surface tension at interface, which is responsible for the hexagonal flow pattern. Pearson (1958) was the first to propose a theoretical analysis on the influence of the surface tension $\mathcal{S}$ assumed to depend linearly on the temperature

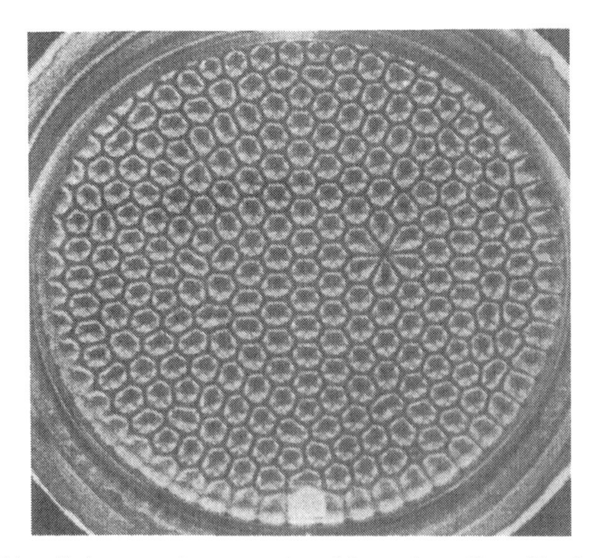

Fig. 6.7 Bénard's hexagonal pattern viewed from above (from Koschmieder 1993)

$$\mathcal{S} = \mathcal{S}_0 + \frac{\partial \mathcal{S}}{\partial T}(T - T_0), \tag{6.37}$$

wherein $\partial \mathcal{S}/\partial T$ is usually a negative quantity. The linear analysis is similar to that of the Rayleigh–Bénard's problem, but the dimensionless Rayleigh number is now replaced by the Marangoni number

$$Ma = -\frac{\partial \mathcal{S}}{\partial T}\Delta T \frac{d}{\rho_0 \nu \chi}, \tag{6.38}$$

where ΔT is the temperature difference applied between the boundaries of the fluid layer. At the upper surface, the boundary conditions read now, in dimensionless form,

$$W = 0, \quad D^2 W = k^2 Ma\, \Theta \quad (D = \mathrm{d}/\mathrm{d}Z). \tag{6.39}$$

The first relation arises from the assumed non-deformability of the surface, whereas the second one expresses the equality between the normal component of the mechanical stress tensor and the horizontal gradient of the surface tension (see Problem 6.8). For a stress-free boundary, the Marangoni number vanishes and one finds back the classical result $D^2W = 0$. It is important to realize that there are significant differences between the Rayleigh and Marangoni numbers, as the former varies as d^3 and the latter as d; this means that surface tension-driven instability will be dominant in thin layers (small d), whereas buoyancy effects will be more important in thick layers (large d). Remembering that in Bénard's original experiments the depth of the layer was about 1 mm, it is not surprising that the instability observed by Bénard was essentially caused by surface tension forces. If the ratio $Ma/Ra \to \infty$, the surface forces dominate the dynamics of the fluid and gravity effects can be neglected. Under terrestrial conditions, $Ma/Ra = 4$ for a layer with a thickness $d = 1$ mm and $Ma/Ra = 4 \times 10^{-4}$ for $d = 10$ cm. Under microgravity conditions ($\boldsymbol{g} \approx 0$), $Ma/Ra = 10^4$ for $d = 1$ mm.

As an illustration, consider a fluid layer (in which buoyancy effects are neglected, i.e. $Ra = 0$) in contact with a perfectly heat conducting bottom plate and whose upper surface is adiabatically isolated but subject to a temperature-dependent surface tension. The marginal stability curve giving $(Ma)_0$ as a function of k is similar to the marginal curve of the Rayleigh–Bénard's problem (Pearson 1958)

$$(Ma)_0 = \frac{8k^2 \cosh k (k - \sinh k \cosh k)}{k^3 \cosh k - \sinh^3 k}, \tag{6.40}$$

and is represented in Fig. 6.8. The corresponding critical values are $(Ma)_c = 79.61$ and $k_c = 1.99$.

The theory has been advanced further by Scriven and Sternling (1964) to allow for deformations of the free surface, in which case the onset of convection may be oscillatory. Nield (1964) extended Pearson's analysis to account for coupled buoyancy and surface tension effects. The numerical results (see

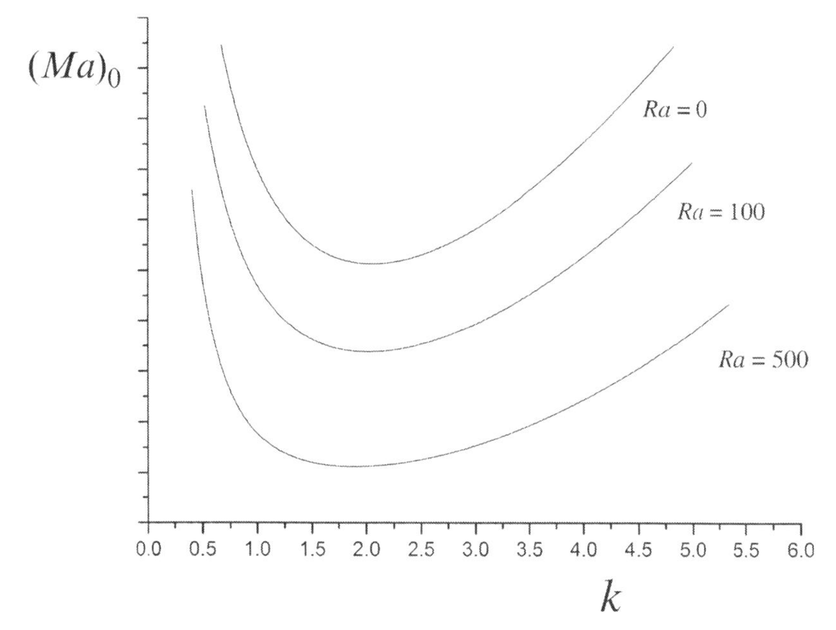

Fig. 6.8 Marginal stability curves for coupled buoyancy and surface tension-driven instability; the curve $Ra = 0$ corresponds to surface tension-driven instability

Fig. 6.8) show that the critical value $(Ma)_c$ decreases upon increasing Ra but that the critical wave number remains practically unchanged. Nield was able to derive the following linear relation between the critical Rayleigh and Marangoni numbers

$$\frac{Ra}{(Ra)_c} + \frac{Ma}{(Ma)_c} = 1, \tag{6.41}$$

which exhibits the strong coupling between the two motors of instability; $(Ra)_c$ and $(Ma)_c$ are the critical values corresponding to the absence of surface tension gradients and gravity, respectively.

All the above theoretical considerations are based on linearized equations; they give interesting information on the onset of convection but are silent about the planform and other characteristics of the flow; to go further a non-linear analysis is required.

Non-linear effects were considered by Cloot and Lebon (1984), who solved the problem with a power series expansion (Shlüter et al. 1965), and by Bragard and Lebon (1993) using the amplitude method. To incorporate possible hexagonal patterns, the velocity disturbance is taken to be of the general form

$$w = W(z)\left[Z(t)\cos ky + Y(t)\cos(\sqrt{3}/2kx)\cos(ky/2)\right], \tag{6.42}$$

which corresponds to rolls for $Y = 0$ and to hexagons for $Y = 2Z$. After some lengthy but rather simple calculations, Bragard and Lebon (1993) obtained under zero-gravity conditions the following set of amplitude equations

$$\begin{aligned} \frac{\mathrm{d}Y}{\mathrm{d}t} &= \varepsilon Y - \gamma YZ - RY^3 - PYZ^2, \\ \frac{\mathrm{d}Z}{\mathrm{d}t} &= \varepsilon Z - \frac{1}{4}\gamma Y^2 - R_1 Z^3 - \frac{1}{2}PY^2 Z, \end{aligned} \tag{6.43}$$

where $\varepsilon = [Ma - (Ma)_c]/(Ma)_c$ and $\gamma, P = 4R - R_1, R$, and R_1 are constant coefficients determined from the solvability condition. Observe that the symmetry property $f(Y) = f(-Y)$ and $f(Z) = f(-Z)$ is broken in (6.43). To obtain a hexagonal structure, the condition $Y = 2Z$ must be fulfilled and it is directly checked that the steady versions of (6.43) possess a solution verifying this condition. By performing a linear stability analysis of the steady solutions, it was shown (Bragard and Lebon 1993) that hexagons are stable in the range $-0.0056 < \varepsilon < 1.8$. By setting $Y = 2Z$ in the stationary solution of (6.43), one obtains a second-order equation in Z, namely

$$\varepsilon - \gamma Z - (R_1 + 2P)Z^2 = 0, \tag{6.44}$$

which is represented in Fig. 6.9 and displays an inverted bifurcation, confirming the presence of a region of *subcritical instability*, characterized by the occurrence of motion at $\varepsilon < 0$.

Experiments by Koschmieder (1993) indicate that the pattern of the subcritical instability consists of ill-defined small-scale hexagonal cells. For the sake of completeness, we have also reported in Fig. 6.9 the amplitude curve for the roll solution. The solid portions of the curves refer to stable motions and the broken lines to unstable solutions. For $\varepsilon < \varepsilon_-(= -0.0056)$,

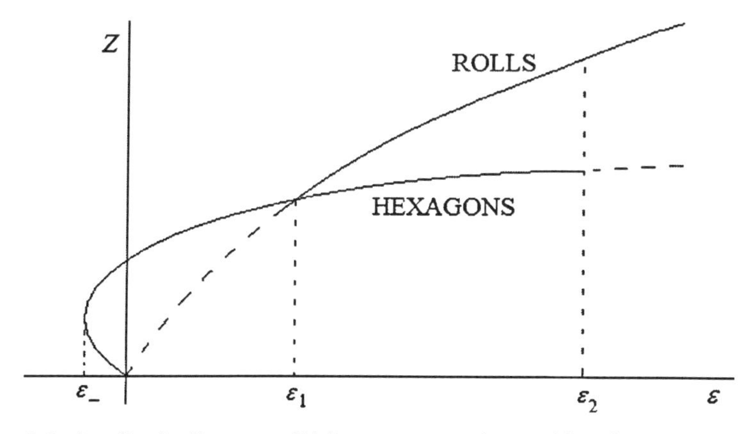

Fig. 6.9 Amplitude Z vs. ε: *solid lines* correspond to stable solutions and *broken lines* correspond to unstable solutions

the fluid remains at rest, as soon as ε is larger than ε_- but less than zero, subcritical stable hexagons are predicted. Moreover, hexagons are the only stable solutions in the interval $\varepsilon_- < \varepsilon < \varepsilon_1 (= 0.53)$ while in the range $\varepsilon_1 < \varepsilon < \varepsilon_2 (= 1.8)$, both hexagonal cells and rolls are stable; finally for $\varepsilon > \varepsilon_2$, only stable rolls can be found.

A behaviour similar to Bénard–Marangoni's instability is observed in Rayleigh–Bénard's problem when Boussinesq approximation is given up. In the particular case of a temperature-dependent viscosity, it was found (Palm 1975) that the hexagonal pattern is the preferred structure, even for a fluid enclosed between two rigid plates. Experiments have also revealed that the motion in the hexagonal platform may be either ascending or descending at the centre of the cell: the motion is downward in the cell centres of gases and upward in liquids. This change in circulation is associated to the property that, in gases, the viscosity increases with temperature while it decreases in liquids.

Other interesting aspects of buoyancy and surface tension-driven instabilities have been treated in the specialized literature (e.g. Colinet et al. 2002). Among them, let us mention the problem of fluid layers submitted to horizontal temperature gradients. This is important in crystal growth processes (Davis 1987; Parmentier et al. 1993; Madruga et al. 2003), convection in porous media, which is of interest in the mineral oil industry, thermal convection in rotating systems (Busse 1978; Davis 1987), and double diffusive convection under the mutual action of temperature and salinity gradients (thermohaline convection and salt fingers). All these phenomena play a central role in geophysics, astrophysics, and oceanography. Let us also point out the effects resulting from surface deformations (Regnier et al. 2000), particularly relevant in small boxes and in presence of lateral walls, unavoidable in practical experiments (Rosenblat et al. 1982; Dauby and Lebon 1996; Dauby et al. 1997). Bénard's convection has also been used to model the Earth's mantle motions (Turcotte 1992); moreover, it offers an attractive basis for the interpretation of the transition to turbulence (Bergé et al. 1984). The above list, although being not exhaustive, reflects the impressively wide range of applications of Rayleigh–Bénard–Marangoni's instability.

6.4 Taylor's Instability

Taylor's instability is observed in a viscous incompressible fluid column contained between two vertical cylinders rotating at different angular velocities, the temperature being assumed to remain uniform. There occurs a competition between centrifugal forces and viscosity: if we assume that the inner cylinder moves at a higher velocity than the outer one, the fluid close to the inside wall will move outward and replace there the slow moving fluid. At small angular velocity, one has an ordinary Couette flow where angular

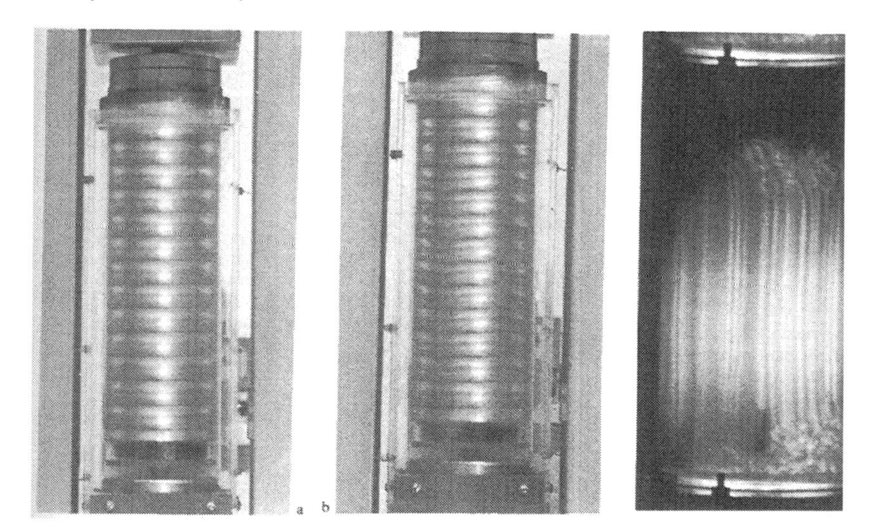

Fig. 6.10 (**a**) Taylor's vortices, (**b**) wavy flow, and (**c**) turbulence (from Coles 1965)

momentum is transported from the inner to the outer cylinder by viscosity in the form of stationary annular convective cells. When the angular velocity is gradually increased, this state becomes unstable and toroidal vortices, also called *Taylor's vortices*, are formed. At higher angular velocities, wavy deformations appear and finally a transition to turbulent motion occurs (see Fig. 6.10).

At low angular velocity, the motion is laminar with a velocity field given by

$$v_r = 0, \quad v_\theta = Ar + \frac{B}{r}, \quad v_z = 0, \tag{6.45}$$

when expressed in cylindrical coordinates r, θ, z, where r is the radial distance to the axis of the cylinders, θ the azimuthal angle, and z the coordinate parallel to the vertical axis, A and B are constants to be determined from the boundary conditions. With Ω_1 and Ω_2 the angular velocities of the inner and outer cylinders, respectively, and R_1 and R_2 the corresponding radii, the non-slip condition imposes the following conditions at the walls

$$v_\theta = R_1\Omega_1 \text{ at } r = R_1, \quad v_\theta = R_2\Omega_2 \text{ at } r = R_2; \tag{6.46}$$

this allows us to determine the values of the constants A and B, namely

$$A = -\Omega_1 \frac{\Omega^2 - \mu}{1 - \Omega^2}, \quad B = \Omega_1 \frac{R_1^2(1-\mu)}{1-\Omega^2}, \tag{6.47}$$

where $\Omega = \Omega_2/\Omega_1$ and $\mu = R_1/R_2$.

We first proceed with a *linear stability analysis*. Denote the velocity disturbance by u', v', w' and the pressure disturbance by p'; assuming that the disturbances are small and axisymmetric, i.e. θ independent, it is easily checked that they obey the following linearized continuity and Navier–Stokes equations

$$\frac{1}{r}\frac{\partial}{\partial r}(ru') + \frac{\partial w'}{\partial z} = 0, \tag{6.48a}$$

$$\frac{\partial u'}{\partial t} - 2v_\theta \frac{v'}{r} = -\frac{\partial}{\partial r}\left(\frac{p'}{\rho}\right) - \nu\left(\nabla^2 u' - \frac{u'}{r^2}\right), \tag{6.48b}$$

$$\frac{\partial v'}{\partial t} + \left(\frac{\partial v_\theta}{\partial r} + \frac{v_\theta}{r}\right)u' = \nu\left(\nabla^2 v' - \frac{v'}{r^2}\right), \tag{6.48c}$$

$$\frac{\partial w'}{\partial t} = -\frac{\partial}{\partial z}\left(\frac{p'}{\rho}\right) + \nu\nabla^2 w', \tag{6.48d}$$

where $\nabla^2 \equiv (1/r)(\partial/\partial r) + \partial^2/\partial r^2 + \partial^2/\partial z^2$. The pressure disturbance p' is easily eliminated from relations (6.48b) and (6.48d). Moreover, by supposing that the fluid column is of infinite length, we need only the boundary conditions at R_1 and R_2, namely

$$u' = v' = w' = 0. \tag{6.49}$$

According to the normal mode technique, we seek for solutions of the form

$$(u', v', w') = [U(r), V(r), W(r)]\exp(\mathrm{i}kz + \sigma t). \tag{6.50}$$

To simplify the calculations, we shall make the narrow-gap approximation, which implies that the distance $d = R_2 - R_1$ between the cylinders is much smaller than R_1. This permits us to put $\mu = 1$ and to omit $1/r$ compared to $\partial/\partial r$, moreover within the small-gap limit, $A + B/r^2 \approx \Omega_1$ and $A \approx -\Omega_1$. Substituting (6.50) in (6.48), making all the quantities dimensionless, and proceeding by analogy with the Rayleigh–Bénard's problem, we obtain the following relations for the radial and azimuthal velocity amplitudes at the marginal state $\sigma = 0$

$$(D^2 - k^2)^2 U = [1 + (\Omega - 1)Z]V, \tag{6.51}$$

$$(D^2 - k^2)V = -Ta\, k^2 U, \tag{6.52}$$

where we have introduced the notation $D = \mathrm{d}/\mathrm{d}Z$ with

$$Z = (r - R_1)/d \tag{6.53}$$

and

$$Ta = 4\Omega_1^2 d^4/\nu^2, \tag{6.54}$$

the relevant boundary conditions are

$$U = DU = V = 0 \text{ at } Z = 0 \text{ and } Z = 1. \tag{6.55}$$

The quantity Ta is the dimensionless Taylor number, which is the ratio between the centrifugal forces and the viscous dissipation. It is worth to stress that relations (6.51) and (6.52) are similar to (6.1.13) and (6.1.14) describing Rayleigh–Bénard's instability. For each particular value of $\Omega = \Omega_2/\Omega_1$, the eigenvalue problem (6.51) and (6.52) results in a marginal stability curve (see Fig. 6.11), which is similar to the curves for the Rayleigh–Bénard's problem in the rigid–rigid configuration. The minimum of the curve gives the critical value $(Ta)_c$ determining the minimal angular velocity at which toroidal axisymmetric vortices will set in. For $Ta < (Ta)_c$, all the disturbances are dampened and the flow is independent of the z coordinate; at $Ta = (Ta)_c$, there is a supercritical transition from Couette flow to Taylor's vortices characterized by a critical wave number k_c or a critical wavelength $\lambda_c = 2\pi/k_c$ which is easily detectable on the photographs of Fig. 6.10. In the small-gap limit, the critical values are given by

$$(Ta)_c = 3,430/(1+\Omega), \quad k_c = 3.12, \quad \text{for } 0 < \Omega < 1. \tag{6.56}$$

The critical Taylor number is a function of Ω and it is interesting to note that, for $\Omega = 1$, one has $(Ta)_c = 1,715$, which is very close to the value 1,708 obtained for the critical Rayleigh number in the rigid–rigid convection case. The same remark applies to the critical wave number whose value $k_c = 3.12$ is comparable to the value $k_c = 3.117$ of the Rayleigh–Bénard's problem. It is therefore clear that the linear stability problems for Rayleigh–Bénard's and Taylor's instabilities in the narrow-gap approximation are formally identical.

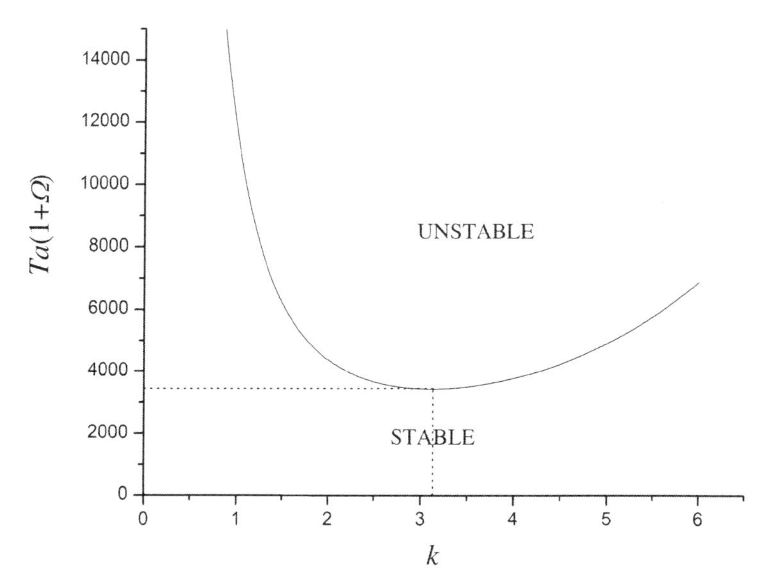

Fig. 6.11 Marginal stability curve expressing the Taylor number Ta as a function of the wave number k

For wide distances between the cylinders, it is useful to work with a Taylor number (Chandrasekhar 1961), which is an explicit function of the radius ratio. The results remain qualitatively unchanged with respect to the small-gap approximation; the main features are that the critical Taylor number increases with decreasing values of the radius ratio whereas the critical wave number remains practically unchanged.

To obtain the form of the fluid flow after the onset of instability, a *non-linear analysis* is required. It is not our purpose to enter into the details of the non-linear developments as they are rather intricate and therefore we advice the interested reader to refer to the specialized literature for details. A wide discussion can be found in Drazin and Reid (1981) and Koschmieder (1993). Starting from an amplitude equation of the form (6.28), it is found that, in the weakly supercritical non-linear regime, the amplitude of the flow is proportional to $[Ta - (Ta)_c]^{1/2}$ and the motion takes the form of toroidal vortices, the interval of stable wave numbers lies, however, in a narrower band than the one predicted by the linear approach. The stability of the Taylor's vortices was studied experimentally by Coles (1965). By increasing the Taylor number, a sequence of bifurcations is displayed, which may be summarized as follows. For $Ta < (Ta)_c$, one has a pure Couette flow that becomes unstable at the critical Taylor number at which Taylor's vortices are observed. At higher values of Ta, wavy vortices with a characteristic frequency are displayed (see Fig. 6.10b), the next step is the appearance of a quasi-periodic regime with two characteristic frequencies. Later on, a transition towards weak turbulence characterized by the disappearance of one of the frequencies takes place, still further, the second characteristic frequency vanishes and a state of full turbulence emerges (see Fig. 6.10c). One of the interests of the study of Taylor's instability is that one can follow the transition from laminar flow to turbulence in great detail through a limited number of rather characteristic stages, it offers therefore an attractive route to an exhaustive study of turbulence at the same footing as Rayleigh–Bénard's convection.

In the forthcoming sections, we will discuss qualitatively several examples of patterns occurring in such various domains as chemistry, biology, oceanography, and electricity. Unlike the presentations of the Bénard–Marangoni's and Taylor's instabilities, we shall not enter into detailed mathematical formulations but rather outline the main features, referring the interested reader to specialized books and articles.

6.5 Chemical Instabilities

Dissipative patterns are not exclusive of hydrodynamics but are also found in many other processes and particularly in chemistry. Some chemical reactions give rise to temporal and spatial variations of the mass concentrations of the active substances. In the course of time, some substances may undergo periodic oscillations whereas some spatial patterns may be formed in the reaction vessel. It should be stressed that these phenomena are underlying the activities of life and are therefore far from being purely academic games. The central feature governing the appearance of self-organization in chemistry is the *autocatalytic* reaction: it is a reaction in which the products of some step take part to a subsequent step. It looks like a feedback process wherein the presence of one particular species stimulates the production of more of that species.

A representative example of chemical reaction exhibiting dissipative patterns is the Belousov–Zhabotinsky (BZ) reaction, which is discussed later on. When chemical reactions are coupled with diffusion, as it occurs in spatially inhomogeneous mixtures, the corresponding spatio-temporal instabilities are designated as *Turing's instabilities*. In a remarkable paper, Turing (1952) showed that a steady and uniform state in a reaction–diffusion system might become unstable after some control parameter has overcome some critical value.

Maintenance of non-equilibrium states for a long period demands that chemical reagents are added to the system. But a result of such a process is that the concentrations of the various components are made very inhomogeneous and it is necessary to stir the mixture with the consequence that spatial patterns will be destroyed. Only time oscillations can be observed in such stirred homogeneous reactions. As illustrations, we will discuss in the next sections the Lotka–Volterra and the Brusselator models (Prigogine and Lefever 1968; Nicolis and Prigogine 1977). To maintain non-equilibrium conditions in absence of stirring has for a long time been considered as a difficult problem. It is only recently that experimental methods using molecular diffusion have been developed which allow for the presence of both spatial and temporal patterns.

6.5.1 Temporal Organization in Spatially Homogeneous Systems

6.5.1.1 The Lotka–Volterra Model

A simple example of temporal organization is the Lotka–Volterra model, which is well known in ecology. It describes the growth and death of a population of X individuals of a species of prey and Y individuals of a species of predators, X and Y represent for instance two populations of fishes where Y is subsisting exclusively on X. This prey–predator system is governed by the non-linear equations

$$\frac{\mathrm{d}X}{\mathrm{d}t} = X(k_1 - k_2 Y), \tag{6.57}$$

$$\frac{\mathrm{d}Y}{\mathrm{d}t} = -Y(k_2 - k_3 X), \tag{6.58}$$

where k_1, k_2, and k_3 are positive constants.

Relations (6.57) and (6.58) admit two trivial stationary solutions $X_\mathrm{s} = Y_\mathrm{s} = 0$ and two non-trivial solutions

$$X_\mathrm{s} = k_2/k_3, \quad Y_\mathrm{s} = k_1/k_2. \tag{6.59}$$

Their linear stability is investigated by writing $X = X_\mathrm{s} + x'$, $Y = Y_\mathrm{s} + y'$ and by substituting these expressions in (6.57) and (6.58); making use of (6.59), it is checked that the disturbances x' and y' obey the following linear equations

$$\frac{\mathrm{d}x'}{\mathrm{d}t} = -k_2 X_\mathrm{s} y', \quad \frac{\mathrm{d}y'}{\mathrm{d}t} = k_3 Y_\mathrm{s} x', \tag{6.60}$$

and, after elimination of y',

$$\frac{\mathrm{d}^2 x'}{\mathrm{d}t^2} = -k_1 k_2 x', \tag{6.61}$$

whose solution is of the form

$$x' = x_0 \cos(\omega t) \quad \text{with} \quad \omega = \sqrt{k_1 k_2}. \tag{6.62}$$

It follows that the stationary state is not asymptotically stable in the sense that the perturbations do not vanish in the limit of $t \to \infty$, instead they are periodic in time with an angular frequency ω depending on the parameters characterizing the system. This periodic behaviour is easy to interpret: when the population of preys X is increased, the predators Y have more food at their disposal and their number is increasing. Since Y consumes more individuals X, their population diminishes and less food is available for the species Y, which in turn decreases. This reduction of the number of predators allows a larger number of preys to survive with the consequence that their population will increase and a new cycle is initiated. The ratio of (6.58) and (6.57) can be written as

$$\frac{\mathrm{d}Y}{\mathrm{d}X} = -\frac{Y(k_2 - k_3 X)}{X(k_1 - k_2 Y)}, \tag{6.63}$$

which, after integration, yields the following relation

$$k_3 X + k_2 Y - k_2 \log X - k_1 \log Y = \text{constant}, \tag{6.64}$$

with the constant depending on the initial conditions, and playing the same role as total energy in classical mechanics. Clearly, (6.64) defines an infinity of trajectories in the phase space $X - Y$ corresponding to different initial conditions (see Fig. 6.12).

These trajectories take the form of concentric curves surrounding the stationary state S with a period depending on the initial conditions. This behaviour is typical of conservative systems and has to be contrasted with limit cycle oscillations as exhibited by dissipative systems. This gives also a strong argument against the use of such models to describe oscillations observed in nature, and particularly in some classes of chemical reactions.

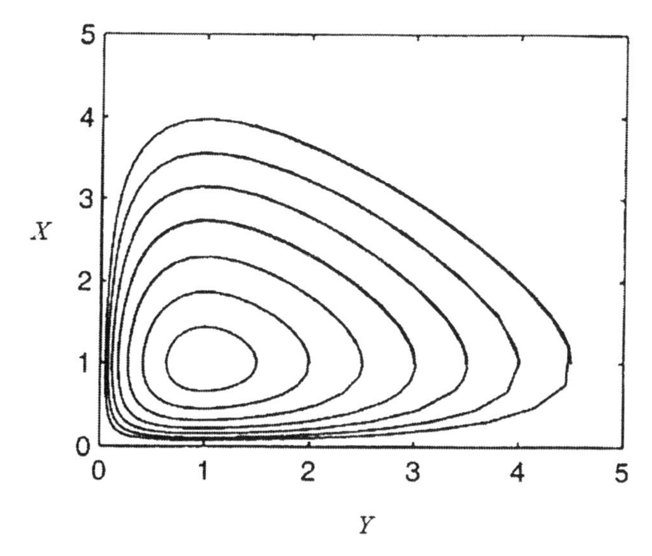

Fig. 6.12 Phase space trajectories: the Lotka–Volterra cycles

6.5.1.2 The Brusselator

A more appropriate modelling of the oscillatory behaviour of chemical reactions is provided by the Brusselator (Prigogine and Lefever 1968; Nicolis and Prigogine 1977) which consists in a simple and instructive example of autocatalytic scheme. The sequence of chemical reactions in the Brusselator is the following

$$A \xrightarrow{k_1} X, \tag{6.65a}$$

$$2X + Y \xrightarrow{k_2} 3X, \tag{6.65b}$$

$$B + X \xrightarrow{k_3} Y + D, \tag{6.65c}$$

$$X \xrightarrow{k_4} E, \tag{6.65d}$$

the four steps are assumed to be irreversible which is achieved by taking all reverse reaction constants equal to zero; the global reaction of the above scheme is

$$A + B \rightarrow D + E. \tag{6.66}$$

The concentrations of the reactants A and B are maintained at a fixed and uniform non-equilibrium value, and the final products D and E are removed as soon as they are formed; the autocatalytic step is the second one, which involves the intermediate species X and Y. It is assumed that the reactions take place under isothermal and well stirred, i.e. spatially homogeneous conditions. According to the laws of chemical kinetics, we have the following rate equations for the species X and Y

$$\frac{\mathrm{d}X}{\mathrm{d}t} = k_1 A + k_2 X^2 Y - k_3 B X - k_4 X, \tag{6.67}$$

$$\frac{\mathrm{d}Y}{\mathrm{d}t} = -k_2 X^2 Y + k_3 B X, \tag{6.68}$$

whose steady solutions are

$$X_{\mathrm{s}} = \frac{k_1}{k_4} A, \quad Y_{\mathrm{s}} = \frac{k_3 k_4}{k_1 k_2} \frac{B}{A}. \tag{6.69}$$

To examine the stability of this solution, let us set $X = X_{\mathrm{s}} + x'$ and $Y = Y_{\mathrm{s}} + y'$ where x' and y' are small disturbances; it is easy to verify that the perturbations are obeying a linear system of the form

$$\frac{\mathrm{d}}{\mathrm{d}t}\begin{pmatrix} x' \\ y' \end{pmatrix} = \mathcal{L}\begin{pmatrix} x' \\ y' \end{pmatrix}, \tag{6.70}$$

where the matrix $\mathcal{L}$ is given by

$$\mathcal{L} = \begin{pmatrix} k_3 B - k_4 & \frac{k_1^2 k_2}{k_4^2} A^2 \\ -k_3 B & -\frac{k_1^2 k_2}{k_4^2} A^2 \end{pmatrix}, \tag{6.71}$$

whose eigenvalues $\sigma_1, \sigma_2, \ldots$ determine the stability of the system. It is left as an exercise to show that the stationary state of (6.70) becomes unstable ($\mathrm{Re}\,\sigma > 0$) when B is larger than a critical concentration given by

$$B_{\mathrm{c}} = \frac{k_4}{k_3} + \frac{k_1^2 k_2}{k_3 k_4^2} A^2. \tag{6.72}$$

For $B < B_{\mathrm{c}}$, the system remains homogeneous; whereas at $B = B_{\mathrm{c}}$, the eigenvalues are purely imaginary, which leads to undamped oscillations of the perturbations, just like in the Lotka–Volterra model; finally for $B > B_{\mathrm{c}}$, the system is unstable and the concentrations X and Y undergo periodic oscillations of the limit cycle type, independently of the initial values of X and Y. Such oscillations have indeed been detected in the Belousov–Zhabotinsky reaction, where fascinating geometric patterns as concentric circles and spirals propagating through the medium have been observed. BZ reaction is basically a catalytic oxidation by potassium bromate KBrO of a organic compound such as malonic acid $CH_2(COOH)_2$ catalysed by the cerium Ce^{3+}–Ce^{4+} ion couple.

After some transitory period, the oscillatory behaviour is evidenced by the variation of the concentration of the Ce^{3+} and Ce^{4+} ions (Fig. 6.13). First, a blue colour indicating an excess of Ce^{4+} is spreading into the mixture, a few minutes later the blue colour disappears and is replaced by a red one

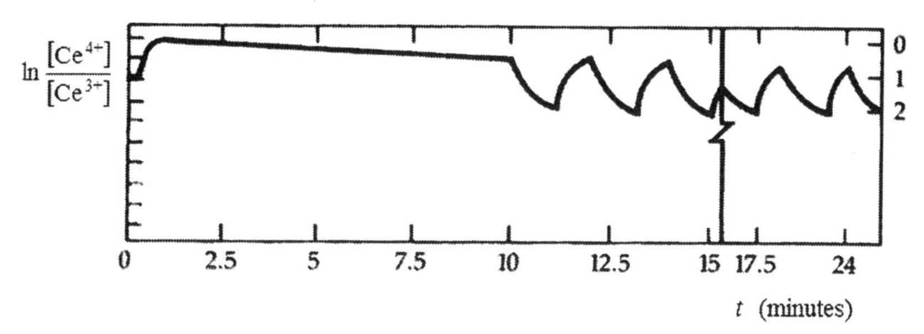

Fig. 6.13 Experimentally observed oscillations of Ce^{4+} and Ce^{3+} ions in Belousov–Zhabotinsky reaction

that indicates an excess of Ce^{3+} ions. The process goes on with a succession of blue, red, blue colours with a perfectly regular period, which to a certain sense constitutes a chemical clock.

When the substances necessary for the reaction are exhausted, the oscillations will die out and the system goes back to its equilibrium state. In contrast with hydrodynamics where complexity is generally characterized by non-homogeneities in space, in chemistry even a spatially homogeneous system may exhibit a complex behaviour in time. More recent and detailed analyses of BZ reaction have shown the occurrence of aperiodic and even chaotic behaviours. There are several analogies of BZ periodic behaviour in living organisms: heart beats, circadian rhythms, and menstrual cycles are a few examples.

6.5.2 Spatial Organization in Spatially Heterogeneous Systems

In absence of stirring, BZ reaction exhibits some non-trivial spatial patterns, which arise from the interplay of the chemical reaction and the diffusion process. When BZ reaction is performed in a long thin vertical tube so that the problem consists of a single spatial dimension, one observes (Fig. 6.14) a superposition of steady horizontal bands of different colours corresponding to low and high concentration region. This structure is analogous to the Rayleigh–Bénard's and Taylor's patterns.

6.5.3 Spatio-Temporal Patterns in Heterogeneous Systems: Turing Structures

The great variety of patterns present in nature, both in the animate and inanimate world, like the captivating beauty of a butterfly, the blobs of a

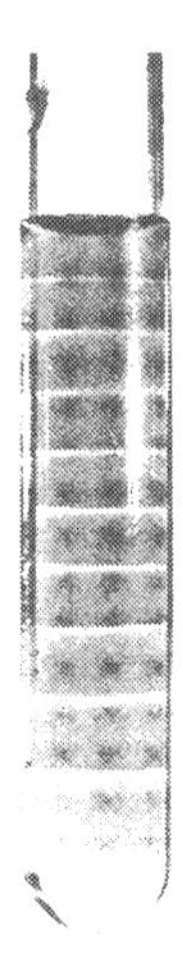

Fig. 6.14 Spatial structure in Belousov–Zhabotinsky reaction

leopard tail, or biological morphogenesis has been a subject of surprise and interrogation for several generations of scientists. The question soon arises about the mechanism behind them. In 1952, Turing proposed an answer based on the coupling between (chemical) reactions and diffusion. As an example of a Turing structure, let us still consider the Brusselator but we suppose now that the chemical reaction takes place in a unstirred thin layer or a usual vessel, so that spatial non-homogeneities are allowed. The basic relations are the kinetic equations (6.67) and (6.68) to which are added the diffusion terms $D_x\nabla^2 X$ and $D_y\nabla^2 Y$, respectively, it is assumed that the diffusion coefficients D_x and D_y are constant. By repeating the analysis of Sect. 6.5.1, it is shown that the stationary homogeneous state becomes unstable for a concentration B larger than the critical value

$$B_{\mathrm{c}} = \frac{k_4}{k_3} + \frac{k_1^2 k_2}{k_3 k_4^2} A^2 + \frac{k_4}{k_3}\frac{j^2\pi^2}{l^2}(D_x + D_y), \quad j = 0, 1, 2 \ldots \tag{6.73}$$

at the condition that the diffusion coefficients are unequal, if $D_x = D_y$ diffusion will not generate an instability, l is a characteristic length. Non-homogeneities will begin to grow and stationary spatial patterns will emerge in two-dimensional configurations. A rather successful reaction for observing Turing's patterns is the CIMA (chlorite–iodide–malonic acid) redox reaction, which was proposed as an alternative to BZ reaction. The oscillatory and space-forming behaviours in CIMA are made apparent through the presence of coloured spots with a hexagonal symmetry. By changing the concentrations, new patterns consisting of parallel narrow stripes are formed instead of the spots (see Fig. 6.15).

Although it is intuitively believed that diffusion tends to homogenize the concentrations, we have seen that, when coupled with an autocatalytic

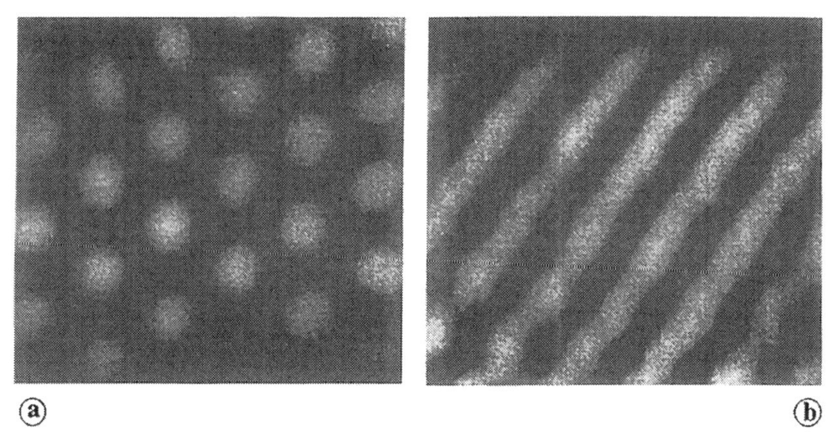

Fig. 6.15 Examples of Turing two-dimensional structures (from Vidal et al. 1994)

reaction under far from equilibrium conditions, it actually gives rise to spatial structures. Turing's stationary patterns are obtained when the eigenvalues of the $\mathcal{L}$ matrix given by (6.71) are real; for complex conjugate eigenvalues, the unstable disturbances are time periodic and one observes spatio-temporal structures taking the form of propagating waves.

The existence of spatio-temporal patterns is not exclusive to fluid mechanics and chemistry. A multitude of self-organizations has been observed in biology and living organisms, which are the most organized and complex examples found in the nature. It has been conjectured that most of the properties of biological systems are the result of transitions induced by far from equilibrium conditions and destabilizing mechanisms similar to autocatalytic reactions. Because of their complexity, these topics will not be analysed here but to further convince the reader about the universality of pattern formation, we prefer to discuss shortly three more examples of dissipative patterns as observed in oceanography, electricity, and materials science.

6.6 Miscellaneous Examples of Pattern Formation

As recalled earlier, one observes many kinds of pattern formation in many different systems. In this section, we give a concise overview of some of them.

6.6.1 Salt Fingers

In *double diffusion* convection, the flow instability is due to the coupling of two diffusive processes, say heat and mass transport. In the case of the *salt fingers*, which are of special interest in oceanography, the basic fields are temperature and salinity, i.e. the concentration of salt in an aqueous solution; in other situations, they are for example the mass concentrations of two solutes in a ternary mixture at uniform temperature or the concentrations of two polymers in polymeric solutions.

If we consider diffusion in a binary mixture submitted to a temperature gradient, new features appear with respect to the simple Rayleigh–Bénard's convection. Apart from heat convection and mass diffusion, the Soret and Dufour cross-effects are present and should be included in the expressions of the constitutive equations, both for the heat flux $\boldsymbol{q}$ and the mass flow $\boldsymbol{J}$. Compared to the Rayleigh–Bénard's problem, the usual balance equations of mass, momentum, and energy must be complemented by a balance equation for the mass concentration of one of the constituents while the mass density is modified as follows to account for the mass concentration

$$\rho = \rho_0[1 - \alpha(T - T_0) + \beta(c - c_0)], \tag{6.74}$$

where β stands for $\beta = \rho_0^{-1}(\partial\rho/\partial c)$. The equations for the disturbances are then linearized in the same way as in Rayleigh–Bénard's problem, and finite amplitude solutions have also been analysed.

As a result of the very different values of the molecular D and heat χ diffusivity coefficients (in salt sea waters $D/\chi = 10^{-2}$), some puzzling phenomena are occurring. Instabilities arise when a layer of cold and pure water is lying under a layer of hot and salty water with densities being such that the cold water is less dense than the warm water above it; convection takes then place in the form of thin fingers of up- and down-going fluids (Brenner 1970; McDougall and Turner 1982).

The mechanism responsible for the onset of instability is easily understood. Imagine that, under the action of a disturbance, a particle from the lower fresh cold water is moving upward. As heat conductivity is much larger than diffusivity, the particle takes the temperature of its neighbouring but as it is less dense than the saltier water outside it, the particle will rise upwards under the action of an upward buoyancy force. Likewise if a particle from the hot salt upper layer is sinking under the action of a perturbation, it will be quickly cooled and, becoming denser than its surroundings, it creates a downward buoyancy force accelerating the downward motion. This example illustrates clearly the property that concentration non-homogeneities are dangerous for the hydrodynamic stability of mixtures when their relaxation time is much larger than that of temperature non-homogeneities.

The resulting finite amplitude motions have been called *salt fingers* because of their elongated structures (see Fig. 6.16). They have been observed in a variety of laboratory experiments with heat-salt and sugar-salt mixtures and in subtropical oceans.

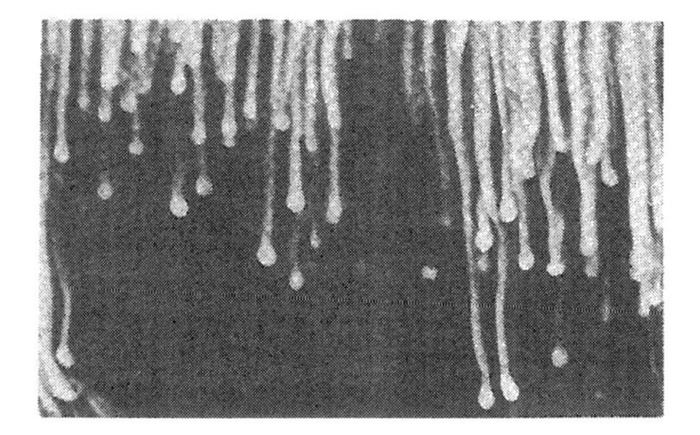

Fig. 6.16 Salt fingers (from Vidal et al. 1994)

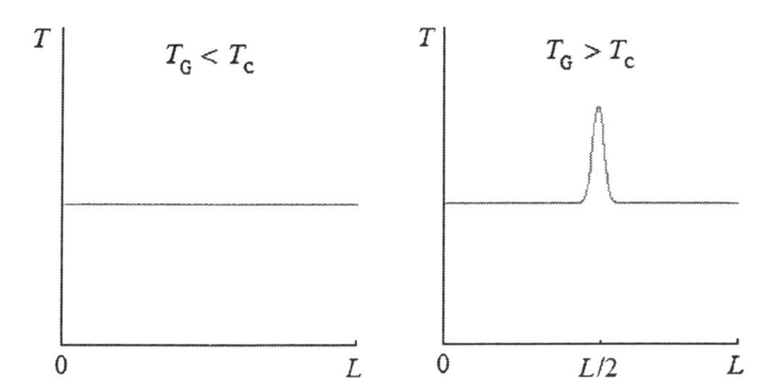

Fig. 6.17 Temperature distributions in the ballast resistor

6.6.2 Patterns in Electricity

6.6.2.1 The Ballast Resistor

Let us first address some attention to the *ballast resistor* (Bedeaux et al. 1977; Pasmanter et al. 1978; Elmer 1992); it is an interesting example because it can be described by a one-dimensional model allowing for explicit analytic treatments and, in addition, it presents useful technological aspects. The device consists of an electrical wire traversing through a vessel of length L filled with a gas at temperature T_G. The control parameters are the temperature T_G and the electric current I crossing the wire. As much as the temperature T_G is lower than a critical value T_c, the temperature of the wire remains uniform but for $T_G > T_c$ there is a bifurcation in the temperature profile, which is no longer homogeneous, but instead is characterized by a peak located at the middle of the electric wire (see Fig. 6.17).

It is worth to stress that, for $T_G > T_c$, the value of the electric current I is insensitive to the variations of the electrical potential which indicates that the ballast resistor can be used as a current stabilizer device.

6.6.2.2 The Laser

When the laser is pumped only weakly, one observes that the emitted light waves have random phases with the result that the electric field strength is a superposition of random waves. Above a critical pump strength, the laser light becomes coherent (Haken 1977), meaning that the waves are now ordered in a well-defined temporal organization (Fig. 6.18).

By increasing the external pumping, a sequence of more and more complicated structures is displayed, just like in hydrodynamic and chemical instabilities. In particular by pumping the laser strength above a second threshold, the continuous wave emission is transformed into ultra-short pulses.

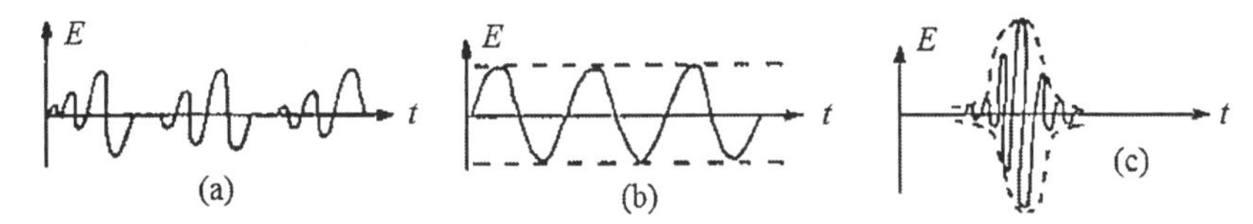

Fig. 6.18 Laser instability: the electrical field strength E is given as a function of time: (**a**) disordered state, (**b**) ordered state, and (**c**) the same above the second threshold

6.6.3 Dendritic Pattern Formation

Formation of dendrites, i.e. tree-like or snowflake-like structures as shown in Fig. 6.19, is a much-investigated subject in the area of pattern formation.

Fig. 6.19 Dendritic xenon crystal growth (from Gollub and Langer 1999)

Research on dendritic crystal growth has been motivated by the necessity to better understand and control metallurgical microstructures.

The process which determines the formation of dendrites is essentially the degree of undercooling that is the degree to which the liquid is colder than its freezing temperature. The fundamental rate-controlling mechanism is diffusion, either diffusion of latent heat away from the liquid–solid interface, or diffusion of chemical species toward and away from this solidification front. These diffusion processes lead to shape instabilities, which trigger the formation of patterns in solidification. In a typical sequence of events, the initially crystalline seed immersed in its liquid phase grows out rapidly in a cascade of branches whose tips move outwards at a given speed. These primary arms become unstable against side branching and the new side branching are in turn unstable with respect to further side branching, ending in a final complicated dendritic structure. The speed at which the dendrites grow, the regularity, and the distances between the side branches determine most of the properties of the solidified material, like its response to heating and mechanical deformation.

To summarize, we have tried in this chapter to convince the reader of the *universality* of pattern-forming phenomena. We have stressed that similar patterns are observed in apparently very different systems, as illustrated by examples drawn from hydrodynamics (Rayleigh–Bénard's and Bénard–Marangoni's convections, Taylor's vortices), chemistry (Belousov–Zhabotinsky's reaction and Turing's instability), electricity (ballast resistor and laser instability), and materials science (dendritic formation). Of course, this list is far from being exhaustive and further applications have been worked out in a great variety of areas. Figure 6.20 displays some examples like a quasi-crystalline standing-wave pattern produced by forcing a layer of silicone oil at two frequencies (a), a standing-wave pattern of granular material-forming stripes (b), a typical mammalian coat as the leopard (c).

Two last remarks are in form. That a great number of particles, of the order of 10^{23}, will behave in a coherent matter despite their random thermal

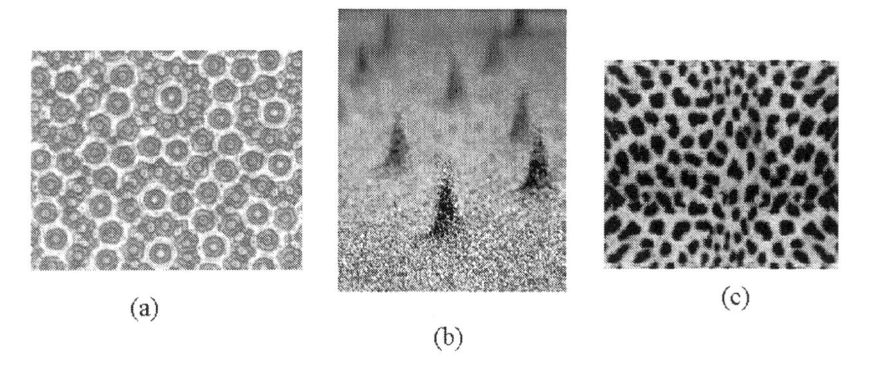

Fig. 6.20 Examples of patterns in quasi-crystalline pattern (**a**), granular material (**b**), and typical leopard's coat (**c**)

agitation is the main feature of pattern formation. As pointed out throughout this chapter, self-organization finds its origin in two causes: non-linear dynamics and external non-equilibrium constraints. Fluctuations arising from the great number of particles and their random motion are no longer damped as in equilibrium but may be amplified with the effect to drive the system towards more and more order. This occurs when the control parameter, like the temperature gradient in Bénard's experiment, crosses a critical point at which the system undergoes a transition to a new state, characterized by regular patterns in space and/or in time.

It may also be asked why appearance of order is not in contradiction with the second law of thermodynamics which states that the universe is evolving towards more and more disorder. There is of course no contradiction, because the second principle, as enounced here, refers to an isolated system while pattern forming can only occur in closed and/or open systems with exchange of energy and matter with the surrounding. The decrease of entropy in individual open or closed cells is therefore consistent with the entropy increase of the total universe and the validity of the second law is not to be questioned.

6.7 Problems

6.1. *Non-linear Landau equation.* Show that the solution of the non-linear Landau equation

$$\frac{\mathrm{d}}{\mathrm{d}t}|A|^2 = 2(\mathrm{Re}\,\sigma)\,|A|^2 - 2l\,|A|^4$$

is given by (6.12).

6.2. *Landau equation and Rayleigh–Bénard's instability.* The Landau equation describing Rayleigh–Bénard's instability can be cast in the form $\mathrm{d}A/\mathrm{d}t = \sigma A - lA^3$ whose steady solution is $A_\mathrm{s} = (\sigma/l)^{1/2}$. Expanding σ around the critical Rayleigh number, one has $\sigma = \alpha[Ra - (Ra)_\mathrm{c}]$, where α is a positive constant from which follows the well-known result

$$A_\mathrm{s} = (\alpha/l)^{1/2}[Ra - (Ra_\mathrm{c})]^{1/2}.$$

Study the stability of this steady solution by superposing to it an infinitesimally small disturbance A' and show that the steady non-linear solution is stable for $Ra > (Ra)_\mathrm{c}$.

6.3. *Third-order Landau equation.* Consider the following third-order Landau equation

$$\frac{\mathrm{d}A}{\mathrm{d}t} = \sigma A + \alpha A^2 + \beta A^3,$$

when $A > 0, \sigma > 0, \alpha < 0$ and $\beta^2 > 4\alpha\sigma$. For sufficiently small values of A at $t = 0$, show that A tends to the equilibrium value $A_\mathrm{e} = -\sigma/\alpha + O(\sigma^2)$ as $t \to \infty$.

6.4. *Rayleigh–Bénard's instability.* Consider an incompressible Boussinesq fluid layer between two rigid horizontal plates of infinite extent. The two plates are perfectly heat conducting and the fluid is heated from below. (a) Establish the amplitude equations in the case of infinitesimally small disturbances (linear approximation). (b) Determine the marginal instability curve $Ra(k)$ between the dimensionless Rayleigh number Ra and wave number k. (c) Calculate the critical values $(Ra)_c$ and k_c corresponding to onset of convection.

6.5. *Rotating Rayleigh–Bénard's problem.* Two rigid horizontal plates extending to infinity bound a thin layer of fluid of thickness d. The system, subject to gravity forces, is heated from below and is rotating around a vertical axis with a constant angular velocity Ω. Determine the marginal curve $Ra(k)$ as a function of the dimensionless Taylor number $Ta = 4\Omega^2 d^4/\nu^2$, where ν is the kinematic viscosity of the fluid. Does rotation play a stabilizing or a destabilizing role?

6.6. *Rayleigh–Bénard's problem with a solute.* Consider a two-constituent mixture (solvent + solute) encapsulated between two free horizontal surfaces and subject to a vertical temperature gradient β. Denoting by $c(\boldsymbol{r}, t)$ the concentration of the solute, assume that the density of the mixture is given by

$$\rho = \rho_0[1 + \alpha(T - T_0) + \gamma(c - c_0)].$$

Neglecting the diffusion of the solute so that $\mathrm{d}c/\mathrm{d}t = 0$, determine the marginal curve $Ra(k)$ when the basic reference state is at rest with a given concentration $c_r(z)$ and a temperature field $T_r = T_0 - \beta z$.

6.7. *Non-linear Rayleigh–Bénard's instability.* A thin incompressible fluid layer of thickness d is bounded by two horizontal stress-free boundaries ($Ma = 0$) of infinite horizontal extent. The latter are perfectly heat conducting and gravity forces are acting on the fluid.

(a) Show that the convective motion can be described by the following non-linear relation

$$\frac{\partial w}{\partial t} + \frac{1}{2}\frac{\partial^3 w^2}{\partial z^3} = \Delta^3 w - Ra\,\Delta_1 w \quad (0 < z < 1),$$

here w is the dimensionless vertical velocity component, Ra the Rayleigh number and

$$\Delta = \Delta_1 + \partial^2/\partial z^2 = \partial^2/\partial x^2 + \partial^2/\partial y^2 + \partial^2/\partial z^2;$$

the corresponding boundary conditions (at $z = 0$ and $z = 1$) are

$$w = \frac{\partial^2 w}{\partial z^2} = \frac{\partial^4 w}{\partial z^4} = 0.$$

(b) Find the solution of the corresponding linear problem with normal mode solutions of the form $w = W(z)f(x,y)\exp(\sigma t)$; to be explicit, determine the expressions of $W(z)$ and σ and the equation satisfied by $f(x,y)$.

(c) By assuming a non-linear solution of the roll type, i.e. $w = A\cos(kx)g(z)$, show that the Landau equation associated to this problem can be written as

$$\frac{\mathrm{d}A}{\mathrm{d}t} = \alpha A - \beta A^3,$$

where α and β are two constants to be determined in terms of the data Ra, $(Ra)_c$, and k_c.

6.8. *Boundary condition with surface tension gradient.* In dimensional form, the kinematic boundary conditions at a horizontal surface normal to the z-axis and subject to a surface tension gradient may be written as $\boldsymbol{\sigma}\cdot\boldsymbol{n} = \nabla S$, where $\boldsymbol{\sigma}$ is the stress tensor or more explicitly $\sigma_{xz} + \partial S/\partial x = 0$ and $\sigma_{yz} + \partial S/\partial y = 0$. Show that, in non-dimensional form, the corresponding boundary condition is given by (6.39) $D^2W = k^2 Ma\theta$. *Hint*: After differentiating the first relation with respect to x, the second with respect to y, make use of Newton's constitutive relation $\boldsymbol{\sigma} = \eta[\nabla\boldsymbol{v} + (\nabla\boldsymbol{v})^{\mathrm{T}}]$ and the continuity relation $\nabla\cdot\boldsymbol{v} = 0$.

6.9. *Bénard–Marangoni's instability.* (a) Show that, in an incompressible liquid layer whose lower boundary is in contact with a rigid plate while the upper boundary is open to air and subject to a surface tension depending linearly on the temperature, the marginal curve relating the Marangoni number Ma to the wave number k is given by (6.40), i.e.

$$(Ma)_0 = \frac{8k^2\cosh k(k - \sinh k\cosh k)}{k^3\cosh k - \sinh^3 k},$$

both boundaries are assumed to be perfectly heat conducting and gravity acceleration is neglected. (b) Find the corresponding critical values $(Ma)_c = 79.6$ and $k_c = 1.99$.

6.10. *Bénard–Marangoni's instability.* The same problem as in 6.9, but now with heat transfer at the upper surface governed by Newton's cooling law

$$-\lambda\frac{\partial T}{\partial z} = h(T - T_\infty),$$

where λ is the heat conductivity of the fluid, h the heat transfer coefficient, and T_∞ the temperature of the outside world, say the laboratory. In dimensionless form, the previous law reads as $D\Theta = -Bi\,\Theta$, ($D = \mathrm{d}/\mathrm{d}Z$) with $Bi = hd/\lambda$ the so-called *Biot number*. The limiting case $Bi = 0$ corresponds to an adiabatically isolated surface while $Bi = \infty$ describes a perfectly heat conductor. (a) Determine the dependence of Ma with respect to k and Bi. (b) Draw the marginal instability curves for $Bi = 0, 1, 10$. (c) Sketch the curves $(Ma)_c(Bi)$ and $k_c(Bi)$.

6.11. *The Rayleigh–Bénard–Marangoni's instability.* Show that, for the coupled buoyancy–surface tension-driven instability, the Ra and Ma numbers obey the relation (6.41)

$$\frac{Ra}{(Ra)_\mathrm{c}} + \frac{Ma}{(Ma)_\mathrm{c}} = 1,$$

where $(Ra)_\mathrm{c}$ is the critical Rayleigh number without Marangoni effect and $(Ma)_\mathrm{c}$ the critical Marangoni number in absence of gravity.

6.12. *The Lorenz model.* (a) Show that the steady solutions of (6.25) corresponding to supercritical convection are given (6.35) by $A_1 = A_2 = \sqrt{b(r-1)}$ for $r > 1$. (b) Prove that this solution becomes unstable at $r = Pr(Pr\,b+3)/(Pr - b - 1)$. (c) Solve numerically the Lorenz equations for $Pr = 10$, $b = 8/3$, and $r = 28$ (Sparrow 1982).

6.13. *Couette flow between two rotating cylinders.* Consider a non-viscous fluid contained between two coaxial rotating cylinders. The reference state is stationary with $u_r = u_z = 0$, $u_\theta(r) = r\Omega(r)$, the quantity $\Omega(r)$ is an arbitrary function of the distance r to the axis of rotation and is related to the reference pressure by $p_\mathrm{ref} = \rho \int r\Omega^2(r)\mathrm{d}r$. (a) Show that the latter result is directly obtained from the radial component of the momentum equation. (b) Using the normal mode technique, show that, for axisymmetric disturbances $(\partial/\partial\theta = 0)$, the amplitude equation is given by

$$(DD^* - k^2)U_\mathrm{r} - \frac{k^2}{\sigma}\phi(r)U_\mathrm{r} = 0,$$

where $D = \mathrm{d}/\mathrm{d}r$, $D^* = D + 1/r$, and $\phi(r) = (1/r^3)\mathrm{d}[(r^2\Omega)^2]/\mathrm{d}r$ is the so-called *Rayleigh discriminant.* It is interesting to observe that the quantity $(r^2\Omega)$ in the Rayleigh discriminant is related to the circulation along a circle of radius r by

$$\int_0^{2\pi} u_\theta(r) r\,\mathrm{d}\theta = 2\pi r^2\Omega(r).$$

(c) Show further that the flow is stable with respect to axisymmetric disturbances if $\phi \geq 0$. This result reflects the celebrated Rayleigh circulation criterion stating that a necessary and sufficient condition of stability is that the square of the circulation does not decrease anywhere.

6.14. *Lotka chemical reactions.* Show that the following sequence of chemical autocatalytic reactions

$$\mathrm{A + X \to 2X}$$
$$\mathrm{X + Y \to 2Y}$$
$$\mathrm{Y + B \to E + D}$$

where the concentrations of substances A and B are maintained fixed, correspond to the Lotka–Volterra model.

Chapter 7
Extended Irreversible Thermodynamics

Thermodynamics of Fluxes: Memory and Non-Local Effects

With this chapter, we begin a panoramic overview of non-equilibrium thermodynamic theories that go beyond the local equilibrium hypothesis, which is the cornerstone of classical irreversible thermodynamics (CIT). We hope that this presentation, covering Chaps. 7–11, will convince the reader that non-equilibrium thermodynamics is a fully alive and modern field of research, combining practical motivations and conceptual questions. Indeed, such basic topics as the definition and meaning of temperature and entropy, the formulation of the second law, and its consequences on the admissible transport equations are still open questions nowadays.

Modern technology strives towards miniaturized devices and high-frequency processes, whose length and timescales are comparable to the mean free path of the particles and to the internal relaxation times of the devices, thus requiring extensions of the classical transport laws studied in the previous chapters. Indeed, these laws assume an instantaneous response of the fluxes to the imposed thermodynamic forces, whereas, actually, it takes some time for the fluxes to reach the values predicted by the classical laws. As a consequence, when working at short timescales or high frequencies, and correspondingly at short length scales or short wavelengths, the generalized transport laws must include memory and non-local effects. The analysis of these generalized transport laws is one of the main topics in modern non-equilibrium thermodynamics, statistical mechanics, and engineering. Such transport laws are generally not compatible with the local equilibrium hypothesis and a more general thermodynamic framework must be looked for.

Going beyond CIT and exploring new frontiers are the driving impetus for the development of recent non-equilibrium thermodynamic theories. In that respect, we will successively analyse extended irreversible thermodynamics (EIT), theories with internal variables, rational thermodynamics, Hamiltonian formulation, and mesoscopic theories.

This overview starts with EIT, because of its formal simplicity and its proximity to the methods of CIT to which the reader is already acquainted. EIT provides a macroscopic and causal description of non-equilibrium processes and is based on conceptually new ideas, like the introduction of the fluxes as additional non-equilibrium independent variables, and the search for general transport laws taking the form of evolution equations for these fluxes. Such equations will be generally obtained by considering the restrictions imposed by the second law of thermodynamics.

To be explicit, in EIT the space $\mathcal{V}$ of state variables is formed by the union of the space $\mathcal{C}$ of classical variables like mass, momentum, energy, and composition, and the space $\mathcal{F}$ of the corresponding fluxes, i.e. $\mathcal{V} = \mathcal{C} \cup \mathcal{F}$. The physical nature of the $\mathcal{F}$-variables is different from that of the $\mathcal{C}$-variables. The latter are slow and conserved with their behaviour governed by the classical balance laws. In contrast, the $\mathcal{F}$-variables are fast and non-conserved: they do not satisfy conservation laws and their rate of decay is generally very short. In dilute gases, it is of the order of magnitude of the collision time between molecules, i.e. 10^{-12} s. This means that, for time intervals much larger than this value, fast variables can be ignored. This is no longer true in high-frequency phenomena, when the relaxation time of the fluxes is comparable to the inverse of the frequency of the process, or in some materials, like polymers, dielectrics, or superfluids, characterized by rather large relaxation times of the order of seconds or minutes. The independent character of the fluxes is also made evident when the mean free path of heat or charge carriers becomes comparable to the dimensions of the sample, as in nano-systems. Other motivations for elevating the fluxes to the rank of variables are given at the end of this chapter.

The domain of application of EIT enlarges the frontiers of CIT, whose range of validity is limited to small values of the relaxation times τ of the fluxes, i.e. to small values of the Deborah number $De \equiv \tau/t_{\mathrm{M}}$, with t_{M} a macroscopic timescale, and to small values of the Knudsen number $Kn \equiv \ell/L$, where ℓ is the mean free path and L a macroscopic length. The transport equations derived from EIT reduce to the CIT expressions for $De \ll 1$ and $Kn \ll 1$, but are applicable to describe a wider range of situations encompassing $De > 1$ and $Kn > 1$. Examples of situations for which $De \geq 1$ are processes where the macroscopic timescale becomes short enough to be comparable to the microscopic timescale, as in ultrasound propagation in dilute gases or neutron scattering in liquids, or when the relaxation time becomes long enough to be comparable to the macroscopic timescale, as in polymer solutions, suspensions, superfluids, or superconductors. The property $Kn \geq 1$ is characteristic of micro- and nano-systems as thin films, superlattices, submicronic electronic devices, porous media, and shock waves.

A simple way, although not unique, to obtain the time evolution equations of the fluxes on a macroscopic basis is to assume the existence of a generalized entropy and to follow the same procedure as in CIT. In EIT, it is taken for granted that there exists a non-equilibrium entropy s to which the following properties are assigned:

- It is an additive quantity.
- It is a function of the whole set of variables $s = s(\mathcal{V})$.

- It is a concave function of the state variables.
- Its rate of production is locally positive.

Once the expression of s is known, it is an easy matter to derive generalized equations of state, which are of interest in the description of non-equilibrium steady states.

The scope of this chapter is to give a general presentation of what EIT is, how it works, and what can be expected from it. EIT has been the object of several monographs by Jou et al. (2000, 2001), Lebon (1992) and Müller and Ruggeri (1998). For a more microscopic perspective, we refer to the books by Eu (1992, 1998) and Luzzi et al. (2001, 2002); see also two collections of contributions by several authors (Casas-Vázquez et al. 1984; Sieniutycz and Salamon 1992) or reviews by Garcia-Colin (1991, 1995) and Nettleton and Sobolev (1995).

As an introductory example, we will study heat conduction in a rigid body with memory effects; in this problem, only the heat flux is introduced as extra variable. Afterwards, we shall discuss more complicated situations, such as viscous fluids, polymer solutions, and electric transport in microelectronic devices, where other fluxes, as the viscous pressure tensor and the electric current, are selected as supplementary independent variables.

7.1 Heat Conduction

After outlining the shortcomings of the classical approach, we shall motivate the choice of the heat flux as independent variable in the prototype problem of heat conduction in a non-deformable solid at rest, and we will take advantage of this particular example to introduce the main tenets of the general formalism.

7.1.1 Fourier's vs. Cattaneo's Law

The best-known model for heat conduction is Fourier's law, which relates linearly the temperature gradient ∇T to the heat flux $\boldsymbol{q}$ according to

$$\boldsymbol{q} = -\lambda \nabla T, \tag{7.1}$$

where λ is the heat conductivity, depending generally on the temperature. By substitution of (7.1) in the energy balance equation, written in absence of source terms as

$$\rho \frac{\partial u}{\partial t} = -\nabla \cdot \boldsymbol{q}, \tag{7.2}$$

and relating the specific internal energy u to the temperature by means of $\mathrm{d}u = c_v \mathrm{d}T$, with c_v being the heat capacity per unit mass at constant volume, one obtains

$$\rho c_v \frac{\partial T}{\partial t} = \nabla \cdot (\lambda \nabla T). \tag{7.3}$$

The material time derivative has been replaced here by the partial time derivative because the body is at rest. From a mathematical point of view, (7.3) is a parabolic differential equation. Although this equation is well tested for most practical problems, it fails to describe the transient temperature field in situations involving short times, high frequencies, and small wavelengths. For example, it was shown by Maurer and Thomson (1973) that, by submitting a thin slab to an intense thermal shock, its surface temperature is 300°C larger than the value predicted by (7.3). The reasons for this failure must be found in the physical statement of Fourier's law, according to which a sudden application of a temperature difference gives *instantaneously* rise to a heat flux everywhere in the system. In other terms, any temperature disturbance will propagate at infinite velocity. Physically, it is expected, and it is experimentally observed, that a change in the temperature gradient should be felt after some build-up or relaxation time, and that disturbances travel at finite velocity. From a microscopic point of view, Fourier's law is valid in the collision-dominated regime, where there are many collisions amongst the particles, but it loses its validity when one approaches the ballistic regime, in which the dominant collisions are those of the particles with the boundaries of the system rather than the collisions amongst particles themselves.

To eliminate these anomalies, Cattaneo (1948) proposed a damped version of Fourier's law by introducing a heat flux relaxation term, namely

$$\tau \frac{\partial \boldsymbol{q}}{\partial t} = -(\boldsymbol{q} + \lambda \nabla T). \tag{7.4}$$

The term containing the time τ represents the heat flux relaxation. When the relaxation time τ of the heat flux is negligible or when the time variation of the heat flux is slow, this equation reduces to Fourier's law. For homogeneous solids, τ describes molecular-scale energy transfer by either phonons or electrons, and it is very small, of the order of time between two successive collisions at the microscopic level. Therefore, in most practical heat transfer problems, infinite propagation is not relevant as those parts of the signals with infinite velocity are strongly damped at room temperature. However, when slow internal degrees of freedom are involved, as in polymers, superfluids, porous media, or organic tissues, τ reflects the time required to transfer energy between different degrees of freedom and it may be relatively large, of the order or larger than 1 s.

Relaxational effects on heat conductors were already discussed by Maxwell (1867), Cattaneo (1948), Vernotte (1958) and Grad (1958), but without referring to its thermodynamic implications, which will be analysed in Sect. 7.1.2 in the frame of EIT.

Assuming constant values of c_v and λ and introducing (7.4) in (7.2) results in the following hyperbolic equation

$$\tau \frac{\partial^2 T}{\partial t^2} + \frac{\partial T}{\partial t} - \chi \nabla^2 T = 0, \tag{7.5}$$

where $\chi = \lambda/\rho c_v$ designates the heat diffusivity. Equation (7.5) is sometimes called *telegrapher's equation* because it is similar to the one describing propagation of electrical signals along a wire. For small values of the time $t \ll \tau$, the first term of (7.5) is dominant, so that it reduces to

$$\tau \frac{\partial^2 T}{\partial t^2} = \chi \nabla^2 T. \tag{7.6}$$

This is a wave equation with a wave propagating at the velocity $(\chi/\tau)^{1/2}$; it describes a reversible process, as it is invariant with respect to time inversion. In contrast, for timescales much longer than $\tau (t \gg \tau)$, the first term of (7.5) is negligible and one obtains a partial differential equation of the form

$$\frac{\partial T}{\partial t} = \chi \nabla^2 T, \tag{7.7}$$

which is associated with diffusion of heat, as shown in Chap. 2. Diffusion is typically an irreversible process, as (7.7) is not invariant when t is changed into $-t$. To summarize, at short times the transport equation (7.5) is reversible and heat propagates as a wave with a well-defined speed (which may be microscopically interpreted as a ballistic motion of heat carriers), whereas at longer times the process becomes irreversible and heat is diffused throughout the system. It is therefore clear that τ can be interpreted as the characteristic time for the crossover between ballistic motion and the onset of diffusion. In the context of chaotic deterministic systems, τ is interpreted as the Lyapunov time beyond which predictivity is lost (Nicolis and Prigogine 1989).

The dynamical properties of (7.5) have been thoroughly analysed. By assuming that there exists a solution of the form $T(x,t) = T_0 \exp[\mathrm{i}(kx - \omega t)]$, where ω is a (real) frequency and k is a (complex) wave number, it is found (see Problem 7.2) that the solution is characterized by a phase speed v_p and an attenuation length α, respectively, given by

$$v_\mathrm{p} = \frac{\omega}{\mathrm{Re}\, k} = \frac{\sqrt{2\chi\omega}}{\sqrt{\tau\omega + \sqrt{1+\tau^2\omega^2}}}, \quad \alpha = \frac{1}{\mathrm{Im}\, k} = \frac{2\chi}{v_\mathrm{p}}. \tag{7.8}$$

In the low-frequency limit ($\tau\omega \ll 1$), it is found that

$$v_{\mathrm{p},0} = \sqrt{2\chi\omega}, \quad \alpha_0 = \sqrt{2\chi/\omega}, \tag{7.9}$$

which are the results obtained directly from Fourier's law. In the high-frequency limit ($\tau\omega \gg 1$), the phase speed and attenuation length are

$$v_{\mathrm{p},\infty} \equiv U = (\chi/\tau)^{1/2}, \quad \alpha_\infty = 2(\chi\tau)^{1/2}. \tag{7.10}$$

The velocity U corresponds to the so-called *second sound*, which is a temperature wave, not to be confused with the *first sound*, which is a pressure wave with velocity $c_0 = \sqrt{(\partial p/\partial \rho)_\mathrm{s}}$. The value of $v_{\mathrm{p},\infty}$ diverges when the relaxation time vanishes, thus leading to an infinite speed of propagation. The presence of the relaxation time represents much more than a minor correction to the classical results, as in the high-frequency regime it leads to a completely different behaviour than that predicted by the classical Fourier's law. In Sect. 7.1.3, we discuss other physical consequences of (7.5). The analysis of thermal waves has been the topic of much research (Joseph and Preziosi 1989; Dreyer and Struchtrup 1993; Tzou 1997; Jou et al. 2001).

The differences between parabolic and hyperbolic equations are shown in Box 7.1. Figure 7.1 provides a qualitative comparison between the parabolic ($\tau = 0$) and hyperbolic ($\tau \neq 0$) solutions of (7.5), at a short time after application of a thermal pulse in a one-dimensional rod. According to Cattaneo's

Box 7.1 Differences Between Hyperbolic and Parabolic Heat Transport

As shown in Box 2.3, the response to a delta pulse of temperature is given, according to the classical diffusion equation, by

$$\Delta T(x,t) = \frac{g_0}{2(\pi\chi t)^{1/2}} \exp\left(-\frac{x^2}{4\chi t}\right), \tag{7.1.1}$$

where g_0 is a value which depends on the intensity of the initial pulse. The response to a delta pulse predicted by the telegrapher's equation is much more complicated, but has been derived in many textbooks (see, in particular, Morse and Feshbach 1953; Jou et al. 2001; Tzou 1997). It is found that

$$\Delta T(x,t) = \frac{g_0}{4} \exp\left(-\frac{t}{2\tau}\right) \left\{ 2I_0(\xi) U\tau\delta(Ut - |x|) + \left[I_0(\xi) + \frac{t}{\tau\xi}\frac{\mathrm{d}I_0(\xi)}{\mathrm{d}\xi} \right] H(Ut - |x|) \right\}, \tag{7.1.2}$$

where $I_0(x)$ is the modified Bessel function of the first kind, $H(x)$ the Heaviside function, U the second sound velocity given by $U \equiv (\chi/\tau)^{1/2}$ and $\xi \equiv \frac{1}{2}[(t/\tau)^2 - (x/U\tau)^2]^{1/2}$. The main differences with the solution of the classical diffusion equation (7.1.1) are shown in Fig. 7.1. They are essentially:

1. In the classical description, $\Delta T(x,t)$ is non-zero at any position in space for $t > 0$. In contrast, $\Delta T(x,t)$ given by (7.1.2) vanishes for $x > Ut$, U being the speed of the front.
2. In (7.1.2) there are two main contributions: the first, in $\delta(Ut - x)$, describes the propagation of the initial pulse with speed U, damped by the exponential term. The second term, in $H(Ut - x)$, is related to the wake, which, for sufficiently long times $Ut \gg x$, tends to the diffusive solution (7.1.1).

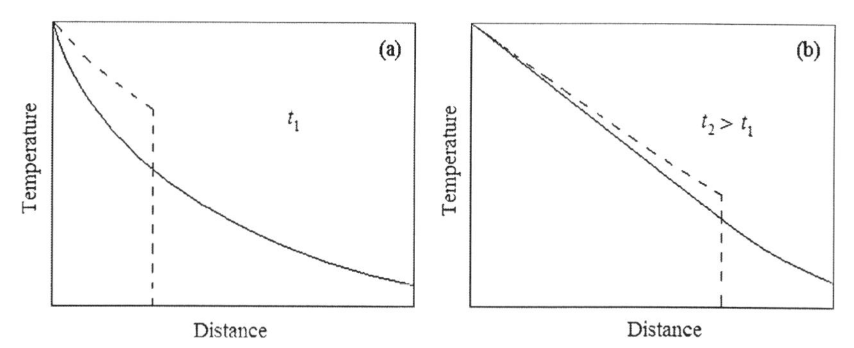

Fig. 7.1 Fourier (*solid line*) and Cattaneo (*dashed line*) temperature distributions in one-dimensional rod: (**a**) corresponds to a given time t_1 and (**b**) corresponds to a time $t_2 > t_1$

law, the wave propagates with a speed $v_{\mathrm{p},\infty} = (\chi/\tau)^{1/2}$ through the system; at the right of the wave front, located at $\ell = v_{\mathrm{p},\infty} t = (\chi/\tau)^{1/2} t$, the temperature perturbation is not felt. The quantity ℓ is called the *penetration depth*, which in Fourier's law ($\tau = 0$) extends to infinity, making that the disturbance is instantaneously felt throughout the system (see Fig. 7.1a). The temperature behind the wave front will be higher than the one corresponding to Fourier's law, because the same amount of energy is confined into a smaller volume. When the time is increased, the diffusion of energy over a larger volume has a damping effect on the wave, and the temperature distribution becomes closer to the Fourier prediction (Fig. 7.1b). As time increases, the difference between the hyperbolic and the parabolic temperature distributions becomes smaller and smaller.

Despite the success of Cattaneo's model and its generalizations to describe heat transfer at high frequencies, its thermodynamic consequences are less known and are worth examining. First, it should be stressed that Cattaneo's equation is not compatible with CIT. Indeed, after substitution of Cattaneo's equation (7.4) in the classical expression for the entropy production as derived in Chap. 2, namely $\sigma^s = \boldsymbol{q} \cdot \nabla T^{-1}$, it is found that

$$\sigma^s = \frac{\lambda}{T^2}(\nabla T)^2 + \frac{\tau}{T^2}\frac{\partial \boldsymbol{q}}{\partial t} \cdot \nabla T, \tag{7.11}$$

which is no longer positive definite, because of the presence of the second term of the right-hand side. The problem finds its origin in the local equilibrium hypothesis, which must therefore be revisited. This is better seen by examining the time evolution of entropy in an isolated rigid body. Making use of the local equilibrium assumption, one obtains for the total entropy of the system a non-monotonic increase, while use of EIT yields a monotonic increase as it is shown by the solid curve in Fig. 7.2 (Criado-Sancho and Llebot 1993).

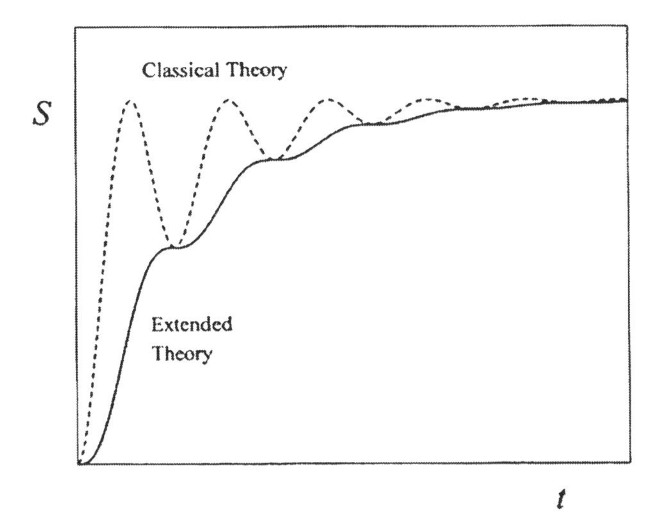

Fig. 7.2 The evolution of the classical entropy S_{CIT} during the equilibration of an isolated system when use is made of Cattaneo's equation is given by the *dashed curve*. The evolution of the extended entropy S_{EIT}, obtained from (7.2.15), is represented by the *solid curve*, which, in contrast with that of S_{CIT}, increases monotonically

Instead of being monotonically increasing, the classical entropy behaves in an oscillatory way, as shown in Box 7.2. Strictly speaking, this result is not incompatible with the Clausius' formulation of the second law, which states that the entropy of the final equilibrium state must be higher than the entropy of the initial equilibrium state. However, the non-monotonic behaviour of the entropy is in contradiction with the local equilibrium formulation of the second law, which requires that the entropy production must be positive everywhere at any time. Note, finally, that Cattaneo's model does not preclude that heat flows from lower to higher temperature. Nevertheless, this is not in contradiction with the Clausius' statement of the second law, about the impossibility of constructing a cyclic heat engine whose only effect is to transport heat from a colder to a hotter reservoir, because in (7.4) the mentioned unusual behaviour lasts only over time intervals of the order of the relaxation time during which Clausius' claim is at least questionable. Notwithstanding, this behaviour is in opposition with the local equilibrium formulation of the second law, requiring that heat should always move from hotter to colder regions, in absence of any other effect.

Box 7.2 Evolution of Entropy in a Discrete Isolated System: CIT vs. EIT Behaviour

We study heat transfer between two rigid bodies at temperatures T_1 and $T_2(< T_1)$ put into contact. Let the total system be isolated from the outside world and let $\dot{Q}$ be the amount of heat exchanged between the two bodies per unit time. In the local equilibrium formulation, one has $S(U_1, U_2) = S_1(U_1) + S_2(U_2)$ and therefore

$$\frac{\mathrm{d}S}{\mathrm{d}t} = \frac{\mathrm{d}S_1}{\mathrm{d}t} + \frac{\mathrm{d}S_2}{\mathrm{d}t} = T_1^{-1}\frac{\mathrm{d}U_1}{\mathrm{d}t} + T_2^{-1}\frac{\mathrm{d}U_2}{\mathrm{d}t}. \tag{7.2.1}$$

Since the global system is isolated, $\mathrm{d}U_1 + \mathrm{d}U_2 = 0$, and from the first law of thermodynamics, one has

$$\frac{\mathrm{d}U_1}{\mathrm{d}t} = C_1\frac{\mathrm{d}T_1}{\mathrm{d}t} = -\dot{Q}, \quad \frac{\mathrm{d}U_2}{\mathrm{d}t} = C_2\frac{\mathrm{d}T_2}{\mathrm{d}t} = \dot{Q}, \tag{7.2.2}$$

where use has been made of the equations of state $\mathrm{d}U_i = C_i\mathrm{d}T_i$, and C_i are the respective heat capacities of the systems.

For an isolated system, (7.2.1) represents the rate of entropy produced inside the system, and in virtue of (7.2.2), it can be written as

$$\frac{\mathrm{d}S}{\mathrm{d}t} = -(T_1^{-1} - T_2^{-1})\dot{Q}, \tag{7.2.3}$$

or assuming that $T_1 - T_2 \equiv \varepsilon$ is small,

$$\frac{\mathrm{d}S}{\mathrm{d}t} = \frac{\varepsilon}{T^2}\dot{Q}, \tag{7.2.4}$$

where T is an intermediate temperature between T_1 and T_2. According to the second law of thermodynamics, (7.2.3) must be non-negative. This implies that heat only flows from the region of highest temperature to the region of lowest temperature, according to the original Clausius' formulation of the second law.

The simplest hypothesis ensuring the positiveness of (7.2.4) is to assume, as in Fourier's law, that the heat flux $\dot{Q}$ is proportional to the driving force ε/T^2, so that

$$\dot{Q} = K\frac{\varepsilon}{T^2}, \tag{7.2.5}$$

with K being a positive coefficient. Combining (7.2.2) with (7.2.5) one obtains for the evolution of ε

$$\frac{\mathrm{d}\varepsilon}{\mathrm{d}t} = -C^{-1}\dot{Q}, \tag{7.2.6}$$

where C is defined as $C^{-1} = C_1^{-1} + C_2^{-1}$. When (7.2.5) is introduced into (7.2.6), one finds that

$$\frac{\mathrm{d}\varepsilon}{\mathrm{d}t} = -\frac{K}{CT^2}\varepsilon. \tag{7.2.7}$$

Thus ε decays exponentially as $\varepsilon = \varepsilon_0 \exp[-Kt/(CT^2)]$. Now, substitution of (7.2.4) in (7.2.3) leads to

$$\frac{\mathrm{d}S}{\mathrm{d}t} = K\frac{\varepsilon^2}{T^2} \geq 0, \tag{7.2.8}$$

from which follows that the entropy is a monotonically increasing function of time.

Consider now the more general situation in which heat transfer is described by the Cattaneo's law

$$\tau \frac{\mathrm{d}\dot{Q}}{\mathrm{d}t} + \dot{Q} = K \frac{\varepsilon}{T^2}, \tag{7.2.9}$$

where τ is the relaxation time of $\dot{Q}$. After combining (7.2.6) and (7.2.9), one finds that the evolution of ε is governed by

$$\tau \frac{\mathrm{d}^2\varepsilon}{\mathrm{d}t^2} + \frac{\mathrm{d}\varepsilon}{\mathrm{d}t} + \frac{K}{CT^2}\varepsilon = 0, \tag{7.2.10}$$

which is similar to the equation of motion of a damped pendulum. The decay of ε is no longer exponential but will exhibit oscillatory behaviour for $4K\tau/(CT^2) > 1$. The term in $\mathrm{d}\varepsilon/\mathrm{d}t$ corresponds to diffusion of heat, whereas the term in $\mathrm{d}^2\varepsilon/\mathrm{d}t^2$ describes the propagation of the heat wave. Substituting the solution of (7.2.10) in (7.2.3) with $\dot{Q}$ given by (7.2.5) leads to an expression for the entropy S which exhibits a non-monotonic dependence with respect to time, as indicated by the dashed curve in Fig. 7.2.

In EIT, the quantity $\dot{Q}$ is viewed as an independent variable of a generalized entropy S_{EIT}, whose rate of variation is, up to the second order in $\dot{Q}$,

$$\frac{\mathrm{d}S_{\mathrm{EIT}}}{\mathrm{d}t} = T_1^{-1}\frac{\mathrm{d}U_1}{\mathrm{d}t} + T_2^{-1}\frac{\mathrm{d}U_2}{\mathrm{d}t} - a\dot{Q}\frac{\mathrm{d}\dot{Q}}{\mathrm{d}t} \tag{7.2.11}$$

instead of (7.2.1). Using the conservation of energy (7.2.2), one may write (7.2.11) as

$$\frac{\mathrm{d}S_{\mathrm{EIT}}}{\mathrm{d}t} = -\left(T_1^{-1} - T_2^{-1}\right)\dot{Q} - a\dot{Q}\frac{\mathrm{d}\dot{Q}}{\mathrm{d}t}, \tag{7.2.12}$$

which, for small temperature differences, simplifies to

$$\frac{\mathrm{d}S_{\mathrm{EIT}}}{\mathrm{d}t} = \left(\frac{\varepsilon}{T^2} - a\frac{\mathrm{d}\dot{Q}}{\mathrm{d}t}\right)\dot{Q}. \tag{7.2.13}$$

Assuming a linear relation between the flux $\dot{Q}$ and the force, given by the quantity inside brackets, one gets

$$\dot{Q} = K\left(\frac{\varepsilon}{T^2} - a\frac{\mathrm{d}\dot{Q}}{\mathrm{d}t}\right) \tag{7.2.14}$$

wherein positiveness of the entropy production requires that $K > 0$. Making use of (7.2.6) and replacing (7.2.14) by (7.2.13), one obtains

$$\frac{\mathrm{d}S_{\mathrm{EIT}}}{\mathrm{d}t} = \frac{C^2}{K}\left(\frac{\mathrm{d}\varepsilon}{\mathrm{d}t}\right)^2. \tag{7.2.15}$$

This expression is either positive or zero, but never negative. It is therefore concluded that S_{EIT}, whose evolution is plotted by the continuous curve in Fig. 7.2, increases monotonically in the course of time and is thus compatible with evolution equations of the Cattaneo type. For an explicit analysis of the continuous problem, see Criado-Sancho and Llebot (1993).

7.1.2 Extended Entropy

Going back to Cattaneo's equation (7.4), one observes that it represents truly an evolution equation for the heat flux, which can therefore be considered as an independent variable. This is indeed the key of EIT, where it is assumed that the entropy $s = s(u, \boldsymbol{q})$ does depend not only on the classical variable, namely the specific internal energy u, but also on the heat flux $\boldsymbol{q}$.

In differential form, the generalized entropy is written as follows:

$$\mathrm{d}s = \left(\frac{\partial s}{\partial u}\right)\mathrm{d}u + \left(\frac{\partial s}{\partial \boldsymbol{q}}\right)\cdot \mathrm{d}\boldsymbol{q}. \tag{7.12}$$

In analogy with the classical theory, one can define a *non-equilibrium* temperature $\theta(u, q^2)$ by means of the reciprocal of $(\partial s/\partial u)_{\boldsymbol{q}}$. This possibility is examined in Box 7.3. Here, we will neglect second- and higher-order contributions in $\boldsymbol{q}$ and simply identify $\partial s/\partial u$ with the inverse of the local equilibrium temperature $T(u)$, so that

$$\frac{\partial s}{\partial u} = \frac{1}{T(u)}. \tag{7.13}$$

The remaining partial derivative in (7.12) will be denoted as

$$\left(\frac{\partial s}{\partial \boldsymbol{q}}\right) = -T^{-1}v\alpha_1(u)\boldsymbol{q}, \tag{7.14}$$

wherein the minus sign and the factor $T^{-1}v$ are introduced for convenience, with v the specific volume (namely, the reciprocal of the mass density); α_1 is a undetermined function of u alone because, like temperature, its dependence on q^2 is assumed to be negligible. Accordingly, the final expression of the generalized Gibbs' equation (7.12) is

$$\mathrm{d}s = T^{-1}\mathrm{d}u - T^{-1}v\alpha_1\boldsymbol{q}\cdot \mathrm{d}\boldsymbol{q}. \tag{7.15}$$

From this expression and the internal energy balance law (7.2), one obtains, for the material time derivative $\dot{s}$ of the entropy,

$$\rho\dot{s} = -T^{-1}\nabla\cdot\boldsymbol{q} - T^{-1}\alpha_1\boldsymbol{q}\cdot\dot{\boldsymbol{q}}, \tag{7.16}$$

or, equivalently,

$$\rho \dot{s} = -\nabla \cdot (T^{-1}\boldsymbol{q}) + \boldsymbol{q} \cdot (\nabla T^{-1} - T^{-1}\alpha_1 \boldsymbol{q} \cdot \dot{\boldsymbol{q}}). \tag{7.17}$$

This equation can be cast in the general form of a balance equation

$$\rho \dot{s} = -\nabla \cdot \boldsymbol{J}^s + \sigma^s, \tag{7.18}$$

with the entropy flux $\boldsymbol{J}^s$ and the (positive) entropy production σ^s given, respectively, by

$$\boldsymbol{J}^s = T^{-1}\boldsymbol{q} \tag{7.19}$$

and

$$\sigma^s = \boldsymbol{q} \cdot (\nabla T^{-1} - T^{-1}\alpha_1 \dot{\boldsymbol{q}}) \geq 0. \tag{7.20}$$

Relation (7.20) has the structure of a bilinear form $\sigma^s = \boldsymbol{q} \cdot \boldsymbol{X}$ in the flux $\boldsymbol{q}$ and the force $\boldsymbol{X}$, identified as the quantity within parentheses in (7.20); it differs from the classical thermodynamic force, which is simply ∇T^{-1}, by the presence of a term in the time derivative of the heat flux. The simplest way to obtain an evolution equation for $\boldsymbol{q}$ compatible with the positiveness of σ^s is to assume that the force $\boldsymbol{X}$ is linear in $\boldsymbol{q}$, namely

$$\nabla T^{-1} - T^{-1}\alpha_1 \dot{\boldsymbol{q}} = \mu_1 \boldsymbol{q}, \tag{7.21}$$

where the phenomenological coefficient μ_1 may depend on u but not on $\boldsymbol{q}$ because, as previously, third-order contributions in $\boldsymbol{q}$ are omitted. Introduction of (7.21) into (7.20) results in $\sigma^s = \mu_1 \boldsymbol{q} \cdot \boldsymbol{q} \geq 0$, from which is inferred that $\mu_1 > 0$.

Expression (7.21) contains two non-defined coefficients α_1 and μ_1, which must be identified on physical grounds. Under steady state conditions, (7.21) simplifies to

$$\boldsymbol{q} = -\frac{1}{\mu_1 T^2}\nabla T. \tag{7.22}$$

A comparison with Fourier's law $\boldsymbol{q} = -\lambda \nabla T$ yields then $\mu_1 = (\lambda T^2)^{-1}$ from which it results $\lambda \geq 0$. Next, by comparing (7.21) with Cattaneo's equation (7.4), one is led to

$$\alpha_1 = \tau / \lambda T. \tag{7.23}$$

Box 7.3 Non-Equilibrium Temperature vs. Local Equilibrium Temperature

In analogy with the classical theory, we define the non-equilibrium temperature θ by

$$\theta^{-1}(u, \boldsymbol{q}) = \left(\frac{\partial s}{\partial u}\right)_{\boldsymbol{q}}. \tag{7.3.1}$$

Expanding θ^{-1} around $\boldsymbol{q} = 0$, and omitting terms of order higher than q^2, one can write

$$\theta^{-1}(u, \boldsymbol{q}) = T^{-1}(u) + \alpha(u) q^2, \tag{7.3.2}$$

where T designates the local equilibrium temperature; the coefficient $\alpha(u)$ depends only on u at this order of approximation.

With the identification (7.23), the generalized Gibbs' equation (7.15) takes the form

$$ds = \theta^{-1}du - \frac{\tau}{\rho\lambda T^2}\boldsymbol{q}\cdot d\boldsymbol{q}. \tag{7.3.3}$$

The integrability condition of (7.3.3) demands that

$$\frac{\partial\theta^{-1}}{\partial\boldsymbol{q}} = -\frac{\partial}{\partial u}\left(\frac{\tau}{\rho\lambda T^2}\right)\boldsymbol{q}, \tag{7.3.4}$$

and, after integration,

$$\theta^{-1}(u,\boldsymbol{q}) = T^{-1}(u) - \frac{1}{2}\frac{\partial}{\partial u}\left(\frac{\tau}{\rho\lambda T^2}\right)\boldsymbol{q}\cdot\boldsymbol{q}. \tag{7.3.5}$$

This result is interesting because it provides an explicit expression of the lowest order correction of the temperature in systems out of equilibrium. Note that θ can be identified with T when the quantity $\tau/\rho\lambda T^2$ is constant. For an overview of the current discussions on the meaning and consequences of temperature in non-equilibrium systems, see Casas-Vázquez and Jou (2003).

With the above identifications of μ_1 and α_1, the generalized Gibbs' equation (7.15) takes the form

$$ds = T^{-1}du - \frac{\tau}{\rho\lambda T^2}\,\boldsymbol{q}\cdot d\boldsymbol{q}, \tag{7.24}$$

wherein it is important to observe that the coefficient of the new term in $d\boldsymbol{q}$ is completely identified in terms of physical quantities, namely the relaxation time τ and the heat conductivity λ. After integration of (7.24), the explicit expression for the entropy outside (local) equilibrium up to second-order terms in $\boldsymbol{q}$ is

$$\rho s(u,\boldsymbol{q}) = \rho s_{\text{eq}}(u) - \frac{1}{2}\frac{\tau}{\lambda T^2}\boldsymbol{q}\cdot\boldsymbol{q}. \tag{7.25}$$

The monotonic increase of extended entropy as shown in Fig. 7.2 is due to the last term on the right-hand side of (7.25).

7.1.3 Non-Local Terms: From Collision-Dominated Regime to Ballistic Regime

However, all is not well with the Cattaneo's equation. Although it is qualitatively satisfactory, as it predicts that heat pulses and high-frequency thermal waves will propagate at finite speed, some quantitativepredictions are not sufficiently accurate. For instance, the theory of solids predicts that $\chi/\tau = c_0^2/3$ in the so-called *Debye approximation*, with c_0 the sound velocity. After introducing this value in (7.10), it is found that the second sound is given by the constant value $U = c_0/\sqrt{3}$, in contradiction with experiments showing that U depends on temperature. Similarly, the kinetic theory of gases predicts that, for monatomic gases of mass m, $\chi/\tau = 5k_\mathrm{B}T/3m$ from which it follows that $U = \sqrt{5k_\mathrm{B}T/3m}$, but this value deviates by more than 20% from experimental data. One may then ask how to make EIT compatible with the above experimental evidences. It appeared rather soon that the above discrepancies find their origin in non-local effects, which were not incorporated in Cattaneo's original equation.

Non-local effects become important when the mean free path ℓ of the heat carriers becomes comparable to the wavelength of the external perturbation or to the dimensions L of the system. In this case, the Knudsen number $Kn = \ell/L$ is of the order or larger than 1 and the transport laws change from a collision-dominated regime, i.e. a regime with many collisions amongst the particles, to a ballistic regime, where there are few collisions amongst the particles and the predominant interactions are collisions of the particles with the walls. The difference of behaviour of transport in both situations is remarkable. For instance, in heat pulse experiments, one distinguishes several modes of transport: conduction, thermal waves, and ballistic fronts, and therefore transport equations combining these factors are desirable.

The question then arises how to include these effects in the formalism. A way out is to introduce a new extra variable, the flux of the heat flux, described by a second-order tensor $\mathbf{Q}$ and to write, instead of the Cattaneo's equation (7.4), the following expression

$$\tau_1 \frac{\partial \boldsymbol{q}}{\partial t} = -(\boldsymbol{q} + \lambda \nabla T) + \nabla \cdot \mathbf{Q}. \tag{7.26}$$

The tensor $\mathbf{Q}$, assumed to be symmetric as confirmed by the kinetic theory of phonons (Dreyer and Struchtrup 1993), may be split in the usual form $\mathbf{Q} = Q\mathbf{I} + \overset{0}{\mathbf{Q}}$, the scalar Q being one-third of its trace and $\overset{0}{\mathbf{Q}}$ the deviatoric part. In a relaxational approach, the evolution equations for Q and $\overset{0}{\mathbf{Q}}$ may be written as

$$\tau_0 \frac{\partial Q}{\partial t} = -Q + \beta' \nabla \cdot \boldsymbol{q}, \tag{7.27}$$

$$\tau_2 \frac{\partial \overset{0}{\mathbf{Q}}}{\partial t} = -\overset{0}{\mathbf{Q}} + 2\beta'' (\nabla \overset{0}{\boldsymbol{q}})^\mathrm{s}, \tag{7.28}$$

where superscript s refers to the symmetric part of the tensor. Note that, when the relaxation times are neglected, these expressions parallel the Newton–Stokes' equations, with $\boldsymbol{q}$ playing the role of the fluid velocity and $\mathbf{Q}$

that of the viscous pressure tensor. This is characteristic of the hydrodynamic regime of phonon's flow, which is indeed described by these equations.

Assuming that the relaxation times τ_0 and τ_2 are negligibly small, substituting (7.27) and (7.28) into (7.26) and restricting the analysis to the linear approximation, one obtains, for the evolution equation of the heat flux,

$$\tau_1 \frac{\partial \boldsymbol{q}}{\partial t} = -(\boldsymbol{q} + \lambda \nabla T) + \beta'' \nabla^2 \boldsymbol{q} + \left(\beta' + \frac{1}{3}\beta''\right) \nabla(\nabla \cdot \boldsymbol{q}). \tag{7.29}$$

Expression (7.29) is comparable with that obtained by Guyer and Krumhansl (1966) from phonon kinetic theory. In this approach, heat transport is the result of energy and momentum exchanges between massless colliding carriers, called *phonons*. To be more explicit, the phonons undergo two types of collisions: resistive R collisions which are collisions with defects of the lattice and/or the boundaries of the crystal, and so-called *Umklapp phonon–phonon processes*; they conserve energy but not momentum and have a characteristic time τ_{R}. The second type of collisions are referred to as normal N processes; they conserve energy and momentum, and their characteristic time will be denoted by τ_{N}. Starting from the Boltzmann's equation, Guyer and Krumhansl were able to derive the following equation for the heat flux:

$$\frac{\partial \boldsymbol{q}}{\partial t} + \frac{1}{\tau_{\mathrm{R}}} \boldsymbol{q} + \frac{1}{3} \rho c_v c_0^2 \nabla T = \frac{1}{5} c_0^2 \tau_{\mathrm{N}} \left[\nabla^2 \boldsymbol{q} + 2\nabla(\nabla \cdot \boldsymbol{q})\right]. \tag{7.30}$$

Comparison between (7.29) and (7.30) yields the following identifications

$$\tau_1 = \tau_{\mathrm{R}}, \quad \frac{\lambda}{\tau_{\mathrm{R}}} = \frac{1}{3} \rho c_v c_0^2, \quad \beta'' = \frac{1}{5} c_0^2 \tau_{\mathrm{N}} \tau_{\mathrm{R}}, \quad \beta' = \frac{1}{3} c_0^2 \tau_{\mathrm{N}} \tau_{\mathrm{R}}, \tag{7.31}$$

from which follows that τ_1 is associated with resistive collisions while the non-local contributions find their origin in the normal collisions. The result (7.29) contains as particular cases the Fourier's and the Cattaneo's equations for heat transport. Indeed:

1. For $1/\tau_{\mathrm{R}}$ and $1/\tau_{\mathrm{N}} \to \infty$, which corresponds to very high frequency of R and N collisions, one recovers Fourier's law.
2. For $1/\tau_{\mathrm{R}}$ finite, $1/\tau_{\mathrm{N}} \to \infty$, the frequency of normal N phonon–phonon collisions is large compared to that of R collisions and Cattaneo's equation is well suited.
3. In the general case where the frequencies of R and N collisions are comparable, non-local effects are important and Guyer–Krumhansl equation is more adequate. It becomes also clear that the flux of the heat flux $\mathbf{Q}$ accounts for the presence of the momentum-preserving N phonon–phonon collisions.

Combining (7.29) with the energy balance equation and formulating the final result in the Fourier's space (i.e. in terms of frequency ω and wave number k), the dispersion relation between ω and k for thermal waves is (see Problem 7.5)

$$\mathrm{i}\omega = \frac{-\chi k^2}{1 + \mathrm{i}\omega\tau_\mathrm{R} + \frac{l^2 k^2}{1+\mathrm{i}\omega\tau_\mathrm{N}}}, \tag{7.32}$$

with $l^2 = \frac{4}{3}\beta'' + \beta'$ for longitudinal waves and $l^2 = \beta''$ for transverse waves. For $\omega\tau_\mathrm{R} \gg 1 \gg \omega\tau_\mathrm{N}$ and ignoring non-local effects ($\ell = 0$), (7.32) leads to the phase velocity $v_\mathrm{p} = (\chi/\tau_\mathrm{R})^{1/2} = c_0/\sqrt{3}$. This is the regime referred to as second sound, and is attained when τ_N and τ_R are of different orders of magnitude. At higher frequencies, when both $\omega\tau_\mathrm{N} \gg 1$ and $\omega\tau_\mathrm{R} \gg 1$, the phase velocity obtained from (7.32) for longitudinal waves is (Dreyer and Struchtrup 1993, Dedeurwaerdere et al. 1994)

$$v_\mathrm{p} = \left(\frac{\chi}{\tau_\mathrm{R}} + \frac{l^2}{\tau_\mathrm{R}\tau_\mathrm{N}}\right)^{1/2} = \sqrt{\frac{14}{15}}c_0. \tag{7.33}$$

This regime is typical of ballistic propagation of phonons, i.e. a flow in which phonons travel through the whole crystal without suffering any collision. In fact, the theoretical result (7.33) is not fully satisfactory because experiments assign to the ballistic speed, a value equal to c_0.

Better agreement with experiments may be achieved when higher-order fluxes $\mathbf{Q}_1, \mathbf{Q}_2, \ldots, \mathbf{Q}_\mathrm{N}$ are incorporated as independent variables in the formalism. Indeed, in some situations, as in diluted gases, the relaxation times of the different higher-order fluxes are of the same order as that of the heat flux itself, in such a way that when $\boldsymbol{q}$ is taken as an independent variable, all higher-order fluxes should also be taken into account. In this way, one obtains for the different fluxes a hierarchy of evolution equations for the infinite number of higher-order fluxes of the form

$$\tau_n \dot{\boldsymbol{Q}}_n = -\boldsymbol{Q}_n + \frac{l_{n-1}^2}{\tau_{n-1}}\nabla \boldsymbol{Q}_{n-1} - \nabla \cdot \boldsymbol{Q}_{n+1}. \tag{7.34}$$

Writing (7.34) in Fourier's space ω k and solving the corresponding expression for the heat flux, one obtains a generalized Fourier's law with a (ω, k)-dependent thermal conductivity

$$\tilde{q}(\omega, k) = -\mathrm{i}k\lambda(\omega, k)\tilde{T}(\omega, k), \tag{7.35}$$

where $\tilde{q}(\omega, k)$ and $\tilde{T}(\omega, k)$ are the Fourier transforms of $\boldsymbol{q}(\boldsymbol{r}, t)$ and $T(\boldsymbol{r}, t)$, respectively. It is found that the generalized heat conductivity $\lambda(\omega, k)$ is given by (Dedeurwaerdere et al. 1996; Jou et al. 2001)

$$\lambda(\omega, k) = \frac{\lambda(T)}{1 + \mathrm{i}\omega\tau_1 + \frac{k^2 l_1^2}{1+\mathrm{i}\omega\tau_2 + \frac{k^2 l_2^2}{1+\mathrm{i}\omega\tau_3 + \frac{k^2 l_3^2}{1+\mathrm{i}\omega\tau_4} + \cdots}}}, \tag{7.36}$$

where l_n are characteristic lengths of the order of the mean free path, and $\tau_1, \tau_2, \ldots \tau_\mathrm{N}$ are the relaxation times of the respective higher-order fluxes.

This result will be exploited in Sect. 7.1.4. Two types of truncations of the continued-fraction expansion (7.36) have been considered. The most usual is to assume that $l_i = 0$ with $i > n$, for some given n. But a better approach is to suppose that $l_i = l_n$ for $i > n$. Asymptotic developments have been performed, with the advantage that the whole formalism can be expressed in terms of one single "effective" relaxation time, which incorporates in a compact way the effect resulting from the introduction of an infinite number of higher-order fluxes (see Jou et al. 2001).

7.1.4 Application to Steady Heat Transport in Nano-Systems

We have mentioned that in nano-systems or submicronic electronic devices, whose dimensions are comparable to the mean free path of phonons, ballistic transport is dominant and, consequently, the validity of Fourier's law is questioned. Here we propose to revisit this law when the Knudsen number $Kn \gg 1$. To formulate the problem in simple terms, consider a one-dimensional system of length L, whose opposite boundaries are at temperature T and $T + \Delta T$, and transient effects are ignored. Depending on the values of Kn, the heat flux q takes the following limiting forms

$$q = \lambda \frac{\Delta T}{L} \quad (Kn \ll 1,\ \text{diffusive transport}), \tag{7.37a}$$

$$q = \Lambda \Delta T \quad (Kn \gg 1,\ \text{ballistic transport}), \tag{7.37b}$$

where λ denotes the thermal conductivity and Λ is a heat conduction transport coefficient. In the diffusive limit, the heat flux is proportional to the temperature gradient, according to Fourier's law; in contrast, in the ballistic regime, the heat flux depends only on the temperature difference, but not on the length L of the system. The values of λ and Λ have been derived in the kinetic theory. For a monatomic ideal gas in the diffusive regime, it is found that $\lambda = \frac{5}{2} n k_B (k_B T/m)^{1/2} \ell$, with n the particle number density, m the mass of the particles, k_B the Boltzmann constant, and ℓ the mean free path given by $\ell = (k_B T/m)^{1/2} \tau$, with τ the average time between successive collisions; in the rarefied gas regime, the heat conduction coefficient Λ is given by $\Lambda = \frac{1}{2} n k_B (k_B T/m)^{1/2}$. Note that the thermal conductivity λ is proportional to ℓ, whereas Λ does not depend on it. It is also worth to mention that computer simulations of heat transport in one-dimensional systems suggest that in some situations Fourier's law must be generalized in the form $q = \lambda(T) \Delta T / L^{\alpha}$, with α an exponent whose value depends on the details of the system. For instance, it is found that $\alpha = 0.63$ for some anharmonic chains or one-dimensional gases, $\alpha = 0.5$ for disordered harmonic chains with free boundaries, or $\alpha = 1.5$ for disordered harmonic chains with fixed boundaries (Lepri et al. 2003).

A simple phenomenological modelling of the transition between the diffusive and the ballistic regimes can be achieved by introducing a heat conductivity $\lambda(T, \ell/L)$, in such a way that in whole generality

$$q = \lambda(T, \ell/L)\frac{\Delta T}{L}. \tag{7.38}$$

The limiting values of this generalized conductivity should be

$$\lambda(T, \ell/L) \to \lambda(T) \quad \text{for} \quad \ell/L \to 0,$$
$$\lambda(T, \ell/L) \to \frac{\lambda(T)}{a}\frac{L}{\ell} \equiv \Lambda(T)L \quad \text{for} \quad \ell/L \to \infty,$$

where a is a constant depending on the system.

This may be achieved, for instance, through the continued-fraction approach (7.36), which in the steady state ($\omega = 0$), allows us to define a k-dependent thermal conductivity $\lambda(T, k)$. Since the system is characterized by one single length scale, the length L, it seems natural to identify k with $2\pi/L$. This yields an expression for $\lambda(T, \ell/L)$, which in the simplest case where all the l_n take the same value, for instance, $l_n^2 = \frac{1}{4}\ell^2$, and using the asymptotic expression corresponding to (7.36) (Problem 7.6) yields

$$\lambda(T, \ell/L) = \frac{\lambda(T)L^2}{2\pi^2\ell^2}\left[\left(1 + 4(\pi\ell/L)^2\right)^{1/2} - 1\right], \tag{7.39}$$

(see Jou et al. 2005). It is directly checked that (7.39) reduces to $\lambda(T)$ for $\ell/L \to 0$ and to $\lambda(T)(L/\ell)$ for $\ell/L \to \infty$; these are the required asymptotic behaviours. These considerations confirm that EIT provides a consistent modelling for heat transfer processes taking place not only at short times but also at micro- and nano-scales. Despite its simplicity, (7.39) satisfactorily fits experimental results in silicon thin layers and nano-wires (Alvarez and Jou 2007). Another possibility, which has been explored recently (Chen 2004; Lebon et al. 2006b), is to split the heat flux into a diffusive and a ballistic component, the latter being associated with fast particles and a long mean free path, the former with slow particles and a short mean free path.

7.2 One-Component Viscous Heat Conducting Fluids

Up to now, we have only considered heat transfer in rigid bodies at rest. Let us now proceed to the more general case of a compressible one-component isotropic viscous fluid in motion subject to a temperature gradient. According to EIT, the space of the thermodynamic state variables is the union of the classical one (internal energy u and specific volume v) and the space of the fluxes, namely the heat flux $\boldsymbol{q}$, the bulk viscous pressure p^{v}, and the shear viscous pressure $\overset{0}{\mathbf{P}}{}^{\mathrm{v}}$. The generalized Gibbs' equation takes the form

$$\mathrm{d}s = T^{-1}\,\mathrm{d}u + T^{-1}p\,\mathrm{d}v - T^{-1}v\alpha_0 p^{\mathrm{v}}\,\mathrm{d}p^{\mathrm{v}} - T^{-1}v\alpha_1\boldsymbol{q}\cdot\mathrm{d}\boldsymbol{q} - T^{-1}v\alpha_2\overset{0}{\mathbf{P}}{}^{\mathrm{v}} : \mathrm{d}\overset{0}{\mathbf{P}}{}^{\mathrm{v}}, \tag{7.40}$$

where we have approximated the non-equilibrium temperature by the local equilibrium temperature (see Box 7.3), similarly the non-equilibrium pressure has been identified with the equilibrium pressure p, the coefficients α_0, α_1, and α_2 are unknown scalar functions of u and v, their dependence with respect to the fluxes is assumed to be negligible. Compared to CIT, the last three terms are new. Our objective is to determine the evolution equations for the fluxes satisfying the thermodynamic restrictions imposed by the second law. We shall proceed in parallel with the procedure followed in Chap. 2, which implies that we first determine the expression of the entropy production σ^s. To this end, we combine (7.40) with the expressions of $\dot{u}$ and $\dot{v}$ as derived from the balance laws of mass (2.16) and energy (2.18). Omitting the energy source term, it is easily found that

$$\begin{aligned}\rho\dot{s} = &-T^{-1}\nabla\cdot\boldsymbol{q} - T^{-1}p^{\mathrm{v}}\nabla\cdot\boldsymbol{v} - T^{-1}\overset{0}{\mathbf{P}}{}^{\mathrm{v}} : \overset{0}{\mathbf{V}} - T^{-1}\alpha_0 p^{\mathrm{v}}\dot{p}^{\mathrm{v}}\\ &-T^{-1}\alpha_1\boldsymbol{q}\cdot\dot{\boldsymbol{q}} - T^{-1}\alpha_2\overset{0}{\mathbf{P}}{}^{\mathrm{v}} : (\overset{0}{\mathbf{P}}{}^{\mathrm{v}})^{\bullet}.\end{aligned} \tag{7.41}$$

Before deriving the entropy production

$$\sigma^s = \rho\dot{s} + \nabla\cdot\boldsymbol{J}^s, \tag{7.42}$$

we need an expression for the entropy flux $\boldsymbol{J}^s$. For isotropic systems, the most general vector depending on $u, v, \boldsymbol{q}, \overset{0}{\mathbf{P}}{}^{\mathrm{v}}$, and p^{v} is, up to second order in the fluxes,

$$\boldsymbol{J}^s = T^{-1}\boldsymbol{q} + \beta' p^{\mathrm{v}}\boldsymbol{q} + \beta''\overset{0}{\mathbf{P}}{}^{\mathrm{v}}\cdot\boldsymbol{q}. \tag{7.43}$$

We assume for simplicity that the coefficients β' and β'' are constant; (7.43) is supported by the kinetic theory of gases and may be viewed as a generalization of the classical result $\boldsymbol{J}^s = \boldsymbol{q}/T$. The entropy production is easily derived from (7.42) by replacing $\rho\dot{s}$ and $\boldsymbol{J}^s$, respectively, by their expressions in (7.41) and (7.43). The final result is

$$\begin{aligned}\sigma^s = &\,\boldsymbol{q}\cdot(\nabla T^{-1} + \beta''\nabla\cdot\overset{0}{\mathbf{P}}{}^{\mathrm{v}} + \beta'\nabla p^{\mathrm{v}} - T^{-1}\alpha_1\dot{\boldsymbol{q}})\\ &+p^{\mathrm{v}}(-T^{-1}\nabla\cdot\boldsymbol{v} - T^{-1}\alpha_0\dot{p}^{\mathrm{v}} + \beta'\nabla\cdot\boldsymbol{q})\\ &+\overset{0}{\mathbf{P}}{}^{\mathrm{v}} : [-T^{-1}\overset{0}{\mathbf{V}} - T^{-1}\alpha_2(\overset{0}{\mathbf{P}}{}^{\mathrm{v}})^{\bullet} + \beta''(\nabla\overset{0}{\boldsymbol{q}})^s].\end{aligned} \tag{7.44}$$

Note that (7.44) has the structure of a bilinear form

$$\sigma^s = \boldsymbol{q} \cdot \boldsymbol{X}_1 + p^{\mathrm{v}} X_0 + \overset{0}{\mathbf{P}}{}^{\mathrm{v}} : \overset{0}{\mathbf{X}}_2 \geq 0, \tag{7.45}$$

consisting of a sum of products of the fluxes $\boldsymbol{q}, p^{\mathrm{v}}$, and $\overset{0}{\mathbf{P}}{}^{\mathrm{v}}$ and their conjugate generalized forces $\boldsymbol{X}_1, X_0$, and $\overset{0}{\mathbf{X}}_2$. The latter are the respective terms between the brackets. They are similar to the expressions obtained in CIT, but they contain additional terms depending on the time and space derivatives of the fluxes.

The simplest flux–force relations ensuring the positiveness of σ^s are

$$\boldsymbol{X}_1 = \mu_1 \boldsymbol{q}, \quad X_0 = \mu_0 p^{\mathrm{v}}, \quad \overset{0}{\mathbf{X}}_2 = \mu_2 \overset{0}{\mathbf{P}}{}^{\mathrm{v}}, \tag{7.46}$$

with $\mu_0 \geq 0$, $\mu_1 \geq 0$, and $\mu_2 \geq 0$. Writing explicitly the expressions for the generalized forces, one obtains in the linear approximation the following set of evolution equations for the fluxes:

$$\nabla T^{-1} - T^{-1}\alpha_1 \dot{\boldsymbol{q}} = \mu_1 \boldsymbol{q} - \beta'' \nabla \cdot \overset{0}{\mathbf{P}}{}^{\mathrm{v}} - \beta' \nabla p^{\mathrm{v}}, \tag{7.47}$$

$$-T^{-1}\nabla \cdot \boldsymbol{v} - T^{-1}\alpha_0 \dot{p}^{\mathrm{v}} = \mu_0 p^{\mathrm{v}} - \beta' \nabla \cdot \boldsymbol{q}, \tag{7.48}$$

$$-T^{-1}\overset{0}{\mathbf{V}} - T^{-1}\alpha_2 (\overset{0}{\mathbf{P}}{}^{\mathrm{v}})^{\bullet} = \mu_2 \overset{0}{\mathbf{P}}{}^{\mathrm{v}} - \beta'' (\nabla \overset{0}{\boldsymbol{q}})^s. \tag{7.49}$$

Note that in (7.49) $(\overset{0}{\mathbf{P}}{}^{\mathrm{v}})^{\bullet}$ stands for the time derivative of $\overset{0}{\mathbf{P}}{}^{\mathrm{v}}$.

The main features issued from the above thermodynamic formalism are:

- The positiveness of the coefficients $\mu_0 \geq 0$, $\mu_1 \geq 0$, and $\mu_2 \geq 0$.
- The equality of the cross-coefficients relates $\boldsymbol{q}$ with $\nabla \cdot \overset{0}{\mathbf{P}}{}^{\mathrm{v}}$ and $\overset{0}{\mathbf{P}}{}^{\mathrm{v}}$ with $(\nabla \overset{0}{\boldsymbol{q}})^s$ on the one side, $\boldsymbol{q}$ with ∇p^{v} and p^{v} with $\nabla \cdot \boldsymbol{q}$ on the other side. This is confirmed by the kinetic theory, and belongs to a class of higher-order Onsager's relations.
- The coefficients β' and β'' appearing in the second-order terms of the entropy flux (7.43) are the same as the coefficients of the cross-terms in the evolution equations (7.47)–(7.49). This result is also in agreement with the kinetic theory of gases. These coefficients describe the coupling between thermal and mechanical effects; setting $\beta' = \beta'' = 0$ means absence of coupling in (7.47)–(7.49) and similarly in (7.43) of $\boldsymbol{J}^s$, which boils down to the classical result $\boldsymbol{J}^s = \boldsymbol{q}/T$.

To identify the coefficients appearing in (7.47)–(7.49) in physical terms, consider the particular case where the space derivatives of the fluxes are negligible. Equations (7.47)–(7.49) then reduce to

$$\nabla T^{-1} - T^{-1}\alpha_1 \dot{\boldsymbol{q}} = (\lambda T^2)^{-1} \boldsymbol{q}, \tag{7.50}$$

$$-T^{-1}\nabla \cdot \boldsymbol{v} - T^{-1}\alpha_0 \dot{p}^{\mathrm{v}} = (\zeta T)^{-1} p^{\mathrm{v}}, \tag{7.51}$$

$$-T^{-1}\overset{0}{\mathbf{V}} - T^{-1}\alpha_2 (\overset{0}{\mathbf{P}}{}^{\mathrm{v}})^{\cdot} = (2\eta T)^{-1} \overset{0}{\mathbf{P}}{}^{\mathrm{v}}. \tag{7.52}$$

By setting

$$\alpha_1 = \tau_1(\lambda T)^{-1}, \quad \alpha_0 = \tau_0\zeta^{-1}, \quad \alpha_2 = \tau_2(2\eta)^{-1}, \tag{7.53}$$

$$\mu_1 = (\lambda T^2)^{-1}, \quad \mu_0 = (\zeta T)^{-1}, \quad \mu_2 = (2\eta T)^{-1}, \tag{7.54}$$

(7.50)–(7.52) can be identified with the so-called *Maxwell–Cattaneo's laws*

$$\tau_1 \dot{\boldsymbol{q}} + \boldsymbol{q} = -\lambda \nabla T, \tag{7.55}$$

$$\tau_0 \dot{p}^{\mathrm{v}} + p^{\mathrm{v}} = -\zeta \nabla \cdot \boldsymbol{v}, \tag{7.56}$$

$$\tau_2 (\overset{0}{\mathbf{P}}{}^{\mathrm{v}})^{\bullet} + \overset{0}{\mathbf{P}}{}^{\mathrm{v}} = -2\eta \overset{0}{\mathbf{V}}, \tag{7.57}$$

with $\lambda > 0$, $\zeta > 0$, and $\eta > 0$ being the positive thermal conductivity, bulk viscosity, and shear viscosity, respectively, and where τ_1, τ_0, and τ_2 are the relaxation times of the respective fluxes. In terms of λ, ζ, η, and the relaxation times τ_1, τ_0, and τ_2, the linearized evolution equations (7.50)–(7.52) take the following form:

$$\tau_1 \dot{\boldsymbol{q}} = -(\boldsymbol{q} + \lambda \nabla T) + \beta'' \lambda T^2 \nabla \cdot \overset{0}{\mathbf{P}}{}^{\mathrm{v}} + \beta' \lambda T^2 \nabla p^{\mathrm{v}}, \tag{7.58}$$

$$\tau_0 \dot{p}^{\mathrm{v}} = -(p^{\mathrm{v}} + \zeta \nabla \cdot \boldsymbol{v}) + \beta' \zeta T \nabla \cdot \boldsymbol{q}, \tag{7.59}$$

$$\tau_2 (\overset{0}{\mathbf{P}}{}^{\mathrm{v}})^{\bullet} = -(\overset{0}{\mathbf{P}}{}^{\mathrm{v}} + 2\eta \overset{0}{\mathbf{V}}) + 2\beta'' \eta T (\nabla \overset{0}{\boldsymbol{q}})^s. \tag{7.60}$$

Extension of the present description to the problem of thermodiffusion in multi-component fluids has been achieved by Lebon et al. (2003). Despite in ordinary liquids, the relaxation times are small, of the order of 10^{-12} s, their effects may be observed in neutron scattering experiments; in contrast, in dilute gases or polymer solutions, the relaxation times may be rather large and directly perceptible in light scattering experiments and in ultrasound propagation. To illustrate the above analysis, let us mention that Carrassi and Morro (1972) studied the problem of ultrasound propagation in monatomic gases and compared the results provided by Maxwell–Cattaneo's equations (7.55)–(7.57) and the classical Fourier–Newton–Stokes' laws.

In Table 7.1, the numerical values of the ultrasound phase velocities c_0/v_{p} (c_0 designating the sound speed) vs. the non-dimensional mean free path ℓ of the particles are reported. It is observed that the classical theory deviates appreciably from the experimental results as ℓ is increased, whereas the Maxwell–Cattaneo's theory agrees fairly well with the experimental data.

Table 7.1 Numerical values of c_0/v_{p} as a function of ℓ (Carrassi and Morro 1972)

ℓ	0.25	0.50	1.00	2.00	4.00	7.00
$(c_0/v_{\mathrm{p}})_{(\text{Fourier–Newton–Stokes})}$	0.40	0.26	0.19	0.13	0.10	0.07
$(c_0/v_{\mathrm{p}})_{(\text{Maxwell–Cattaneo})}$	0.52	0.43	0.44	0.47	0.48	0.49
$(c_0/v_{\mathrm{p}})_{(\text{experimental})}$	0.51	0.46	0.50	0.46	0.46	0.46

Taking into account the identifications (7.53), the explicit expression of the entropy, after integration of (7.40), is

$$s_{\mathrm{EIT}} = s_{\mathrm{eq}}(u, v) - \frac{\tau_1 v}{2\lambda T^2} \boldsymbol{q} \cdot \boldsymbol{q} - \frac{\tau_0 v}{2\zeta T} p^{\mathrm{v}} p^{\mathrm{v}} - \frac{\tau_2 v}{4\eta T} \overset{0}{\mathbf{P}}{}^{\mathrm{v}} : \overset{0}{\mathbf{P}}{}^{\mathrm{v}}. \tag{7.61}$$

This expression reduces to the local equilibrium entropy $s_{\mathrm{eq}}(u, v)$ for zero values of the fluxes. Since the entropy must be a maximum at equilibrium, it follows that the relaxation times must be positive. Indeed, stability of equilibrium demands that entropy be a concave function, which implies that the second-order derivatives of s_{EIT} with respect to its state variables are negative; in particular,

$$\frac{\partial^2 s_{\mathrm{EIT}}}{\partial \boldsymbol{q} \cdot \partial \boldsymbol{q}} = -\frac{v\tau_1}{\lambda T^2} < 0, \quad \frac{\partial^2 s_{\mathrm{EIT}}}{\partial \overset{0}{\mathbf{P}}{}^{\mathrm{v}} : \partial \overset{0}{\mathbf{P}}{}^{\mathrm{v}}} = -\frac{v\tau_2}{2\eta T^2} < 0, \quad \frac{\partial^2 s_{\mathrm{EIT}}}{(\partial p^{\mathrm{v}})^2} = -\frac{v\tau_0}{\zeta T^2} < 0. \tag{7.62}$$

Since the transport coefficients λ, ζ, and η must be positive as a consequence of the second law, it turns out from (7.62) that the relaxation times are positive, for stability reasons. The property of concavity of entropy is equivalent to the requirement that the field equations constitute a hyperbolic set. Hyperbolicity of evolution equations is characteristic of EIT and it is sometimes imposed from the start as in Rational Extended Thermodynamics (Müller and Ruggeri 1998).

7.3 Rheological Fluids

Section 7.2 is dedicated to general considerations about the EIT description of viscous fluid flows. As mentioned above, for most ordinary fluids, the relaxation times are generally very small so that for main problems the role of relaxation effects is minute and can be omitted. However, this is no longer true with rheological fluids, like polymer solutions, because the relaxation times of the macromolecules are much longer than for small molecules, and they may be of the order of 1 s and even larger. The study of rheological fluids has been the concern of several thermodynamic approaches like rational thermodynamics, internal variables theories, and Hamiltonian formalisms (see Chaps. 8–10). The main difference between EIT and other theories is that in the former the shear viscous pressure is selected as variable, whereas in other theories, variables related with the internal structure of the fluid, as for instance, the so-called *configuration tensor*, are preferred. The EIT variables are especially useful in macroscopic analyses while internal variables are generally more suitable for a microscopic understanding.

In the simple Maxwell model, which is a particular case of (7.60), the viscous pressure tensor obeys the evolution equation

$$\frac{\mathrm{d}\mathbf{P}^{\mathrm{v}}}{\mathrm{d}t} = -\frac{1}{\tau}\mathbf{P}^{\mathrm{v}} - 2\frac{\eta}{\tau}\mathbf{V}. \tag{7.63}$$

This model captures the essential idea of viscoelastic models: the response to slow perturbations is that of an ordinary Newtonian viscous fluid, whereas for fast perturbations, with a characteristic time t of the order of the relaxation time τ or less, it behaves as an elastic solid. However, the material time derivative introduced in (7.63) is not very satisfactory, neither for practical predictions nor from a theoretical viewpoint, because it does not satisfy the axiom of frame-indifference (see Chap. 9). This has motivated to replace (7.63) by the so-called *upper-convected Maxwell model*

$$\frac{\mathrm{d}\mathbf{P}^{\mathrm{v}}}{\mathrm{d}t} - (\nabla \boldsymbol{v})^{\mathrm{T}} \cdot \mathbf{P}^{\mathrm{v}} - \mathbf{P}^{\mathrm{v}} \cdot (\nabla \boldsymbol{v}) = -\frac{1}{\tau}\mathbf{P}^{\mathrm{v}} - 2\frac{\eta}{\tau}\mathbf{V}. \tag{7.64}$$

In a steady pure shear flow corresponding to velocity components $\boldsymbol{v} = (\dot{\gamma} y, 0, 0)$, with $\dot{\gamma}$ the shear rate, the upper-convected Maxwell model (7.64) reads as

$$\mathbf{P}^{\mathrm{v}} = \begin{pmatrix} -2\tau\eta\dot{\gamma}^2 & -\eta\dot{\gamma} & 0 \\ -\eta\dot{\gamma} & 0 & 0 \\ 0 & 0 & 0 \end{pmatrix}. \tag{7.65}$$

In contrast with the original Maxwell model, element P_{11}^{v} of tensor (7.65) contains a non-vanishing contribution, corresponding to the so-called *normal stresses.* The upper-convected model agrees rather satisfactorily with a wide variety of experimental data.

In (7.63) and (7.64), we have introduced one single relaxation time but in many cases it is much more realistic to consider $\mathbf{P}^{\mathrm{v}}$ as a sum of several independent contributions, i.e. $\mathbf{P}^{\mathrm{v}} = \sum_j \mathbf{P}_j^{\mathrm{v}}$ with each $\mathbf{P}_j^{\mathrm{v}}$ obeying a linear evolution equation such as (7.63) or (7.64), characterized by its own viscosity η_j and relaxation time τ_j. These independent contributions arise from the different internal degrees of freedom of the macromolecules. In this case, the "extended" entropy should be written as

$$s(u, v, \mathbf{P}_i^{\mathrm{v}}) = s_{\mathrm{eq}}(u, v) - \frac{v}{4T}\sum_i \frac{\tau_i}{\eta_i}\mathbf{P}_i^{\mathrm{v}} : \mathbf{P}_i^{\mathrm{v}} \tag{7.66}$$

and the corresponding model is known as the *generalized Maxwell model.*

In the above descriptions, the viscosity was supposed to be independent of the shear rate. However, there exists a wide class of so-called *non-Newtonian fluids* characterized by shear rate-dependent viscometric functions, like the viscous coefficients. Such a topic is treated at full length in specialized works on rheology and will not be discussed here any more.

The study of polymer solutions is often focused on the search of constitutive laws for the viscous pressure tensor. One of the advantages of EIT is to establish a connection between such constitutive equations and the non-equilibrium equations of state derived directly by differentiating the expression of the extended entropy. These state equations are determinant in the study of flowing polymer solutions, which is important in engineering, since most of polymer processing take place under motion. The phase diagrams established for equilibrium situations cannot be trusted in the presence of

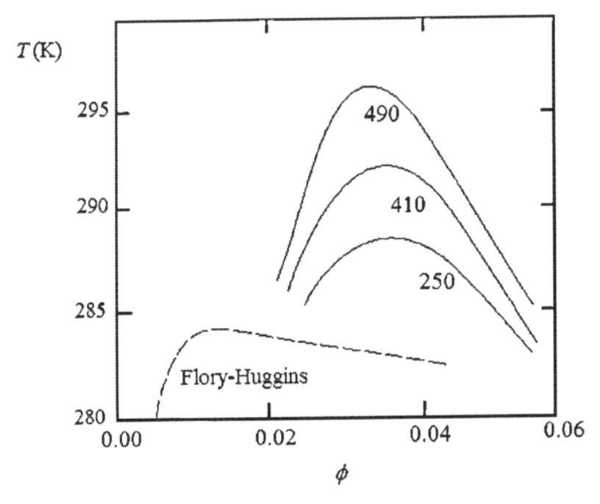

Fig. 7.3 Phase diagram (temperature T vs. volume fraction ϕ) of a polymer solution under shear flow. This is a binary solution of polymer polystyrene in dioctylphtalate solvent for several values of P_{12}^{v} (expressed in $\mathrm{N\,m^{-2}}$). The *dashed curve* is the equilibrium spinodal line (corresponding to a vanishing viscous pressure)

flows as the latter may enhance or reduce the solubility of the polymer and the conditions under which phase separation occurs.

This explains why many efforts have been devoted to the study of flow-induced changes in polymer solutions (Jou et al. 2000, 2001; Onuki 1997, 2002). Classical local equilibrium thermodynamics is clearly not a good candidate because the equations of state should incorporate explicitly the influence of the flow. Moreover, the equilibrium thermodynamic stability conditions cannot be extrapolated to non-equilibrium steady states, unless a justification based on dynamic arguments is provided. According to EIT, the chemical potential will explicitly depend on the thermodynamic fluxes, here the viscous pressure. It follows that the physico-chemical properties related to the chemical potential – as for instance, solubility, chemical reactions, phase diagrams, and so on – will depend on the viscous pressure, and will be different from those obtained in the framework of local equilibrium thermodynamics. This is indeed observed in the practice, see Fig. 7.3, where it is shown that the critical temperature of phase change predicted by the equilibrium theory (281.4 K), as developed by Flory and Huggins, is shifted towards higher values under the action of shear flow. The corrections are far from being negligible when the shear is increased.

7.4 Microelectronic Devices

The classical thermodynamic theory of electric transport has been examined in Chap. 3. Here, we briefly discuss the EIT contribution to the study of charge transport in submicronic electronic devices. Although the carrier transport can always be described by means of the Boltzmann's equation, to solve it is a very difficult task and, furthermore, it contains more information than needed in practical applications. The common attitude is to consider a reduced number of variables (usually expressed as moments of the distribution function), which are directly related with density, charge flux, internal energy, energy flux, and so on, and which are measurable and controllable variables, instead of the full distribution function. This kind of approach is referred to as a *hydrodynamic model* and EIT is very helpful in determining which truncations among the hierarchy of evolution equations are compatible with thermodynamics.

Before considering microelectronic systems, let us first study electric conduction in a rigid metallic sample. We assume that the electric current is due to the motion of electrons with respect to the lattice. In CIT, the independent variables are selected as being the specific internal u and the charge per unit mass, z_{e}; in EIT, the electric current $\boldsymbol{i}$ is selected as an additional independent variable. For the system under study, the balance equations of charge and internal energy may be written as

$$\rho \dot{z}_{\mathrm{e}} = -\nabla \cdot \boldsymbol{i}, \tag{7.67}$$

$$\rho \dot{u} = -\nabla \cdot \boldsymbol{q} + \boldsymbol{i} \cdot \boldsymbol{E}, \tag{7.68}$$

with $\boldsymbol{E}$ the electric field and $\boldsymbol{i} \cdot \boldsymbol{E}$ the Joule heating term. Ignoring heat transport ($\boldsymbol{q} = 0$) for the moment, the generalized Gibbs' equation takes the form

$$\mathrm{d}s = T^{-1}\mathrm{d}u - T^{-1}\mu_{\mathrm{e}}\mathrm{d}z_{\mathrm{e}} - \alpha \boldsymbol{i} \cdot \mathrm{d}\boldsymbol{i}, \tag{7.69}$$

with μ_{e} being the chemical potential of electrons and α a phenomenological coefficient independent of $\boldsymbol{i}$. By following the same procedure as in the previous sections, it is easily checked that the evolution equation for $\boldsymbol{i}$ is

$$\tau_{\mathrm{e}} \frac{\mathrm{d}\boldsymbol{i}}{\mathrm{d}t} = -(\boldsymbol{i} - \sigma_{\mathrm{e}} \boldsymbol{E}'), \tag{7.70}$$

where $\boldsymbol{E}' = \boldsymbol{E} - T\nabla(T^{-1}\mu_{\mathrm{e}})$, τ_{e} is the relaxation time, and σ_{e} is the electrical conductivity, provided that α in (7.69) is identified as $\alpha = \tau_{\mathrm{e}}(\rho\sigma_{\mathrm{e}}T)^{-1}$. The generalized entropy is now given by

$$\rho s = \rho s_{\mathrm{eq}} - \frac{\tau_{\mathrm{e}}}{2\sigma_{\mathrm{e}}T} \boldsymbol{i} \cdot \boldsymbol{i}. \tag{7.71}$$

Equation (7.70), i.e. a generalization of Ohm's law $\boldsymbol{i} = \sigma_{\mathrm{e}}\boldsymbol{E}$, is often used in plasma physics and in the analysis of high-frequency currents but without any reference to its thermodynamic context.

A challenging application is the study of charge transport in submicronic semiconductor devices for its consequences on the optimization of their functioning and design. The evolution equations for the moments are directly obtained from the Boltzmann's equation. Depending on the choice of variables and the level at which the hierarchy is truncated, one obtains different hydrodynamic models. A simple one is the so-called *drift-diffusion model* (Hänsch 1991), where the independent variables are the number density of electrons and holes, but not their energies. More sophisticated is the approach of Baccarani–Wordeman, wherein the energy of electrons and holes is taken as independent variables, but not the heat flux, assumed to be given by the Fourier's law. To optimize the description, a sound analysis of other possible truncations is highly desirable. Application of EIT to submicronic devices has been performed in recent works (Anile and Muscato 1995, Anile et al. 2003) wherein the energy flux rather than the electric flux is raised to the level of independent variables. We will not enter furthermore into the details of the development as they are essentially based on Boltzmann's equation for charged particles, which is outside the scope of this book. Let us simply add that a way to check the quality of the truncation is to compare the predictions of the hydrodynamic models with Monte Carlo simulations.

In particular, for a n^+-n-n^+ silicon diode (Fig. 7.4) at room temperature, the EIT model of Anile et al. (2003) provides results, which are in good agreement with Monte Carlo simulations, as reflected by Fig. 7.5. Compared to a Monte Carlo simulation, the advantage of a hydrodynamic model is its much more economical cost with regard to the computing time consumption.

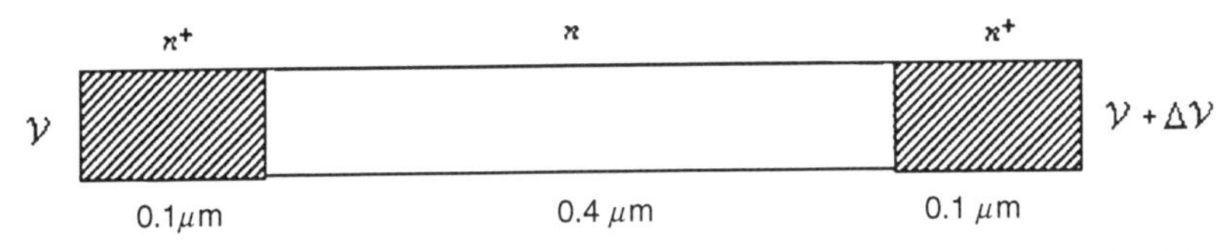

Fig. 7.4 A n^+-n-n^+ silicon diode. The doping density in the region n^+ is higher than in the region n

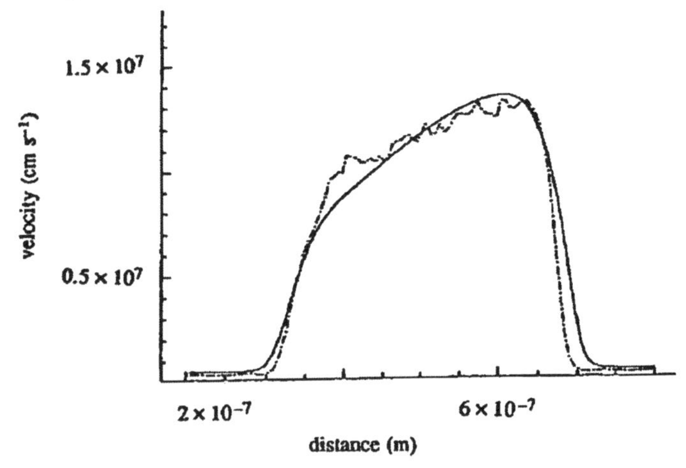

Fig. 7.5 Velocity profiles in the n^+-n-n^+ silicon diode obtained, respectively, by Monte Carlo simulations (*dotted line*) and the hydrodynamical model of Anile and Pennisi (1992) based on EIT

7.5 Final Comments and Perspectives

To shed further light on the scope and perspectives of EIT, some general comments are in form:

1. In EIT, the state variables are the classical hydrodynamic fields supplemented by the fluxes provided by the balance laws, i.e. the fluxes of mass, momentum, energy, electric charge, and so on. This attitude is motivated by the fact that these dissipative fluxes are typically non-equilibrium variables vanishing at equilibrium. The choice of fluxes is natural as the only accessibility to a given system is through its boundaries. Moreover in processes characterized by high frequencies or systems with large relaxation times (polymers, superfluids, etc.) or short-scale dimensions (nano- and microelectronic devices), the fluxes lose their status of fast and negligible variables and find naturally their place among the set of state variables. Other fields where the fluxes may play a leading part are relativity, cosmology, traffic control (flux of cars), economy (flux of money), and world wide web (flux of information). The choice of the fluxes as variables finds its roots in the kinetic theory of gases. Indeed, it amounts to selecting as variables the higher-order moments of the velocity distribution function; in particular, taking the heat flux and the pressure tensor as variables is suggested by Grad's thirteen-moment theory (1958), which therefore provides the natural basis for the development of EIT (Lebon et al. 1992). The main consequence of elevating the fluxes to the rank of variables is that the phenomenological relations of the classical approach (CIT) are replaced by first-order time evolution equations of Maxwell–Cattaneo type. In EIT, the field equations are hyperbolic; note, however, that this property may not be satisfied in the whole space of state variables, especially in the non-linear regime (Müller and Ruggeri 1998; Jou et al. 2001). In CIT, the balance laws are parabolic of the diffusion type with the consequence that signals move at infinite velocity. EIT can be viewed as a generalization of CIT by including inertia in the transport equations.
2. The space of the extra variables is not generally restricted to the above ordinary dissipative fluxes. For instance, to cope with the complexity of some fast non-equilibrium processes and/or non-local effects as in nano-systems, it is necessary to introduce higher-order fluxes, such as the fluxes of the fluxes, as done in Sect. 7.1.3. Moreover, it is conceivable that fluxes may be split into several independent contributions, each with its own evolution equation, as in non-ideal gases (Jou et al. 2001) and polymers (see Sect. 7.3). In some problems, like those involving shock waves (Valenti et al. 2002), it may be more convenient to use as variables combinations of fluxes and transport coefficients.
3. Practically, it is not an easy task to evaluate the fluxes at each instant of time and at every point in space. Nevertheless, for several problems of practical interest, such as heat wave propagation, the fluxes are eliminated

from the final equations. Although the corresponding dispersion relations may still contain the whole set of parameters appearing in the evolution equations of the fluxes, like the relaxation times, the latter may however be evaluated by measuring the wave speed, its attenuation, or shock properties. A direct measurement of the fluxes is therefore not an untwisted condition to check the bases and performances of EIT.

4. There are several reasons that make preferable to select the fluxes rather than the gradients of the classical variables (for instance, temperature gradient or velocity gradient) as independent variables. (a) The fluxes are associated with well-defined microscopic operators, and as such allow for a more direct comparison with non-equilibrium statistical mechanics and the kinetic theory. (b) The fluxes are generally characterized by short relaxation times and therefore are more adequate than the gradients for describing fast processes. Of course, for slow or steady phenomena, the use of both sets of variables is equivalent because under these conditions the former ones are directly related to the latter. (c) Expressing the entropy in terms of the fluxes offers the opportunity to generalize the classical theory of fluctuations and to evaluate the coefficients of the non-classical part of the entropy as will be shown in Chap. 11. This would not be possible by taking the gradients as variables. (d) Finally, the selection of the gradients as extra variables leads to the presence of divergent terms in the formulation of constitutive equations, a well-known result in the kinetic theory.
5. EIT provides a strong connection between thermodynamics and dynamics. In EIT, the fluxes are no longer considered as mere control parameters but as independent variables. The fact that EIT makes a connection between dynamics and thermodynamics should be underlined. EIT enlarges the range of applicability of non-equilibrium thermodynamics to a vast domain of phenomena where memory, non-local, and non-linear effects are relevant. Many of them are finding increasing application in technology, which, in turn, enlarges the experimental possibilities for the observation of non-classical effects in a wider range of non-equilibrium situations.
6. It should also be underlined that EIT is closer to Onsager's original conceptualization than CIT. Indeed, according to Onsager, the fluxes are defined as the time derivative of the state variables a_α, and the forces are given by the derivatives of the entropy with respect to the a_αs

$$J_\alpha = \frac{\mathrm{d}a_\alpha}{\mathrm{d}t}, \quad X_\alpha = \frac{\partial s}{\partial a_\alpha}. \tag{7.72}$$

Following Onsager, the time evolution equations of the a_αs are obtained by assuming linear relations between fluxes and forces

$$\frac{\mathrm{d}a_\alpha}{\mathrm{d}t} = \sum_\beta L_{\alpha\beta} X_\beta. \tag{7.73}$$

Now, the fluxes and forces of CIT are completely unrelated to Onsager's interpretation; clearly, the heat flux and the pressure tensor are not time derivatives of state variables, similarly, the forces ∇T and $\mathbf{V}$, widely used in CIT, cannot be considered as derivatives of s with respect to the variables a_α. Turning now back to EIT, one can define generalized fluxes J_α and forces X_α, respectively, by

$$\boldsymbol{J}_q = \frac{\mathrm{d}\boldsymbol{q}}{\mathrm{d}t}, \qquad \boldsymbol{X}_q = \frac{\partial s}{\partial \boldsymbol{q}} = \alpha \boldsymbol{q}, \tag{7.74}$$

where $\boldsymbol{q}$ is the heat flux or any other flux variable and α is a phenomenological coefficient. Assuming now a linear flux–force relation $\boldsymbol{J}_q = L\boldsymbol{X}_q$, with $L = 1/\alpha\tau$, one obtains an evolution equation for the state variables $\boldsymbol{q}$ of the form

$$\frac{\mathrm{d}\boldsymbol{q}}{\mathrm{d}t} = \frac{1}{\tau}(\boldsymbol{q}_{\mathrm{ss}} - \boldsymbol{q}), \tag{7.75}$$

where $\boldsymbol{q}_{\mathrm{ss}} \equiv -\lambda \nabla T$ is the classical Fourier steady state value of $\boldsymbol{q}$. After recognizing in (7.75) a Cattaneo-type relation, it is clear that the structure of EIT is closer to Onsager's point of view than that of CIT. Moreover, by transposing Onsager's arguments, it can be shown that the phenomenological coefficient L is symmetric (Lebon et al. 1992; Jou et al. 2001).

7. Extended irreversible thermodynamics is the first thermodynamic theory which proposes an explicit expression for non-equilibrium entropy and temperature. In most theories, this problem is even not evoked or the temperature and entropy are selected as their equilibrium values, as for instance in the kinetic theory of gases.

To summarize, the motivations behind the formulation of EIT were the following:

- To go beyond the local equilibrium hypothesis
- To avoid the paradox of propagation of signals with an infinite velocity
- To generalize the Fourier, Fick, Stokes, and Newton laws by including:
  - Memory effects (fast processes and polymers)
  - Non-local effects (micro- and nano-devices)
  - Non-linear effects (high powers)

The main innovations of the theory are:

- To raise the dissipative fluxes to the status of state variables
- To assign a central role to a generalized entropy, assumed to be a given function of the whole set of variables, and whose rate of production is always positive definite

Extended irreversible thermodynamics provides a decisive step towards a general theory of non-equilibrium processes by proposing a unique formulation of seemingly such different systems as dilute and real gases, liquids, polymers, microelectronic devices, nano-systems, etc. EIT is particularly well suited to describe processes characterized by situations where the product of relaxation time and the rate of variation of the fluxes is important, or when

Table 7.2 Examples of application of EIT

High-frequency phenomena	*Short-wavelength phenomena*
Ultrasounds in gases	Light scattering in gases
Light scattering in gases	Neutron scattering in liquids
Neutron scattering in liquids	Heat transport in nano-devices
Second sound in solids	Ballistic phonon propagation
Heating of solids by laser pulses	Phonon hydrodynamics
Nuclear collisions	Submicronic electronic devices
Reaction–diffusion waves in ecosystems	Shock waves
	Fast moving interfaces
Long relaxation times	*Long correlation lengths*
Polyatomic molecules	Rarefied gases
Suspensions, polymer solutions	Transport in harmonic chains
Diffusion in polymers	Cosmological decoupling eras
Propagation of fast crystallization fronts	Transport near critical points
Superfluids, superconductors	

the mean free path multiplied by the gradient of the fluxes is high; these situations may be found when either the relaxation times or the mean free paths are long, or when the rates of change in time and space are high. Table 7.2 provides a list of situations where EIT has found specific applications.

It should nevertheless not be occulted that some problems remain still open like:

1. Concerning the choice of state variables:
   - Are the fluxes the best variables? Should it not be more judicious to select a combination of fluxes or a mixing of fluxes and transport coefficients?
   - Where to stop when the flux of the flux and higher-order fluxes are taken as variables? The answer depends on the timescale you are working on. Shorter is the timescale, larger is the number of variables that are needed.
   - How far is far from equilibrium? In that respect, it should be convenient to introduce small parameters related for instance to Deborah's and Knudsen's numbers, allowing us to stop the expansions at a fixed degree of accurateness.
2. What is the real status of entropy, temperature, and the second law far from equilibrium?
3. Most of the applications concern fluid mechanics, therefore a description of solid materials including polycrystals, plasticity, and viscoplasticity is highly desirable.
4. The introduction of new variables increases the order of the basic differential field equations requiring the formulation of extra initial and boundary conditions.
5. Turbulence remains a challenging problem.

It should also be fair to stress that, during the last decade, many efforts have been spent to bring a partial answer to these acute questions.

7.6 Problems

7.1. *Extended state space.* Assume that the entropy s is a function of a variable α and its time derivative $\eta = \mathrm{d}\alpha/\mathrm{d}t$, and that α satisfies the differential equation

$$M\frac{\mathrm{d}^2\alpha}{\mathrm{d}t^2} + \frac{\mathrm{d}\alpha}{\mathrm{d}t} = L\frac{\partial s}{\partial \alpha}.$$

(a) Show that the positiveness of the entropy production demands that $L > 0$ but does not imply any restriction on the sign of M. *Hint*: Write $\mathrm{d}\alpha/\mathrm{d}t$ and $\mathrm{d}\eta/\mathrm{d}t$ in terms of $\partial s/\partial\alpha$ and $\partial s/\partial\eta$. (b) Assume that $\partial s/\partial\eta = a\eta$, with a being a constant. Show that the stability condition $\mathrm{d}^2 s < 0$ implies that $M > 0$ and $a = -(M/L)$.

7.2. *Phase velocity.* Determine the expressions (7.8) of the phase velocity v_p and the attenuation factor α. *Hint*: Substitute the solution $T(x,t) = T_0 \exp[\mathrm{i}(kx-\omega t)]$ written in the form $T(x,t) = T_0 \exp[\mathrm{i}\,\mathrm{Re}\,k(x-v_\mathrm{p}t)]\exp(-x/\alpha)$ in the hyperbolic equation (7.5) and split the result in real and imaginary parts.

7.3. *Non-local transport.* (a) Check that the entropy and entropy flux of the non-local formalism including the flux of the heat flux presented in Sect. 7.1.3 are given by

$$\mathrm{d}s = \frac{1}{T}\mathrm{d}u - \frac{\tau_1 v}{\lambda T^2}\,\boldsymbol{q}\cdot\mathrm{d}\boldsymbol{q} - \frac{\tau_2 v}{2\lambda T^2\beta''}\overset{0}{\mathbf{Q}} : \mathrm{d}\overset{0}{\mathbf{Q}} - \frac{\tau_0 v}{\lambda T^2\beta'}Q\,\mathrm{d}Q$$

and

$$\boldsymbol{J}^s = \frac{1}{T}\boldsymbol{q} + \frac{1}{\lambda T^2}\overset{0}{\mathbf{Q}}\cdot\boldsymbol{q} + \frac{1}{\lambda T^2}Q\boldsymbol{q}.$$

(b) Prove that the entropy production corresponding to the Guyer–Krumhansl equation (7.30) is

$$T\sigma^s = \frac{3}{\tau_\mathrm{R}\rho c_v c_0^2}\boldsymbol{q}\cdot\boldsymbol{q} + \frac{3\tau_\mathrm{N}}{5\rho c_v}\left[(\nabla\boldsymbol{q}) : (\nabla\boldsymbol{q})^\mathrm{T} + 2(\nabla\cdot\boldsymbol{q})(\nabla\cdot\boldsymbol{q})\right].$$

(c) Show that the stationary heat flux that satisfies the Guyer–Krumhansl equation is the necessary condition for the entropy production to be a minimum, under the constraint $\nabla\cdot\boldsymbol{q} = 0$. In other terms, show that the Euler–Lagrange equations corresponding to the variational equation

$$\delta\int(T\sigma^s - \gamma\nabla\cdot\boldsymbol{q})\mathrm{d}V = 0,$$

with respect to variations of $\boldsymbol{q}$ and γ are the steady state equations $\partial u/\partial t = 0$ and $\partial q/\partial t = 0$ provided one identifies the Lagrange multiplier γ with twice the temperature (Lebon and Dauby 1990).

7.4. *Non-local transport.* Show that (7.26)–(7.28) can be obtained by writing for $\boldsymbol{q}$ and $\boldsymbol{Q}$ general evolution equations of the form

$$\dot{\boldsymbol{q}} = -\nabla \cdot \boldsymbol{\Pi} + \boldsymbol{\sigma}^q,$$
$$\dot{\mathbf{Q}} = \nabla \cdot \boldsymbol{\Xi} + \sigma^Q,$$

at the condition that the flux and source terms are given by the following constitutive equations:

$$\boldsymbol{\Pi} = A(T)\mathbf{I} + Q\mathbf{I} + \overset{0}{\mathbf{Q}}, \quad \boldsymbol{\sigma}^q = -\boldsymbol{q}$$

and

$$\boldsymbol{\Xi} = \boldsymbol{q}\mathbf{I}, \quad \sigma^Q = -\mathbf{Q},$$

wherein $\boldsymbol{\Xi}$ is a third-order tensor.

7.5. *High-frequency wave speeds.* (a) Obtain (7.32) from the energy balance equation (7.2) and Guyer–Krumhansl relation (7.29), for longitudinal thermal waves. (b) Verify that the high-frequency wave speed for longitudinal waves is given by (7.33).

7.6. *Continued-fraction expansions and generalized thermal conductivity.* To clarify the way to obtain the asymptotic expressions used in (7.36) and (7.40) for the generalized thermal conductivity, (a) show that the continued fraction

$$R = \frac{a}{1 + \frac{a}{1+\frac{a}{1+\ldots}}}$$

tends, in the asymptotic limit of an infinite expansion, to

$$R_\infty = \frac{1}{2}(\sqrt{1+4a} - 1).$$

Hint: Note that, in this limit, R_∞ must satisfy $R_\infty = a/(1+R_\infty)$. (b) From this result, and assuming that all correlation lengths are equal, check that (7.36) leads to (7.39) for steady state (namely $\omega = 0$) and for $k = 2\pi/L$.

7.7. *Phonon hydrodynamics.* Poiseuille flow of phonons may be observed in cylindrical heat conductors of radius R when the mean free paths $\ell_{\mathrm{N}} = c_0\tau_{\mathrm{N}}$ and $\ell_{\mathrm{R}} = c_0\tau_{\mathrm{R}}$ satisfy $\ell_{\mathrm{N}}\ell_{\mathrm{R}} \gg R^2$ and $\ell_{\mathrm{N}} \ll R$. In this case, (7.30) reduces in a steady state to

$$\nabla T - \frac{9}{5}\frac{\tau_{\mathrm{N}}}{\rho c_v}\nabla^2 \boldsymbol{q} = 0.$$

(a) Show that the total flux along the cylinder will be given by

$$Q = -\frac{5\pi}{24}\frac{\rho c_v}{\tau_{\mathrm{N}}}R^4\nabla T.$$

(b) Compare the dependence of Q with respect to the radius R with the corresponding expression obtained from Fourier's law. Note that this dependence may be useful to describe the decrease of the effective thermal conductivity in very thin nanowires, in comparison with the usual thermal conductivity of the corresponding bulk material.

7.8. *Double time-lag behaviour.* Instead of the Cattaneo's equation (7.4), some authors (Tzou 1997) use a generalized transport equation with two relaxation times

$$\tau_1 \dot{\boldsymbol{q}} + \boldsymbol{q} = -\lambda \left(\nabla T + \tau_2 \nabla \dot{T} \right).$$

(a) Introduce this equation into the energy balance equation (7.2) and obtain the evolution equation for the temperature. (b) Discuss the limiting behaviour of high-frequency thermal waves.

7.9. *Two-temperature models.* Many systems consist of several subsystems, each with its own temperature, as for instance, the electrons and the lattice in a metal. It has been shown that the evolution equations for the electron and lattice temperatures T_e and T_l are given, respectively, by:

$$c_\mathrm{e} \frac{\partial T_\mathrm{e}}{\partial t} = \nabla \cdot (\lambda \nabla T_\mathrm{e}) - C(T_\mathrm{e} - T_\mathrm{l}),$$
$$c_\mathrm{l} \frac{\partial T_\mathrm{l}}{\partial t} = C(T_\mathrm{e} - T_\mathrm{l}).$$

The constant C describes the electron–phonon coupling, which accounts for the energy transfer from the electrons to the lattice, and c_e and c_l are the specific heats of the electrons and lattice per unit volume, respectively. (a) When the solution of the first equation, namely $T_\mathrm{e} = T_\mathrm{l} + (c_\mathrm{l}/C)\partial T_\mathrm{l}/\partial t$, is introduced into the second one, prove that it leads to

$$\nabla^2 T_\mathrm{l} + \frac{c_\mathrm{l}}{C}\frac{\partial \nabla^2 T_\mathrm{l}}{\partial t} = \frac{c_\mathrm{l} + c_\mathrm{e}}{\lambda}\frac{\partial T_\mathrm{l}}{\partial t} + \frac{c_\mathrm{e} c_\mathrm{l}}{\lambda C}\frac{\partial^2 T_\mathrm{l}}{\partial t^2}.$$

(b) Show that this equation can also be obtained by eliminating $\boldsymbol{q}$ between the energy balance equation and Guyer–Krumhansl's equation, with the suitable identifications of the parameters.

7.10. *Limits of stability of non-equilibrium steady states.* Let us first consider heat conduction in a rigid solid for which $s = s(u, \boldsymbol{q})$. (a) Write the second differential $\delta^2 s$ of the generalized entropy (7.25) around a non-equilibrium state with non-vanishing value of $\boldsymbol{q}$. (b) Show that one of the conditions to be satisfied in order that the matrix of the second differential be negative definite is

$$\frac{\alpha}{c_v T^2} + \left[\frac{\alpha}{2}\frac{\partial^2 \alpha}{\partial u^2} - \left(\frac{\partial \alpha}{\partial u} \right)^2 \right] q^2 \geq 0,$$

where $\alpha \equiv v\tau/\lambda T^2$. (b) If v, τ_1, c_v, and λ are constant, it is found that $\partial\alpha/\partial u = -2\alpha/(c_v T)$ and $\partial^2\alpha/\partial u^2 = 6\alpha(c_v T)^{-2}$. Show that in this case the former inequality is satisfied for values of q such that

$$q \leq \left(\frac{c_v}{\alpha}\right)^{1/2} = \rho c_v T \left(\frac{\lambda}{\rho c_v \tau_1}\right)^{1/2},$$

where $(\lambda/\rho c_v \tau_1)^{1/2} = U$ is the second sound.

7.11. *Development in gradients or in fluxes.* Compare the behaviour of the wave number-dependent viscosity $\eta(k, \omega)$ appearing in the two following second-order expansions (a) the flux expansion

$$\tau_2 \frac{\partial \mathbf{P}^{\mathrm{v}}}{\partial t} + \mathbf{P}^{\mathrm{v}} = -2\eta \mathbf{V} + \nabla \cdot \overset{0}{\mathbf{J}}{}^{\mathrm{v}} \quad \text{and} \quad \overset{0}{\mathbf{J}}{}^{\mathrm{v}} = -\eta'' \langle \nabla \overset{0}{\mathbf{P}}{}^{\mathrm{v}} \rangle,$$

with $\overset{0}{\mathbf{J}}{}^{\mathrm{v}}$ the flux of the viscous pressure tensor, η'' a phenomenological coefficient, and $\langle \ldots \rangle$ the completely symmetrized part of the corresponding third-order tensor and (b) the velocity gradient expansion

$$\tau_2 \frac{\partial \mathbf{P}^{\mathrm{v}}}{\partial t} + \mathbf{P}^{\mathrm{v}} = -2\eta \mathbf{V} + \ell^2 \nabla^2 \mathbf{V}.$$

Note that (b) yields an unstable behaviour for high values of k, since the generalized viscosity becomes negative. (For a discussion of these instabilities arising in kinetic theory approaches to generalized hydrodynamics, see Gorban et al. 2004.)

7.12. *Two-layer model and the telegrapher's equation.* The so-called *two-layer model* consists of a system whose particles jump at random between two states, 1 and 2, with associated velocities $v_1 = v$ and $v_2 = -v$, respectively, along the x-axis. Assume that the rate R of particle exchange between the two states per unit time and length is proportional to the difference of the probability densities P_1 and P_2, i.e. $R = r(P_1 - P_2)$. Show that the evolution equations for the total probability density $P = P_1 + P_2$ and for the probability flux $J = (P_1 - P_2)v$ are, respectively,

$$\frac{\partial P}{\partial t} + \frac{\partial J}{\partial x} = 0,$$
$$\tau \frac{\partial J}{\partial t} + J = -D \frac{\partial P}{\partial x},$$

with $\tau = 1/2r$ and $D = v^2 \tau$ (see van den Broeck 1990; Camacho and Zakari 1994).

7.13. *Electrical system with resistance R and inductance L.* The expression relating the intensity I of the electrical current to the electromotive force ξ is a relaxational equation given by

$$\xi = IR + L \frac{\mathrm{d}I}{\mathrm{d}t},$$

with relaxation time $\tau_e = L/R$. The intensity I is related to the flux of electric current i by $I = iA$, with A the cross section of the conductor. The magnetic energy stored in the inductor is given by $U_m = \frac{1}{2}LI^2$. Consider the total internal energy $U_{tot} = U + U_m$, with U the internal energy of the material. Show that the Gibbs' equation $dS = T^{-1}dU + T^{-1}p\,dV$ may be rewritten as

$$dS = T^{-1}dU + T^{-1}p\,dV - \frac{\tau_e V}{\sigma_e T} i\,di,$$

which is a relation reminiscent of the Gibbs' equation proposed in EIT. *Hint*: Recall that $R = (\sigma_e A)^{-1}l$, with l the length of the circuit.

7.14. *Chemical potential.* According to (7.66), the differential equation for the Gibbs' free energy at constant temperature and pressure reads as

$$dg = \sum_k \mu_k dc_k + \frac{v\tau}{2\eta}\mathbf{P}^v : d\mathbf{P}^v,$$

and, after integration, $g(c_i, \mathbf{P}^v) = g_{eq}(c_i) + \frac{1}{4}vJ\mathbf{P}^v : \mathbf{P}^v$, where τ designates the relaxation time of the viscous pressure tensor and $J = \eta/\tau$, is the so-called *steady state compliance*, a function which is often studied in polymer solutions. (a) Obtain the modification to the chemical potential of the ith component, defined by

$$\mu_i = \left(\frac{\partial g}{\partial n_i}\right)_{T,p,\mathbf{P}^v},$$

if the steady state compliance J is assumed to have the form $J = \alpha M(ck_B T)^{-1}$, with M the molecular mass, c the number of macromolecules per unit volume, and α a constant (e.g. $\alpha = 0.400$ for Rouse model of bead-and-spring macromolecules, and $\alpha = 0.206$ for the Zimm model, the latter including the hydrodynamic interaction amongst the different beads of a macromolecule (Doi and Edwards, 1986; Bird et al., vol. 2, 1987a, 1987b)). (b) Study the influence of the non-equilibrium contribution on the stability of the system; in particular, determine whether the presence of a non-vanishing viscous pressure will reinforce or not the stability with respect to that at equilibrium (Jou et al. 2000).

Chapter 8
Theories with Internal Variables

The Influence of Internal Structure on Dynamics

An alternative approach of non-equilibrium thermodynamics may be carried out through the internal variables theory (IVT). This phenomenological description has been very successful in the study of a wide variety of processes in rheology, deformable bodies, physico-chemistry and electromagnetism. IVT may be considered as a generalization of the classical theory of irreversible processes, just like extended irreversible thermodynamics. In the classical theory, the state of the system is described by the local equilibrium variables, say energy, volume or deformation, but it is well known that many non-equilibrium effects cannot be adequately described by this set of variables. To avoid, among others, the inconvenience of a heavy formalism in terms of functionals, one has, in extended irreversible thermodynamics, supplemented the local equilibrium variables by the corresponding fluxes. In the IVT, one adopts a similar attitude by adding a certain number of *internal* variables to the local equilibrium variables. Such variables are called internal because they are connected either to internal motions or to local microstructures. They should not be confused with order parameters as introduced in Chap. 6, which are describing some broken symmetry related to long-range interactions as in phase transitions. We should also add that unlike the fluxes in extended irreversible thermodynamics, the internal variables are arbitrary extensive quantities, not identified from the outset. It is the freedom in the choice of the internal variables that explains the wide domain of applicability of the formalism.

What is really meant by internal variables? The usual interpretation is that they are macroscopic representations of some microscopic internal structures; they can be of geometrical or physico-chemical nature, but generally there is a priori no reference to their origin. The hope is that at the end, some physical meaning will come out from the theory itself. Mandel (1978) wrote that "a clever physicist will always manage to detect internal variables and measure them". Internal variables are essentially introduced to compensate for our lack of knowledge about the behaviour of the system. The alternative name of *hidden* variables is also found in the literature, as they refer to the internal structure that is hidden from the eyes of an external observer. At the difference of the *external* variables, like temperature or volume, the *internal* parameters are not coupled to external forces which provide a means of control. As they cannot be adjusted to a prescribed value through surface and body forces, internal variables are not controllable. Being not attached to an external force, they cannot take part in a mechanical work and therefore they do not appear explicitly in the balance equations of mass, momentum and energy. Additional equations (taking generally the form of rate equations) are needed to describe their evolution.

To summarize, internal variables are in principle measurable but not controllable and this makes difficult their selection and definition. They are associated to processes that are completely dissipated inside the system instead of being developed against external surface or body forces.

Examples of internal variables are the extent of advancement in a chemical reaction, the density of dislocations in plastic metals, mobile atoms in crystals, the end-to-end distance between macromolecules in polymer solutions, the shape and orientations of microstructures in macromolecules, the magnetization in ferromagnets, etc. In the following, we shall present the general scheme of the thermodynamic theory of internal variables and afterwards we shall discuss some applications both in fluid and solid mechanics.

Following our policy of seeking a presentation as simple as possible, we shall restrict our analysis to rather simple examples. The reader interested by more complete developments is advised to consult the comprehensive review papers by Bampi and Morro (1984), Kestin (1990, 1992), Maugin and Muschik (1994), Lhuillier et al. (2003), and the well-documented book by Maugin (1999) which includes a large bibliography and a wide and systematic presentation.

8.1 General Scheme

8.1.1 Accompanying State Axiom

The hypothesis of local equilibrium that is the basis of classical irreversible thermodynamics is replaced in the IVT by the *accompanying state axiom*. The essence of this postulate is that, to each non-equilibrium state, corresponds an *accompanying equilibrium state*, and to every irreversible process is associated an accompanying "reversible" process. This notion can be traced back to Meixner (1973a, 1973b), Muschik (1990) and mainly to Kestin (1990, 1992): it provides an interesting picture of IVT by associating non-equilibrium states with equilibrium states by means of a projection. The accompanying equilibrium state is a fictitious one; its state space $\mathcal{V}$ consists of the union of the spaces of standard equilibrium variables $\mathcal{C}$ and internal variables $\mathcal{I}$. The ensemble $\mathcal{V}$ chosen for further analysis will consist of the following extensive variables

$$\mathcal{V} := u, \boldsymbol{a}, \xi; \tag{8.1}$$

the space of classical variables includes the specific internal energy u and the other local equilibrium variables, for instance the specific volume v in fluid mechanics or the elastic strain tensor $\boldsymbol{\varepsilon}$ in solid mechanics, which will be generically designed as $\boldsymbol{a}$, and finally $\xi \equiv \xi_1, \xi_2, \ldots, \xi_n$ stands for a finite set of internal variables (either scalars, vectors, or tensors). Although the number of variables needed to specify a non-equilibrium state n is generally much larger than the set (8.1), we can always imagine an accompanying equilibrium state e whose extensive variables $u, \boldsymbol{a}, \boldsymbol{\xi}$ take the same values as in a non-equilibrium state n, i.e. $u(e) = u(n)$, $\boldsymbol{a}(e) = \boldsymbol{a}(n)$, $\boldsymbol{\xi}(e) = \boldsymbol{\xi}(n)$. A way of obtaining these accompanying state variables from the actual physical non-equilibrium state is illustrated by the Gedanken experience of Fig. 8.1.

Consider a volume element n suddenly surrounded by an adiabatic rigid enclosure so that no heat flows and no external work can be performed; as a consequence of the first law, $\dot{Q} = \dot{W} = 0$ implies that $u(n) = u(e)$. Simultaneously, we constrain the internal variables ξ. Since no work and no heat are exchanged, the system relaxes towards an equilibrium-constrained state e with the values of $u, \boldsymbol{a}$, and ξ unmodified. Observe however that other variables, like the temperature T or the entropy s, relax to values which are different from their actual non-equilibrium values: $T(e) \neq T(n)$ and $s(e) \neq s(n)$. The entropy $s(e)$ is larger than the entropy in non-equilibrium $s(n)$ because the former is obtained from the latter by an adiabatic no-work process. At this point it is interesting to emphasize the difference between the axiom of accompanying state and the local equilibrium hypothesis. The gist of this latter statement is to assume that the *non-equilibrium* entropy $s(n)$ and temperature $T(n)$ are the same as in their accompanying equilibrium state $s(e)$ and $T(e)$, respectively, while in IVT, the entropy $s(e)$ is assumed to characterize the *equilibrium* state.

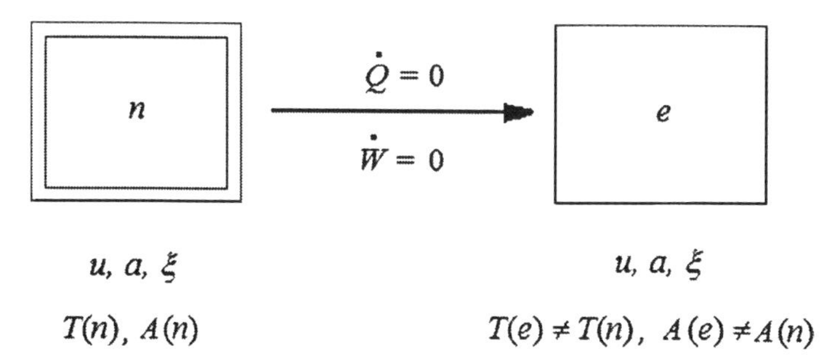

Fig. 8.1 Producing an accompanying equilibrium e state from a non-equilibrium state n

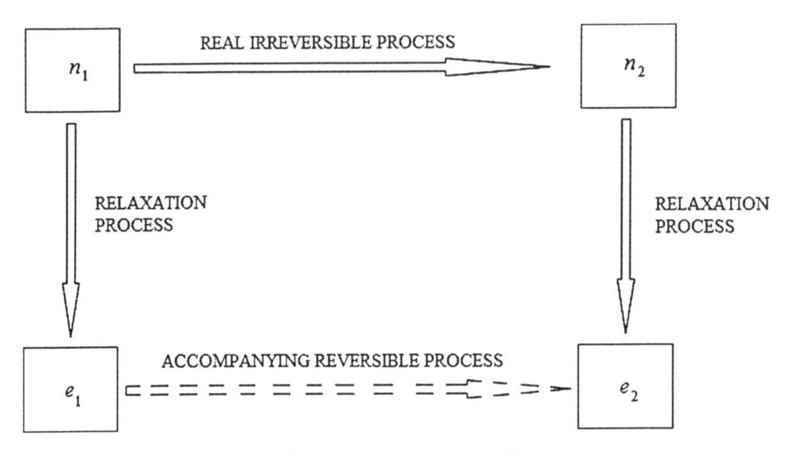

Fig. 8.2 Accompanying reversible process

To clarify the notion of accompanying *reversible process*, consider an irreversible process between states n_1 and n_2. The accompanying states e_1 and e_2 are obtained by projection on the state space $(u, \boldsymbol{a}, \xi)$ as shown in Fig. 8.2 and the sequence of accompanying states between e_1 and e_2 constitutes the "reversible" accompanying process. By contrast with a true reversible process, which takes place infinitely slowly, the former one occurs at finite velocity and, as seen later, is characterized by a non-zero entropy production.

Now the question remains of how many internal variables are needed for a valuable description. Of course it is desirable to have a number of variables as small as possible. There exists a criterion based on the notion of characteristic time that allows deciding whether an internal variable is relevant or not. Let us denote by $\tau_\xi = \dot{\xi}/\xi$ the intrinsic time of the internal mechanism and by $\tau_a = \dot{a}/a$ the characteristic macroscopic time associated with the external variable a; as usual, a upper dot denotes time derivative. If the Deborah number defined by $De = \tau_\xi/\tau_a$ is such that $De \ll 1$, then the corresponding internal variable relaxes quickly towards its equilibrium value; this amounts to say that the internal constraint is more or less removed, thus allowing the system to reach a state of so-called *unconstrained* equilibrium (Kestin 1990, 1992). In contrast for $De \gg 1$, the internal variable will remain frozen during the timescale of the evolution of the variable a, we shall then speak about a state of *constrained* equilibrium. Practically, in IVT only the internal variables for which $De \gg 1$ will be retained. It was also suggested by Kestin (1990, 1992) to use the Deborah number as a measure of the "distance" between a non-equilibrium state and an accompanying equilibrium state: the smaller is De, the shorter is the distance between such states. Situations with nonlocal effects are discussed in Box 8.1.

Box 8.1 Non-Local Effects. Histories of Variables

In presence of non-local effects, as in the cases of damages or non-homogeneous electrical or magnetic fields, it may in principle be necessary to introduce spatial gradients $\nabla\xi$ of the internal variables in the description. The evolution of such internal variables is now determined by complete balance equations, involving both a rate and a divergence term. The difficulty is that such balance laws require appropriate boundary conditions, which cannot in principle be assigned, as the internal variables are not controllable. This problem has not received sufficient attention and only partial answers can be found in the recent literature. The underlying idea is that, although the internal variables are not controllable, the second law of thermodynamics forces them to assume given values on the boundaries (Woods 1975; Valanis 1996; Cimmelli 2002).

Because of the difficulty to identify the nature of the internal variables, it may be tempting to formulate the problem without having recourse to them. To obtain such a description, it suffices to eliminate the internal variables ξ from (8.11) to (8.13). In the case that (8.13) can be solved, we can substitute its solutions in (8.11) and (8.12), leading to a set of constitutive relations, which turn out to be functionals $\mathcal{Q}$ and $\mathcal{F}$ of the histories of u (or T), ∇T, and $\boldsymbol{a}$ (Kestin 1992):

$$\boldsymbol{q} = \mathop{\mathcal{Q}}_{t'\geq 0}\{T(t-t'), \nabla T(t-t'), \boldsymbol{a}(t-t')\}, \tag{8.1.1}$$

$$\boldsymbol{F} = \mathop{\mathcal{F}}_{t'\geq 0}\{T(t-t'), \nabla T(t-t'), \boldsymbol{a}(t-t')\}. \tag{8.1.2}$$

As shown in Chap. 9, it is customary in rational thermodynamics to claim that the systems that it describe are materials with memory. The difficulty is how to measure and to catalogue all these histories and how to solve the subsequent integro-differential equations obtained after substitution of the above functionals in the balance equations.

8.1.2 Entropy and Entropy Production

The axiom of accompanying equilibrium state leads us to accept the existence of an accompanying entropy s that is a function of the whole set $\mathcal{V}$ of variables:

$$s = s(u, \boldsymbol{a}, \xi). \tag{8.3}$$

The corresponding Gibbs' equation in rate form will assume the form

$$\dot{s} = T^{-1}\dot{u} - T^{-1}\rho^{-1}\boldsymbol{F}_{\mathrm{e}} \cdot \dot{\boldsymbol{a}} + T^{-1}\boldsymbol{\mathcal{A}} \cdot \dot{\xi}, \tag{8.4}$$

where T is the temperature, $\boldsymbol{F}_{\mathrm{e}}$ the mechanical force conjugate to the observable $\boldsymbol{a}$, and $\boldsymbol{\mathcal{A}}$ the affinity (also known as configurational or Eshelby force in some branches of solid mechanics) conjugate to ξ. They are defined, respectively, by

$$\frac{1}{T} = \frac{\partial s}{\partial u}, \quad \frac{\boldsymbol{F}_{\mathrm{e}}}{\rho T} = -\frac{\partial s}{\partial \boldsymbol{a}}, \quad \frac{\boldsymbol{\mathcal{A}}}{T} = \frac{\partial s}{\partial \xi}. \tag{8.5}$$

The internal energy obeys the balance law

$$\rho \dot{u} = -\nabla \cdot \boldsymbol{q} + \boldsymbol{F} \cdot \dot{\boldsymbol{a}} \tag{8.6}$$

with $\boldsymbol{q}$ the heat flux vector. Remark in passing that the forces $\boldsymbol{F}(n)$ required to produce work on the volume element in the actual physical space are different from the forces $\boldsymbol{F}_{\mathrm{e}}$ in (8.4) which represent these acting on the fictitious reversible process. After elimination of $\dot{u}$ between the Gibbs' equation (8.4) and the energy balance equation (8.6), one obtains the expression for the evolution of the entropy of the accompanying state during the process:

$$\rho \dot{s} + \nabla \cdot (\boldsymbol{q} T^{-1}) = \boldsymbol{q} \cdot \nabla T^{-1} + T^{-1}(\boldsymbol{F} - \boldsymbol{F}_{\mathrm{e}}) \cdot \dot{\boldsymbol{a}} + \rho T^{-1} \boldsymbol{\mathcal{A}} \cdot \dot{\xi}. \tag{8.7}$$

Suppose that the entropy obeys a balance equation of the standard form

$$\rho \dot{s} + \nabla \cdot (T^{-1} \boldsymbol{q}) = \sigma^s \geq 0, \tag{8.8}$$

wherein it is accepted that the rate of entropy production σ^s is positive definite and the entropy flux given by $T^{-1}\boldsymbol{q}$. By comparison with (8.7), it is then found that

$$\sigma^s = \boldsymbol{q} \cdot \nabla T^{-1} + T^{-1}(\boldsymbol{F} - \boldsymbol{F}_{\mathrm{e}}) \cdot \dot{\boldsymbol{a}} + \rho T^{-1} \boldsymbol{\mathcal{A}} \cdot \dot{\xi} \geq 0. \tag{8.9}$$

With the exception of extended thermodynamics, it is currently admitted that, in absence of diffusion of matter, the entropy flux is the ratio of the heat flux and the temperature. The positiveness of the rate of entropy production is a more disputed question and can be considered as a supplementary postulate of the theory.

Instead of starting from (8.9), some authors prefer to work with the Clausius–Duhem's inequality

$$-\rho(\dot{f} + s\dot{T}) + \boldsymbol{F} \cdot \dot{\boldsymbol{a}} + T^{-1} \boldsymbol{q} \cdot \nabla T \geq 0, \tag{8.10}$$

which is directly obtained from (8.8) when use is made of the definition of the free energy $f = u - Ts$ and the energy balance equation (8.6). In the future, we shall indifferently start from (8.9) or (8.10).

8.1.3 Rate Equations

It is noted that (8.9) has the form of the familiar sum of generalized forces $\boldsymbol{X}$ and fluxes $\boldsymbol{J}$, which, at the exception of the heat flux, are the rate of change of the extensive variables. In classical irreversible thermodynamics (see Chap. 2), the fluxes are linear expressions of the forces but this is not mandatory in the present theory wherein the fluxes and forces are related to each other by general functions containing the variables u, $\boldsymbol{a}$, and ξ, i.e. $\boldsymbol{J} = \boldsymbol{J}(\boldsymbol{X}; u, \boldsymbol{a}, \xi)$. Such flux–force relations take generally the form of rate equations, which conventionally will be written as

$$\boldsymbol{q} = \mathcal{Q}(\nabla T, \dot{\boldsymbol{a}}, \boldsymbol{\mathcal{A}}; u, \boldsymbol{a}, \xi), \tag{8.11}$$

$$\boldsymbol{F} - \boldsymbol{F}_{\mathrm{e}} = \mathcal{F}(\nabla T, \dot{\boldsymbol{a}}, \boldsymbol{\mathcal{A}}; u, \boldsymbol{a}, \xi), \tag{8.12}$$

$$\dot{\xi} = \Xi(\nabla T, \dot{\boldsymbol{a}}, \boldsymbol{\mathcal{A}}; u, \boldsymbol{a}, \xi), \tag{8.13}$$

where $\mathcal{Q}, \mathcal{F}$, and Ξ denote functions of their arguments.

After this short digression, let us go back to the evolution equation (8.13). In the simplest case, we may assume that the "flux" $\dot{\xi}$ is linear with the "force" $\boldsymbol{\mathcal{A}}$

$$\dot{\xi} = l\boldsymbol{\mathcal{A}}(u, \boldsymbol{a}, \xi), \tag{8.13}$$

where l is a phenomenological coefficient. Such a simplified evolution equation is justified when the thermal gradients play a negligible role and when $\boldsymbol{F} \approx \boldsymbol{F}_{\mathrm{e}}$ as often occurring in solid mechanics. The entropy production (8.9) takes now the simple form

$$T\sigma^s = \rho\boldsymbol{\mathcal{A}} \cdot \dot{\xi}, \tag{8.14}$$

and we shall examine three particular cases:

1. $\boldsymbol{\mathcal{A}} = \dot{\xi} = 0$: the system has reached a state of unconstrained equilibrium characterized by $\boldsymbol{\mathcal{A}}(u, \boldsymbol{a}, \xi) = 0$ or $\xi(e) = \xi(u, \boldsymbol{a})$.
2. $\boldsymbol{\mathcal{A}} \neq 0$ with $\dot{\xi} = 0$: this is a case of constrained equilibrium still defined by a zero entropy production.
3. $\boldsymbol{\mathcal{A}} \neq 0$ and $\dot{\xi} \neq 0$: the system relaxes towards the equilibrium surface $\xi(e)$.

As mentioned earlier, there exists a wealth of applications of the internal variable theory, mainly in the domain of material sciences. Here we will present only three of them, and we refer the reader to specialized books and articles for a more exhaustive list of examples.

8.2 Applications

8.2.1 Viscoelastic Solids

Viscoelastic solids are a privileged field of application of the theory of internal variables. Thermodynamics of viscoelasticity in solids was initiated by Bridgman (1941) and Eckart (1948) and developed later on by Eringen (1967), Kluitenberg (1962) and Kestin and Bataille (1980), among others. Our purpose is to discuss qualitatively some essential features of rheological models like those of Maxwell and Kelvin–Voigt, which have played a central role in the theory of viscoelasticity; excessive generality will be deliberately avoided.

We restrict our attention to isothermal effects and make the assumption that the solid is a *one-dimensional* homogeneous and isotropic rod subject to infinitesimally small deformations. Under the action of a stress $\boldsymbol{\sigma}$ the solid undergoes a strain ε. It is remembered from Sect. 8.1 that an intensive parameter such as $\boldsymbol{\sigma}$, which is the stress actually applied in the non-equilibrium state, is not the same as the stress $\boldsymbol{\sigma}_{\mathrm{e}}$ in the accompanying equilibrium state. Moreover, it is assumed that the deformation of the body has, associated with it, an internal work that is completely dissipated in the interior of the system during the actual dissipative process. Its expression is given by $\mathcal{A}\,\mathrm{d}\xi$ where $\mathcal{A}$ is the "affinity" associated to the internal variable ξ. According to IVT, the space of basic variables is constituted by $\mathcal{V} \equiv u, \sigma_{\mathrm{e}}, \xi$ with u the specific internal energy. The entropy associated with the accompanying state satisfies the Gibbs' equation

$$T\,\mathrm{d}s = \mathrm{d}u - \rho^{-1}\sigma_{\mathrm{e}}\mathrm{d}\varepsilon + \mathcal{A}\,\mathrm{d}\xi, \tag{8.15}$$

where ρ is the constant mass density. Since the temperature is a constant, it may be preferable to work with the free energy f for which

$$\mathrm{d}f = \rho^{-1}\sigma_{\mathrm{e}}\mathrm{d}\varepsilon - \mathcal{A}\,\mathrm{d}\xi, \tag{8.16}$$

with

$$\rho^{-1}\sigma_{\mathrm{e}} = \partial f/\partial\varepsilon, \quad \mathcal{A} = -\partial f/\partial\xi. \tag{8.17}$$

The equation of state $f = f(\varepsilon, \xi)$ can be expanded in a Taylor series around a reference state $\varepsilon = \xi = 0$ and put in the form

$$f(\varepsilon, \xi) = \frac{1}{2}E\varepsilon^2 + B\varepsilon\xi + \frac{1}{2}C\xi^2, \tag{8.18}$$

where E, B, and C are constant coefficients with $E > 0$, $C > 0$, and $B^2 \leq EC$ to satisfy the stability property of the equilibrium state. Examples of linear state equations (8.17) are

$$\rho^{-1}\sigma_{\mathrm{e}} = E\varepsilon + B\xi, \tag{8.19}$$

$$-\mathcal{A} = B\varepsilon + C\xi. \tag{8.20}$$

Elimination of $\dot{u}$ from the energy balance equation

$$\rho\dot{u} = \boldsymbol{\sigma} : \dot{\boldsymbol{\varepsilon}}, \tag{8.21}$$

and the Gibbs' equation (8.15) written in rate form leads to the following relation for the evolution of the entropy in the accompanying state:

$$\rho T\dot{s} = (\boldsymbol{\sigma} - \boldsymbol{\sigma}_{\mathrm{e}}) : \dot{\boldsymbol{\varepsilon}} + \rho^{-1}\boldsymbol{\mathcal{A}}\cdot\dot{\boldsymbol{\xi}}. \tag{8.22}$$

Since there is no heat flux, the entropy flux will be zero and therefore the quantity $\rho T\dot{s}$ can be identified with the rate of dissipated energy $T\sigma^s$, with σ^s the rate of entropy production. In very slow processes, it is justified to assume (Kestin and Bataille 1980) that $\sigma = \sigma_e$ so that the dissipated energy acquires the simple form

$$\rho T\sigma^s = \boldsymbol{\mathcal{A}} \cdot \dot{\xi} \geq 0. \tag{8.23}$$

Of course, the simplest rate equation for the internal variable is to put

$$\dot{\xi} = l\boldsymbol{\mathcal{A}}(l > 0). \tag{8.24}$$

Taking the time derivative of (8.19) and using (8.24), one obtains

$$\rho^{-1}\dot{\sigma} = E\dot{\varepsilon} + lB\mathcal{A}, \tag{8.25}$$

where use is made of the assumption $\sigma_e = \sigma$. Substituting now $\mathcal{A}$ from (8.20) and eliminating ξ from (8.19), leads to the following relaxation equation for the stress tensor

$$\tau_\varepsilon\dot{\sigma} + \sigma = E_\infty(\varepsilon + \tau_\sigma\dot{\varepsilon}), \tag{8.26}$$

where $\tau_\varepsilon = 1/lC$, $\tau_\sigma = E(lCE_\infty)^{-1}$, $E_\infty = E - B^2C^{-1}$. It is worth noticing that relation (8.26) is the constitutive equation for a Poynting–Thomson body. By letting τ_ε and τ_σ go to zero, one finds back the Hooke's law of elasticity, and E_∞ is the classical Young modulus. Setting $\tau_\varepsilon = 0$, one recovers the Kelvin–Voigt model

$$\sigma = E_\infty\varepsilon + \eta\dot{\varepsilon}, \tag{8.27}$$

with $\eta = E/lC$, while by putting $E_\infty = 0$ and $E_\infty\tau_\sigma = \eta$ in (8.26), one obtains Maxwell's model:

$$\tau_\varepsilon\dot{\sigma} + \sigma = \eta\dot{\varepsilon}. \tag{8.28}$$

It is therefore claimed that the thermodynamics of internal variables includes the formulation of the standard models of rheology. The above considerations have been generalized to finite strains by Sidoroff (1975, 1976), among others.

IVT has been used with success in several branches of solid mechanics. For instance, an interesting approach of plasticity, in relation with the microscopic theory of dislocations, was developed by Ponter et al. (1978). These authors show that the area swept out by the dislocation lines is a meaningful internal variable. More explicitly, the internal variable selected by Ponter et al. is

$$\xi = \frac{\hat{A}Nb}{\ell^3}, \tag{8.29}$$

where $\hat{A}$ is the area swept out by the dislocation line, N the number of non-interacting so-called Franck–Read's sources, b the Burgers vector associated with dislocations slip, and ℓ^3 the corresponding volume. Another field of application is the theory of fracture, where the internal variable can be chosen as a scalar variable D (D for damage) such that $0 \leq D \leq 1$, $D = 0$ meaning absence of cracks and $D = 1$ corresponding to fracture.

8.2.2 Polymeric Fluids

As a second illustration, we shall examine the behaviour of polymeric fluids, which are one of the great diversity of fluids called complex fluids. Their characteristic property is to possess a certain degree of internal structures or microstructures. Therefore they behave very differently from water or air when forces are exerted on them. The most rigorous method of describing materials with internal microstructures is through the distribution function ψ which depends on the position and velocity vectors of the ensemble of mass points. However, in view of the large number of variables, this approach becomes quickly impracticable, and some mechanical approximations taking the form of rigid rods, dumbbells or more complicated assemblages (see Fig. 8.3) have been introduced.

In the case of dumbbells, the description involves a seven-dimensional function $\psi(\boldsymbol{r}, \boldsymbol{R}, t)$ where $\boldsymbol{r}$ is the position vector and $\boldsymbol{R}$ the end-to-end macromolecular distance. But still more approximations are needed to simplify the model. In that respect, it is frequent to replace the description based on the distribution function by the second-order symmetric tensor

$$\mathbf{C} = \int \boldsymbol{R}\boldsymbol{R}\, \psi \, d\boldsymbol{R}, \tag{8.30}$$

called "configuration" or "conformation" tensor. The advantage gained by this approximation is to solve model equations in a large variety of flow situations. It is the conformation tensor that we shall identify as the relevant internal variable for our forthcoming analysis. We consider a flowing incompressible fluid composed of a carrier, the solvent of mass density ρ_{f}, and long macromolecular chains, the solute, of mass density ρ_{pol}. Because of incompressibility, the total mass $\rho = \rho_{\mathrm{f}} + \rho_{\mathrm{pol}}$ is a constant and $\nabla \cdot \boldsymbol{v} = 0$, where $\boldsymbol{v}$ designates the barycentric velocity defined by means of $\rho \boldsymbol{v} = \rho_{\mathrm{f}} \boldsymbol{v}_{\mathrm{f}} + \rho_{\mathrm{pol}} \boldsymbol{v}_{\mathrm{pol}}$.

The selected basic variables are the mass concentration $c_{\mathrm{pol}} = \rho_{\mathrm{pol}}/\rho$ of the polymeric chain, the barycentric velocity $\boldsymbol{v}$, the specific internal energy u or the temperature T and the conformation tensor $\mathbf{C}$. The corresponding evolution equations are given by

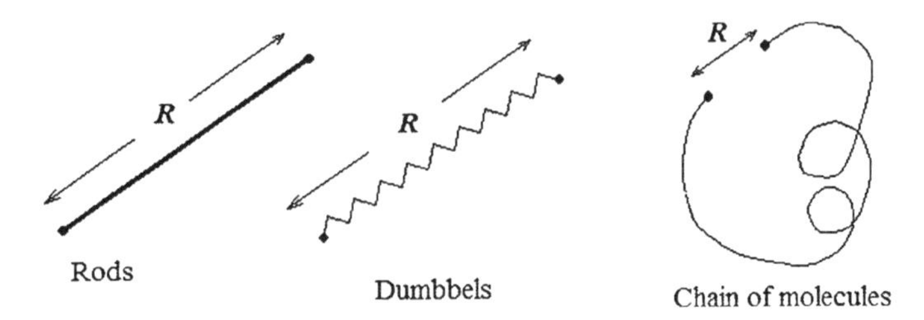

Fig. 8.3 Molecular polymer models

$$\rho \dot{c}_{\text{pol}} = -\nabla \cdot \boldsymbol{J}, \tag{8.31}$$
$$\rho \dot{\boldsymbol{v}} = \nabla \cdot \boldsymbol{\sigma} + \rho \boldsymbol{F}, \tag{8.32}$$
$$\rho \dot{u} = -\nabla \cdot \boldsymbol{q} + \boldsymbol{\sigma}^{\text{v}} : \mathbf{V}, \tag{8.33}$$
$$\dot{\mathbf{C}} = \mathcal{C}(\mathbf{C}, \mathbf{V}, c_{\text{pol}}, u), \tag{8.34}$$

wherein $\boldsymbol{J}$ denotes the flux of matter, $\boldsymbol{F}$ the external body force per unit mass, $\mathbf{V}$ the symmetric velocity gradient tensor, $\mathcal{C}$ a unknown function of the variables $\mathbf{C}$, c_{pol}, u and eventually of $\mathbf{V}$; finally, $\boldsymbol{\sigma}$ is the symmetric stress tensor and $\boldsymbol{\sigma}^{\text{v}}(= \boldsymbol{\sigma} + p\mathbf{I})$ its viscous part with p the hydrostatic pressure. The tensor $\boldsymbol{\sigma}^{\text{v}}$ will be influenced by the presence of the internal variable and therefore will depart from the Newton law by an additional contribution so that in total

$$\boldsymbol{\sigma}^{\text{v}} = 2\eta \mathbf{V} + \boldsymbol{\sigma}_{\text{pol}}(\mathbf{C}, \mathbf{V}, \ldots), \tag{8.35}$$

where η is the usual shear viscosity coefficient. The system of equations (8.31)–(8.34) will be closed by providing constitutive relations for the set $\{\boldsymbol{J}, \boldsymbol{q}, \boldsymbol{\sigma}^{\text{v}} (\text{or}\, \boldsymbol{\sigma}_{\text{pol}}), \mathbf{C}\}$. These are obtained from the axiom of the *local accompanying state* consisting in formulating a generalized Gibbs' equation and calculating the corresponding rate of entropy production inside the material. Let us write the Gibbs' relation in terms of the free energy, i.e.

$$\mathrm{d}f = -s\,\mathrm{d}T + \mu\,\mathrm{d}c_{\text{pol}} - \mathbf{A} : \mathrm{d}\mathbf{C}, \tag{8.36}$$

where $\mu = \mu_{\text{pol}} - \mu_{\text{f}}$ is the relative chemical potential, $\mathbf{A}$ is a symmetric second-order tensor conjugate to the internal variable $\mathbf{C}$ describing the way in which the free energy depends on the internal variable:

$$\mathbf{A} = -\frac{\partial f}{\partial \mathbf{C}}. \tag{8.37}$$

There is no term of the form $p\,\mathrm{d}(\rho^{-1})$ in (8.36) because it is assumed that the mass density is constant. By combining relation (8.36) with the balance equations of mass concentration (8.31) and energy (8.33) and comparing with the general evolution equation for the entropy

$$\rho \dot{s} = -\nabla \cdot \boldsymbol{J}^s + \sigma^s, \tag{8.38}$$

one obtains the following results for the entropy flux and the entropy production, respectively:

$$\boldsymbol{J}^s = \frac{1}{T}(\boldsymbol{q} - \mu \boldsymbol{J}), \tag{8.39}$$
$$T\sigma^s = -\boldsymbol{J}^s \cdot \nabla T - \boldsymbol{J} \cdot \nabla \mu + \boldsymbol{\sigma}^{\text{v}} : \mathbf{V} + \rho \mathbf{A} : \dot{\mathbf{C}} \geq 0. \tag{8.40}$$

Expression (8.39) of the entropy flux is classical while the dissipated energy $T\sigma^s$ contains a supplementary contribution compared to the similar relation for a Newtonian fluid. For the time being, limiting ourselves to expressions of the fluxes $\boldsymbol{J}^s, \boldsymbol{J}, \boldsymbol{\sigma}^{\mathrm{v}}$, and $\dot{\mathbf{C}}$, which are linear and isotropic in the forces ∇T, $\nabla \mu, \mathbf{V}$, and $\mathbf{A}$, one is led to write

$$\boldsymbol{J}^s = -\lambda \nabla T - \chi \nabla \mu, \tag{8.41}$$

$$\boldsymbol{J} = -\chi \nabla T - D \nabla \mu, \tag{8.42}$$

$$\boldsymbol{\sigma}^{\mathrm{v}} = 2\eta \mathbf{V} + l\mathbf{A}, \tag{8.43}$$

$$\dot{\mathbf{C}} = k\mathbf{A} - l\mathbf{V} = -k\frac{\partial f}{\partial \mathbf{C}} - l\mathbf{V}, \tag{8.44}$$

wherein $\lambda, \chi, D, \eta, l$, and k are phenomenological coefficients with $\lambda > 0$, $\eta > 0$, $D > 0$, and $k > 0$. The same coefficient χ (respectively, l) appears in (8.41) and (8.42) (respectively, in (8.43) and (8.44)) because of Onsager's reciprocal property, the change of sign in front of l in (8.43) and (8.44) is a consequence of the property that the fluxes $\boldsymbol{\sigma}^{\mathrm{v}}$ and $\dot{\mathbf{C}}$ are of opposite parity with respect to time reversal. The terms $l\mathbf{A}$ in (8.43) and $-l\mathbf{V}$ in (8.44) do not contribute to the entropy production and have therefore been labelled as *gyroscopic* or *non-dissipative*. In most complex fluids, these non-dissipative contributions may also contain non-linear terms (Lhuillier and Ouibrahim 1980) resulting from the coupling of $\mathbf{A}$ (or $\mathbf{V}$) with $\mathbf{C}$ and (8.43) and (8.44) will more generally be of the form:

$$\boldsymbol{\sigma}^{\mathrm{v}} = 2\eta \mathbf{V} + l\mathbf{A} - \beta(\mathbf{C} \cdot \mathbf{A} + \mathbf{A} \cdot \mathbf{C}), \tag{8.45}$$

$$\dot{\mathbf{C}} = k\frac{\partial f}{\partial \mathbf{C}} - l\mathbf{V} + \beta(\mathbf{C} \cdot \mathbf{V} + \mathbf{V} \cdot \mathbf{C}). \tag{8.46}$$

The main results of the model are contained in (8.45) and (8.46) which show that there is a strong interconnection between the way the molecular deformation $\mathbf{C}$ influences the stress tensor and the way the rate of fluid deformation modifies the molecular deformation. A last remark is in form: if it is wished to satisfy the axiom of material frame-indifference (see Chap. 9), the material time derivative in (8.46), which is non-objective, must be replaced by an objective one, as for instance the Jaumann time derivative

$$D_{\mathrm{J}}\mathbf{C} = \dot{\mathbf{C}} + \mathbf{W} \cdot \mathbf{C} - \mathbf{C} \cdot \mathbf{W}, \tag{8.47}$$

with $\mathbf{W}$ the skew-symmetric velocity gradient tensor. Moreover, there will be no difficulty to generalize (8.45) and (8.46) with terms of third or higher orders in $\mathbf{C}, \mathbf{V}$, and $\mathbf{A}$, but we shall see that the present description offers still enough possibilities (see Box 8.2).

Box 8.2 Simplified Models
Several simplifications of the model (8.45) and (8.46) are of interest:

(1) If in relation (8.46) we set $\beta = 1$ and $\ell = 0$, one recovers a model obtained in the kinetic theory. In this case the additional stress defined by (8.35) is simply $\boldsymbol{\sigma}_{\mathrm{pol}} = 2\mathbf{A} \cdot \mathbf{C}$, but the kinetic theory predicts

instead that $\sigma_{\text{pol}} \approx \mathbf{A}$ from which we conclude that the results of the kinetic theory are in contradiction with the IVT. The problem encountered in the kinetic theory finds probably its origin in the difficulties arising in the evaluation of σ_{pol}.

(2) Let us still assume that $l = 0$ but that the free energy f is quadratic in $\mathbf{C}$ so that $\mathbf{A} = -\partial f/\partial \mathbf{C} = -H\mathbf{C}$ where H is some kind of Hookean spring constant; (8.45) and (8.46) read then as

$$\sigma^{\text{v}} = 2\eta \mathbf{V} + 2\beta H \mathbf{C} \cdot \mathbf{C}, \tag{8.2.1}$$

$$D_{\text{J}}\mathbf{C} = -\frac{1}{\tau}\mathbf{C} + \beta(\mathbf{C} \cdot \mathbf{V} + \mathbf{V} \cdot \mathbf{C}). \tag{8.2.2}$$

This model was proved to be at least as valuable as the model of Oldroyd (Bird et al. 1987) widely used in rheology; (8.2.1) exhibits clearly the dependence of the non-Newtonian stress on the Hookean elasticity of the polymer, whereas (8.2.2) is of the relaxation type with $\tau(= 1/kH)$, a positive relaxation time. The physical interpretation of (8.2.2) is rather evident: the presence of the Jaumann time derivative forces the molecules to rotate with the surrounding fluid, the rate of fluid deformation $\mathbf{V}$ deforms the molecules with an efficiency β and the molecules relax towards an equilibrium configurational state with a relaxation time τ.

(3) Finally, by admitting that $\beta = 0$ and f quadratic in $\mathbf{C}$, relation (8.46) reduces to

$$D_{\text{J}}\mathbf{C} = -\frac{1}{\tau}\mathbf{C} - l\mathbf{V}. \tag{8.2.3}$$

Such an expression presents a striking resemblance with an equation derived by Frankel and Acrivos (1970) to describe emulsions of quasi-spherical drops with $\mathbf{C}$ a second-order tensor, expressing the deviation of drops from sphericity.

8.2.3 Colloidal Suspensions

Colloidal suspensions are another example of complex fluids. A colloidal suspension is a collection of particles suspended in a fluid, called the solvent or the carrier. The driven particles must have a size larger than 10^{-8} m for the particles to be considered as a continuum, but smaller than 10^{-4} m to avoid sedimentation. They may be aerosols (fog, smog, smoke), emulsified drops, ink, paint or asphalt particles. The suspending fluid must be sufficiently dense in order that its mean free path be smaller than the size of the particles, thus excluding rarefied gases. These hypotheses allow us to consider the system as made by two interpenetrating continuous media. To simplify the issue, it is generally admitted that the particles are rigid spheres of the same size moving in a Newtonian fluid so that the system may be considered as isotropic.

In the foregoing analysis, we shall omit the role of velocity fluctuations whose treatment can be found in specialized literature, e.g. Lhuillier (1995, 2001).

It is our aim to show that IVT provides a valuable tool for the study of thermal diffusion in suspensions; the supramolecular size of the particles being taken into account by the introduction of an internal variable. After commenting in Sect. 8.2.3.1 the differences between particles suspensions and molecular mixtures, we will present the evolution equations of the relevant parameters and finally discuss the restrictions placed by the second law of thermodynamics.

8.2.3.1 Suspensions vs. Molecular Mixtures

It is true that the suspensions of particles in a fluid behave in several aspects as a binary molecular mixture of solute molecules in a solvent. However, the particle size is larger than the molecular dimensions and this has two important consequences. First, the volume of a particle is well defined whereas the volume of a molecule is not. A description of suspensions will therefore include not only the particle mass fraction $c(=\mathrm{d}M_\mathrm{p}/\mathrm{d}M)$ but also its volume fraction $\phi(=\mathrm{d}V_\mathrm{p}/\mathrm{d}V)$ which is related to c by

$$\phi = \rho c/\rho_\mathrm{p}, \tag{8.48}$$

where ρ is the total mass per unit volume of the suspension

$$\rho = \phi\rho_\mathrm{p} + (1-\phi)\rho_\mathrm{f}, \tag{8.49}$$

the quantity $\rho_\mathrm{p}(=\mathrm{d}M_\mathrm{p}/\mathrm{d}V_\mathrm{p})$ is the mass per unit volume of the material from which the particles are made and $\rho_\mathrm{f}(=\mathrm{d}M_\mathrm{f}/\mathrm{d}V_\mathrm{f})$ the mass per unit volume of pure fluid; from now on, subscripts p and f will refer systematically to "particle" and "fluid", respectively. Moreover, to describe suspensions it is necessary to introduce a volume-weighted velocity $\boldsymbol{u}$ defined by

$$\boldsymbol{u} = \phi\boldsymbol{v}_\mathrm{p} + (1-\phi)\boldsymbol{v}_\mathrm{f}, \tag{8.50}$$

besides the classical barycentric mass-weighted velocity $\boldsymbol{v}$,

$$\boldsymbol{v} = c\boldsymbol{v}_\mathrm{p} + (1-c)\boldsymbol{v}_\mathrm{f}. \tag{8.51}$$

The second difference between suspensions and molecular mixtures is the large inertia of particles from which follows that the average particle velocity $\boldsymbol{v}_\mathrm{p}$ can be very different from the fluid velocity $\boldsymbol{v}_\mathrm{f}$. Hence the importance of the particle diffusion flux

$$\boldsymbol{J} = \rho c(\boldsymbol{v}_\mathrm{p} - \boldsymbol{v}) = \rho c(1-c)(\boldsymbol{v}_\mathrm{p} - \boldsymbol{v}_\mathrm{f}), \tag{8.52}$$

which is no longer given by the Fick's law but will obey a full evolution equation, as shown below. It is not difficult to convince oneself of the relations

$$\boldsymbol{u} = \boldsymbol{v} + (\rho_{\mathrm{p}}^{-1} - \rho_{\mathrm{f}}^{-1})\boldsymbol{J}, \tag{8.53}$$

$$\phi = c + (1-c)(\rho_{\mathrm{p}}^{-1} - \rho_{\mathrm{f}}^{-1}), \tag{8.54}$$

showing that $\boldsymbol{u}$ and ϕ are generally different from $\boldsymbol{v}$ and c respectively except for the special cases that particles and fluid have similar mass densities.

8.2.3.2 State Variables and Evolution Equations

In presence of thermal gradients, the system will exhibit thermodiffusion. Since we are in presence of two phases, a rather natural choice of variables will be mass densities of phases, the average temperature and the velocities of the particles and fluid phase. However, in analogy with molecular diffusion, a more appropriate selection of variables is the following

$$\rho, c, T, \boldsymbol{v}, \xi (= \boldsymbol{v}_{\mathrm{p}} - \boldsymbol{v}_{\mathrm{f}}), \tag{8.55}$$

where the relative velocity will be considered as an internal variable. More refined descriptions will require the introduction of more internal variables as for instance a structural vector $\boldsymbol{R}$ related to the statistical distribution of the particles (Lhuillier 2001) but, for the present purpose, it is sufficient to restrict our analysis to the set (8.55). The evolution equations for the classical variables ρ, c, T, and $\boldsymbol{v}$ are given by

$$\dot{\rho} = -\rho\nabla\cdot\boldsymbol{v}\,(\text{total mass balance}), \tag{8.56}$$

$$\rho\dot{c} = -\nabla\cdot\boldsymbol{J}\,(\text{mass concentration balance}), \tag{8.57}$$

$$\rho\dot{\boldsymbol{v}} = -\nabla p + \nabla\cdot\boldsymbol{\Pi}^{\mathrm{v}} + \rho\mathbf{g}\,(\text{momentum balance}), \tag{8.58}$$

$$\rho\dot{u} = -\nabla\cdot\mathbf{q} - p\nabla\cdot\mathbf{v} + \boldsymbol{\Pi}^{\mathrm{v}} : (\nabla\mathbf{v})^{\mathbf{s}}\,(\text{internal energy balance}), \tag{8.59}$$

where p is the hydrostatic pressure, $\boldsymbol{\Pi}^{\mathrm{v}}$ is the symmetric viscous stress tensor, $(\nabla\boldsymbol{v})^{s}$ the symmetric part of the tensor $\nabla\boldsymbol{v}$, and u the internal energy per unit mass. It remains to determine the evolution equation of the internal variable ξ, i.e. the relative velocity $\boldsymbol{v}_{\mathrm{p}} - \boldsymbol{v}_{\mathrm{f}}$. The latter is obtained from the two momentum equations for the particulate and fluid phases given, respectively, by Lhuillier (1995)

$$\rho c \frac{\mathrm{d}_{\mathrm{p}}\mathbf{v}_{\mathrm{p}}}{\mathrm{d}t} = -\phi(\nabla p - \nabla\cdot\boldsymbol{\sigma}^{\mathrm{v}}) + \mathbf{f} + \rho c\boldsymbol{g}, \tag{8.60}$$

$$\rho(1-c)\frac{\mathrm{d}_{\mathrm{f}}\mathbf{v}_{\mathrm{f}}}{\mathrm{d}t} = -(1-\phi)(\nabla p - \nabla\cdot\sigma^{\mathrm{v}}) - \mathbf{f} - \rho(1-c)\boldsymbol{g}, \tag{8.61}$$

where two new material time derivatives are introduced $\mathrm{d}_{\mathrm{p}}/\mathrm{d}t = \partial/\partial t + \boldsymbol{v}_{\mathrm{p}}\cdot\nabla$ for the particles and a similar expression for the fluid; $\boldsymbol{f}$ is the mutual interaction force between the particles and the fluid measured per unit volume of the suspension. At this stage, $\boldsymbol{f}$ is an unknown quantity whose constitutive equation will be determined later on, $\boldsymbol{\sigma}^{\mathrm{v}}$ is the so-called suspension viscous stress related to the stress $\boldsymbol{\Pi}^{\mathrm{v}}$ by Lhuillier (1995)

$$\boldsymbol{\Pi}^{\mathrm{v}} = \boldsymbol{\sigma}^{\mathrm{v}} - \xi \boldsymbol{J}, \tag{8.62}$$

besides the viscous term $\boldsymbol{\sigma}^{\mathrm{v}}$, one observes the presence of a stress $\xi \boldsymbol{J}$ of kinetic origin which is introduced to recover the global momentum equation (8.58). Making use of the following identity

$$\dot{\xi} = \frac{\mathrm{d_p}\boldsymbol{v}_\mathrm{p}}{\mathrm{d}t} - \frac{\mathrm{d_f}\boldsymbol{v}_\mathrm{f}}{\mathrm{d}t} + \nabla\left[\frac{1}{2}(2c-1)\xi^2\right] + \xi \cdot \nabla \boldsymbol{v}, \tag{8.63}$$

one obtains the following evolution equation for the internal variable ξ:

$$\dot{\xi} = (\rho_\mathrm{f}^{-1} - \rho_\mathrm{p}^{-1})(\nabla p - \nabla \cdot \boldsymbol{\sigma}^{v}) + \frac{\boldsymbol{f}}{\rho c(1-c)} + \nabla\left[\frac{1}{2}(2c-1)\xi^2\right] - \xi \cdot \nabla \boldsymbol{v}. \tag{8.64}$$

Set (8.56)–(8.59), (8.64) contains some indeterminate unknown quantities like $p, u, \boldsymbol{\sigma}^{\mathrm{v}}, \boldsymbol{q}$, and $\boldsymbol{f}$ whose expressions will be determined from thermodynamics and more particularly from the positive definite property of the entropy production

$$\sigma^s = \rho \dot{s} + \nabla \cdot \boldsymbol{J}^s \geq 0. \tag{8.65}$$

Of course at this stage of the discussion, the entropy s and the entropy flux $\boldsymbol{J}^s$ remain undetermined quantities to be expressed by means of constitutive relations. Let us now examine the consequences issued from the positiveness of σ^s.

8.2.3.3 Restrictions Placed by the Second Law of Thermodynamics

The total kinetic energy per unit mass can be written as

$$\frac{1}{2}cv_\mathrm{p}^2 + \frac{1}{2}(1-c)v_\mathrm{f}^2 = \frac{1}{2}v^2 + \frac{1}{2}c(1-c)\xi^2. \tag{8.66}$$

To keep the usual expression $\frac{1}{2}v^2$ for the kinetic energy per unit mass, we shall admit that the part of the kinetic energy involving the relative velocity pertains to the internal energy so that

$$u(s, \rho, c, \xi) = u_0(s, \rho, c) + \frac{1}{2}c(1-c)\xi^2, \tag{8.67}$$

where u_0 is the local equilibrium energy depending exclusively on the set of "equilibrium" variables. The corresponding Gibbs' equation, written in rate form, will therefore be given by

$$\dot{u} = T\dot{s} + \frac{p}{\rho^2}\dot{\rho} + \mu\dot{c} + \frac{\boldsymbol{J}}{\rho} \cdot \dot{\xi}. \tag{8.68}$$

Note that the chemical potential is now including a kinetic contribution and is related to the local equilibrium chemical potential $\mu_0(=\partial u_0/\partial c)$ by

$$\mu = \mu_0 + \frac{1}{2}(1-2c)\xi^2. \tag{8.69}$$

After elimination of $\dot{u}$, $\dot{\rho}$, $\dot{c}$, and $\dot{\xi}$ by means of the evolution equations (8.56)–(8.59) and (8.64), we obtain the following entropy balance

$$T(\rho\dot{s} + \nabla\cdot\boldsymbol{J}^s) \equiv T\sigma^s = \nabla\cdot[T\boldsymbol{J}^s - \boldsymbol{q} - \mu\boldsymbol{J} + (\boldsymbol{u}-\boldsymbol{v})\cdot\boldsymbol{\sigma}^{\mathrm{v}}] + \boldsymbol{\sigma}^{\mathrm{v}} : (\nabla\boldsymbol{u})^{\mathrm{sym}}$$
$$-\boldsymbol{J}^s\cdot\nabla T - \boldsymbol{J}\cdot\{\nabla\mu_0 - (\rho_{\mathrm{p}}^{-1} - \rho_{\mathrm{f}}^{-1})\nabla p + \boldsymbol{f}[\rho c(1-c)]^{-1}\} \geq 0. \tag{8.70}$$

Positiveness of the dissipated energy $T\sigma^s$ requires that the divergence term in (8.70) vanishes, whence the following expression for the entropy flux:

$$\boldsymbol{J}^s = \frac{1}{T}(\boldsymbol{q} - \mu\boldsymbol{J}) - \frac{1}{T}(\boldsymbol{u}-\boldsymbol{v})\cdot\boldsymbol{\sigma}^{\mathrm{v}}. \tag{8.71}$$

The first two terms in (8.71) are classical but a new term depending on the relative velocity and the mechanical stress tensor is appearing. The remaining terms in (8.70) take the form of bilinear products of thermodynamic fluxes and forces. The simplest way to guarantee the positive definite character of the dissipated energy is to assume that these fluxes and forces are related by means of linear relations, i.e.

$$\boldsymbol{\sigma}^{\mathrm{v}} = \eta(\nabla\boldsymbol{u})^s, \tag{8.72}$$

$$\boldsymbol{J}^s = -\frac{\lambda}{T}\nabla T + \tilde{s}\boldsymbol{J}, \tag{8.73}$$

$$\boldsymbol{f} = -\rho c(1-c)\left[\tilde{s}\nabla T - (\rho_{\mathrm{p}}^{-1} - \rho_{\mathrm{f}}^{-1})\nabla p + \nabla\mu_0 + D^{-1}\boldsymbol{J}\right], \tag{8.74}$$

the phenomenological coefficients $\eta, \lambda, \tilde{s}, D$ depend generally on ρ, c, and T, the same $\tilde{s}$ appears in both (8.73) and (8.74) to satisfy the Onsager reciprocal relations. After substitution of the flux–force relations (8.72)–(8.74) in (8.70) of the dissipated energy, one is led to

$$T\sigma^s = \frac{\lambda}{T}(\nabla T)^2 + \frac{1}{D}J^2 + \frac{1}{\eta}\boldsymbol{\sigma}^{\mathrm{v}} : \boldsymbol{\sigma}^{\mathrm{v}} \geq 0, \tag{8.75}$$

from which follows that $\lambda > 0$, $\eta > 0$, $D > 0$, there is no restriction on the sign of $\tilde{s}$. The above results warrant further comments. It is important to note that the stress $\boldsymbol{\sigma}^{\mathrm{v}}$ is related to the gradient of the volume-weighted velocity $\boldsymbol{u}$ rather than to the gradient of the mass-weighted velocity $\boldsymbol{v}$ as in molecular diffusion. This property has been corroborated by microscopic considerations and is a well-known result in the theory of suspensions. Relation (8.73) can be viewed as an expression of the Soret law stating that a temperature gradient is capable of inducing a flux of matter.

The result (8.74) is important as it provides an explicit relation for the inter-phase force $\boldsymbol{f}$ between the particles and the fluid, and plays, for suspensions, the role of Fick's law for binary mixtures. This interaction force will ultimately appear as a sum of elementary forces involving ∇c (through $(\partial\mu_0/\partial c)_{p,T}$) (concentration-diffusion force), ∇p (baro-diffusion force), ∇T (thermodiffusion force) and the relative velocity $\boldsymbol{v}_\mathrm{p} - \boldsymbol{v}_\mathrm{f}$ (through $D^{-1}\boldsymbol{J}$) (kinematic-diffusion force).

Concerning the concentration-diffusion force, it always drives the particles towards regions of lower particles concentration because $(\partial\mu_0/\partial c)_{p,T} > 0$, which is a consequence of thermodynamic stability. Experimental investigations confirm that the concentration-diffusion force is the most important, that the thermodiffusion force is rather small, and that the baro-diffusion force is negligible.

When the two following conditions are satisfied, $\rho_\mathrm{p} = \rho_\mathrm{f}$, and $\mathrm{d}_\mathrm{p}\boldsymbol{v}_\mathrm{p}/\mathrm{d}t = \mathrm{d}_\mathrm{f}\boldsymbol{v}_\mathrm{f}/\mathrm{d}t$, it is found by subtracting (8.60) from (8.61) that the force $\boldsymbol{f}$ vanishes identically. If in addition the temperature and pressure are kept constant, (8.74) boils down to Fick's law $\boldsymbol{J} = -D\nabla\mu_0$, where D is the positive coefficient of diffusion.

Expression (8.74) of $\boldsymbol{f}$ is sometimes decomposed into a "non-dissipative" force $\boldsymbol{f}^*$ (the three first terms under brackets in (8.74)) and a dissipative contribution, namely

$$\boldsymbol{f} = \boldsymbol{f}^* - \gamma(\boldsymbol{v}_\mathrm{p} - \boldsymbol{v}), \tag{8.76}$$

where use is made of the definition (8.52) of $\boldsymbol{J}$ and where γ, called the friction coefficient, stands for $\gamma = \rho^2 c^2(1-c)/D > 0$. The term "non-dissipative" is justified as it corresponds to situations for which $\gamma = 0$, i.e. $D = \infty$, which is typical of absence of dissipation.

The expression of the heat flux vector $\boldsymbol{q}$ is directly derived by eliminating $\boldsymbol{J}^s$ between (8.71) and (8.73); making use of (8.72) and introducing a pseudo-enthalpy function $\tilde{h} = T\tilde{s} + \mu$, it is found that

$$\boldsymbol{q} = -\lambda\nabla T + \tilde{h}\boldsymbol{J} + \eta(\boldsymbol{u} - \boldsymbol{v})\cdot\nabla\boldsymbol{u}. \tag{8.77}$$

For pure heat conduction, one recovers the classical Fourier law so that the coefficient λ can be identified with the heat conductivity. For a molecular mixture for which $\boldsymbol{u} = \boldsymbol{v}$, the above relation is equivalent to the law of Dufour, expressing that heat can be generated by matter transport.

The above analysis shows that internal variables offer a valuable approach of the theory of suspensions. It is worth noticing that the totality of results obtained in this section was also derived in the framework of extended irreversible thermodynamics (Lebon et al. 2007).

8.3 Final Comments and Comparison with Other Theories

Thermodynamics with IVT provides a rather simple and powerful tool for describing structured materials as polymers, suspensions, viscoelastic bodies, electromagnetic materials, etc. As indicated before, its domain of applicability is very wide, ranging from solid mechanics, hydrodynamics, rheology, electromagnetism to physiology or econometrics sciences. IVT requires only a slight modification of the classical theory of irreversible processes by assuming that the non-equilibrium state space is the union of two subsets. The first one is essentially composed by the same variables as in classical irreversible thermodynamics while the second subset is formed by a more or less large set of internal variables that have two main characteristics: first, they cannot be controlled by an external observer and second, they can be unambiguously measured. Furthermore, it is assumed that to any irreversible process, one can associate a fictitious reversible process referred to as the accompanying state. It was also proved that by eliminating one or several internal variables, one is led to generalized constitutive relations taking the form of functional of the histories of the state variables. In that respect, it can be said that the IVT is equivalent to rational thermodynamics (see next chapter).

The main difficulty with IVT is the selection of the number and the identification of the nature of the internal variables. It is true that for some systems, like polymers or suspensions, the physical meaning of these variables can be guessed from the onset, but this is generally not so. In most cases, the physical nature of the internal variables is only unmasked at the end of the procedure. In some problems, like in suspensions, the dependence of the thermodynamic potentials on these extra variables is a little bit "forced". Referring for instance to (8.67) of the internal energy u, it is not fully justified that the dependence of u on the internal variable ξ is simply the sum of the local internal energy and the diffusive kinetic energy. Another problem is related to the time evolution of the internal variables. Except some particular cases, like diffusion of suspensions, there is no general technique allowing us to derive these evolution equations, in contrast with extended irreversible thermodynamics or GENERIC (see Chap. 10). Moreover, as these variables are in principle not controllable through the boundaries, the evolution equations should not contain terms involving the gradients of the variables. This is a limitation of the theory as it excludes in particular the treatment of nonlocal effects. Some efforts have been recently registered to circumvent this difficulty but the problem is not definitively solved. Unlike extended irreversible thermodynamics, where great efforts have been dedicated to a better understanding of the notion of entropy and temperature outside equilibrium, it seems that such questions are not of great concern in internal variables theories. Here, the entropy that is used is the so-called accompanying entropy and it is acknowledged that its rate of production is positive definite whatever the number and nature of internal variables. The validity of such a hypothesis if questionable and should be corroborated by microscopic theories as the kinetic theory. The temperature is formally defined as the derivative of the internal energy with respect to entropy but questions about the definition of a positive absolute temperature and its measurability in systems far from equilibrium are even not invoked. It is expected that the validity of the results of the IVT become more and more accurate as the number of internal variables is increased and would become rigorously valid when the number of variables is infinite; however, from a practical point of view, this limit is of course impossible to achieve.

8.4 Problems

8.1. *Clausius–Duhem's inequality.* Show that the Clausius–Duhem's inequality (8.10) is equivalent to the dissipation inequality (8.9).

8.2. *Chemical reactions.* Using the degree of advancement of a chemical reaction ξ as an internal variable, formulate the problem of the chemical reaction $\mathrm{A} + \mathrm{B} = \mathrm{C} + \mathrm{D}$ in terms of the internal variable theory.

8.3. *Particle suspensions.* Why is the theory of molecular diffusion not applicable to the description of particle suspensions in fluids?

8.4. *Viscoelastic bodies.* Derive the constitutive relation (8.26) of a Poynting–Thomson body by using Liu's Lagrange multiplier technique developed in Chap. 9.

8.5. *Colloidal suspensions.* Establish the evolution equation (8.64) of the internal variable ξ by using (8.60), (8.61), and (8.63).

8.6. *Colloidal suspensions.* Derive (8.74) of the interaction force $\boldsymbol{f}$ between the particles and the fluid.

8.7. *Colloidal suspensions.* Eliminating the entropy flux between (8.71) and (8.73) show that the heat flux in colloidal suspensions is given by

$$\boldsymbol{q} = -\lambda \nabla T + \tilde{h}\boldsymbol{J} + \eta(\boldsymbol{u} - \boldsymbol{v}) \cdot \nabla \boldsymbol{u}.$$

In the particular problem of pure heat conduction, show that the above expression reduces to Fourier law, while for a mixture for which $\rho_{\mathrm{f}} = \rho_{\mathrm{p}}$ (i.e. $\boldsymbol{u} = \boldsymbol{v}$), it is equivalent to Dufour's law.

8.8. *Superfluids.* Liquid He II is classically described by Landau's two-fluid model (see for instance Khalatnikov 1965). Accordingly, He II is viewed as a binary mixture consisting of a normal fluid with a non-zero viscosity and a superfluid with zero viscosity and zero entropy, the basic variables are ρ_{n}, $\boldsymbol{v}_{\mathrm{n}}$, ρ_{s}, $\boldsymbol{v}_{\mathrm{s}}$, respectively, where ρ denotes the mass density and $\boldsymbol{v}$ the velocity field. Show that an equivalent description may be achieved by selecting the relative velocity $\xi = (\rho_{\mathrm{n}}/\rho)(\boldsymbol{v}_{\mathrm{n}} - \boldsymbol{v}_{\mathrm{s}})$ as an internal variable, with the corresponding Gibbs' equation given by

$$T\,\mathrm{d}(\rho s) = \mathrm{d}(\rho u) - g\,\mathrm{d}\rho - \alpha\xi \cdot \mathrm{d}(\rho\xi),$$

wherein $\rho = \rho_{\mathrm{n}} + \rho_{\mathrm{s}}, \alpha = \rho_{\mathrm{s}}/\rho_{\mathrm{n}}$ while $g = u - Ts + p(1/\rho)$ stands for the specific Gibbs' energy (see Lebon and Jou 1983; Mongiovì 1993, 2001).

8.9. *Superfluids.* Superfluid ^{4}He (see Lhuillier et al. 2003) is an ordered fluid of mass per unit volume ρ and momentum per unit volume $\rho\boldsymbol{v}$; the latter is understood as the sum of two contributions: one from the condensate driving the total mass and moving with velocity $\boldsymbol{v}_\mathrm{s}$, the other from elementary excitations of momentum $\boldsymbol{p}$ and zero mass: $\rho\boldsymbol{v} = \rho\boldsymbol{v}_\mathrm{s} + \boldsymbol{p}$. The other original feature of the superfluid is that it manifests itself by a curl-free velocity:

$$\nabla \times \boldsymbol{v}_\mathrm{s} = 0.$$

Following the reasoning of Sect. 8.2, establish that the evolution equation for $\boldsymbol{p}$, considered as an internal variable, is given by

$$\partial\boldsymbol{p}/\partial t + \nabla \cdot [(\boldsymbol{v} + \boldsymbol{c})\boldsymbol{p}] + [\nabla(\boldsymbol{v} + \boldsymbol{c})] \cdot \boldsymbol{p} = -\rho s \nabla T - \rho \nabla \psi^\mathrm{D} - \nabla \cdot \boldsymbol{\tau}^\mathrm{D},$$

where $\boldsymbol{c}$ is the variable conjugate to $\boldsymbol{p}/\rho$, i.e. $[\boldsymbol{c} = -T\partial s/\partial(\boldsymbol{p}/\rho)]$, ψ^D the dissipative part of Gibbs' function $g = \psi + \psi^\mathrm{D}$, $\boldsymbol{\tau}^\mathrm{D}$ the dissipative part of the mechanical stress tensor.

8.10. *Continuous variable.* The internal variable ξ can also take the form of a continuous variable with a Gibbs' equation written as

$$T\,\mathrm{d}s = \mathrm{d}u - p\,\mathrm{d}v - \int \mu(\xi)\mathrm{d}\rho(\xi)\mathrm{d}\xi.$$

If the rate of change of $\rho(x)$ is governed by a continuity equation $\partial\rho/\partial t = -\partial J(\xi)/\partial\xi$, which defines $J(\xi)$ as a flux in the ξ-space, show that the corresponding entropy production reads as

$$T\sigma^s = -\int J(\xi)\partial\mu(\xi)/\partial\xi\,\mathrm{d}\xi \geq 0,$$

suggesting integral phenomenological relations. However, if the internal variable does not change abruptly, it is sufficient to require that only the integrand of the above expression is positive so that, $J(\xi) = -L\partial\rho(\xi)/\partial\xi$.

8.11. *Application to Brownian motion.* In this problem, the internal variable ξ will be identified as the x-component u of the Brownian particle velocity ($\xi = u$), and the density $\rho(\xi)$ represents the velocity distribution which, at equilibrium is the Maxwellian one,

$$f_\mathrm{eq} = \text{constant} \times \exp(-mu^2/2k_\mathrm{B}T).$$

Assume that the potential $\mu(u)$ is of the form

$$\mu(u) = (k_\mathrm{B}T/m)\ln\rho(u) + A(u),$$

where $\mu_\mathrm{eq} = \mu_0$ is independent of u. Combining the two previous relations, determine the explicit expression of $A(u)$. Show that the phenomenological relation can be cast in the form $J = -L[f(u) - (k_\mathrm{B}T/m)\partial f/\partial u]$, where L

is the friction coefficient of the Brownian particles. Combining this result with the continuity relation, establish the Fokker–Planck equation for the Brownian motion

$$\frac{\partial f(u)}{\partial t} = L\left[\frac{\partial f(u)}{\partial u} + \frac{k_{\mathrm{B}}T}{m}\frac{\partial^2 f(u)}{\partial u^2}\right].$$

8.12. *Magnetizable bodies.* In theories of magnetic solids under strain, it is customary to select magnetization $\boldsymbol{\mathcal{M}} = \boldsymbol{\mathcal{B}} - \boldsymbol{\mathcal{H}}$ (with $\boldsymbol{\mathcal{B}}$ the magnetic induction and $\boldsymbol{\mathcal{H}}$ the magnetic field) as field variable and to split the magnetic variables into a reversible and an irreversible contribution, for instance, $\boldsymbol{\mathcal{M}} = \boldsymbol{\mathcal{M}}_{\mathrm{r}} + \boldsymbol{\mathcal{M}}_{\mathrm{i}}, \boldsymbol{\mathcal{H}}_{\mathrm{r}} + \boldsymbol{\mathcal{H}}_{\mathrm{i}}$. However, to describe the complex relaxation process, some authors (Maugin 1999, p 242) have introduced an extra internal variable $\boldsymbol{\mathcal{M}}_{\mathrm{int}}$. With this choice, the entropy production takes the form

$$T\sigma^s = \boldsymbol{\mathcal{H}}_{\mathrm{r}} \cdot \frac{\mathrm{d}\boldsymbol{\mathcal{M}}_{\mathrm{r}}}{dt} + \boldsymbol{\mathcal{H}}_{\mathrm{i}} \cdot \frac{\mathrm{d}\boldsymbol{\mathcal{M}}_{\mathrm{i}}}{dt} + \boldsymbol{\mathcal{H}}_{\mathrm{int}} \cdot \frac{\mathrm{d}\boldsymbol{\mathcal{M}}_{\mathrm{int}}}{dt}.$$

Show that $\boldsymbol{\mathcal{H}}$ satisfies an evolution equation of the Cattaneo type

$$\tau\frac{\mathrm{d}\boldsymbol{\mathcal{H}}}{\mathrm{d}t} + \boldsymbol{\mathcal{H}} = \frac{\tau}{\chi_{\mathrm{m}}}\frac{\mathrm{d}\boldsymbol{\mathcal{M}}}{\mathrm{d}t},$$

where χ_{m} denotes the magnetic susceptibility.

8.13. *Vectorial internal variable and heat transport.* Assume that the entropy of a rigid heat conductor depends on the internal energy u and a vectorial internal variable $\boldsymbol{j}$, i.e. $s(u, \boldsymbol{j})$. a) Obtain the constitutive equation for the time derivative of $\boldsymbol{j}$. b) From this equation, relate $\boldsymbol{j}$ to the heat flux $\boldsymbol{q}$ and express the evolution equation for $\boldsymbol{q}$, assuming, for simplicity, that all phenomenological coefficients are constant; c) Compare this equation with the double-lag equation presented in Problem 7.8. Which conditions are needed to reduce it to the Maxwell-Cattaneo equation? Which form takes the entropy $s(u, \boldsymbol{j})$ when $\boldsymbol{j}$ is expressed in terms of $\boldsymbol{q}$? Compare it with the extended entropy (7.25).

Chapter 9
Rational Thermodynamics

A Mathematical Theory of Materials with Memory

In Chaps. 7 and 8, it was assumed that the instantaneous local state of the system out of equilibrium was characterized by the union of classical variables and a number of additional variables (fluxes in EIT, internal variables in IVT). Only their instantaneous value at the present time was taken into account and their evolution was described by a set of ordinary differential equations. An alternative attitude, followed in the early developments of rational thermodynamics (RT), is to select a smaller number of variables than necessary for an exhaustive description. The price to be paid is that the state of the material body will be characterized not only by the instantaneous value of the variables, but also by their values taken in the past, namely by their history.

In RT, non-equilibrium thermodynamic concepts are included in a continuum mechanics framework. The roots of RT are found in the developments of the rational mechanics. Emphasis is put on axiomatic aspects with theorems, axioms and lemmas dominating the account. Coleman (1964) published the foundational paper and the name "rational thermodynamics" was coined a few years later by Truesdell (1968). RT deals essentially with deformable solids with memory, but it is also applicable to a wider class of systems including fluids and chemical reactions. Its main objective is to put restrictions on the form of the constitutive equations by application of formal statements of thermodynamics. A typical feature of RT is that its founders consider it as an autonomous branch from which it follows that a justification of the foundations and results must ultimately come from the theory itself. A vast amount of literature has grown up about this theory which is appreciated by the community of pure and applied mathematicians attracted by its axiomatic vision of continuum mechanics.

In the present chapter we present a simplified "idealistic" but nevertheless critical version of RT laying aside, for clarity, the heavy mathematical structure embedding most of the published works on the subject.

9.1 General Structure

The basic tenet of rational thermodynamics is to borrow those notions and definitions introduced in classical thermodynamics to describe equilibrium situations and to admit a priori that they remain applicable even very far from equilibrium. In that respect, temperature and entropy are considered as primitive concepts which are a priori assigned to any state. Quoting Truesdell (1984), it is sufficient to know that "temperature is a measure of how hot a body is, while entropy, sometimes called the caloric, represents how much heat has gone into a body from a body at a given temperature".

Similarly, the second law of thermodynamics written in the form $\Delta S \geq \int đQ/T$ and usually termed the Clausius–Duhem's inequality is always supposed to hold. It is utilized as a constraint restricting the range of acceptable constitutive relations. The consequence of the introduction of the history is that Gibbs' equation is no longer assumed to be valid at the outset as in the classical theory of irreversible thermodynamics. Since the Gibbs equation is abandoned, the distinction between state equations and phenomenological relations disappears, everything will be collected under the encompassing word of constitutive equations. Of course, the latter cannot take any arbitrary forms as they have to satisfy a series of axioms, most of them being elevated to the status of principles in the RT literature.

9.2 The Axioms of Rational Thermodynamics

To each material is associated a set of constitutive equations specifying particular properties of the system under study. In RT, these constitutive relations take generally the form of functionals of the histories of the independent variables and are kept distinct from the balance equations. In the present chapter, the latter turn out to be

$$\dot{\rho} = -\rho \nabla \cdot \boldsymbol{v} \quad \text{(mass balance)}, \tag{9.1a}$$

$$\rho \dot{\boldsymbol{v}} = \nabla \cdot \boldsymbol{\sigma} + \rho \boldsymbol{F} \quad \text{(momentum balance)}, \tag{9.1b}$$

$$\rho \dot{u} = -\nabla \cdot \boldsymbol{q} + \boldsymbol{\sigma} : \nabla \boldsymbol{v} + \rho r \quad \text{(internal energy balance)}. \tag{9.1c}$$

As in the previous chapters, a superimposed dot stands for the material time derivative, ρ is the mass density; u, the specific internal energy; $\boldsymbol{v}$, the velocity; and $\boldsymbol{q}$ is the heat flux vector; in rational thermodynamics, it is preferred to work with the symmetric Cauchy stress tensor $\boldsymbol{\sigma}$ instead of the symmetric pressure tensor $\mathbf{P}(= -\boldsymbol{\sigma})$. It is important to observe the presence of the specific body force $\boldsymbol{F}$ in the momentum equation and the term r in the energy balance, which represents the energy supply due to external sources, for instance the energy lost or absorbed by radiation per unit time and unit mass. It must be realized that the body force and the source term are essentially introduced for the self-consistency of the formalism. Contrary to the classical approach, $\boldsymbol{F}$ and r are not quantities which are assigned a priori, but instead the balance laws will be used to "define" them, quoting the rationalists. In other terms, the balance equations of momentum and energy are always ensured as we have two free parameters at our disposal. The quantities $\boldsymbol{F}$ and r do not modify the behaviour of the body and do not impose constraints on the set of variables, but rather, it is the behaviour of the material, which determines them. This is a perplexing attitude, as $\boldsymbol{F}$ and r, although supplied, will always modify the values of the constitutive response of the system.

The principal aim of RT is to derive the constitutive equations characterizing a given material. Of course, these relations cannot take arbitrary forms, as they are submitted to a series of axioms, which place restrictions on them. Let us briefly present and discuss some of these most relevant axioms.

9.2.1 *Axiom of Admissibility and Clausius–Duhem's Inequality*

By "thermodynamically admissible" is understood a process whose constitutive equations obey the Clausius–Duhem's inequality and are consistent with the balance equations. As will see later, the Clausius–Duhem's inequality plays a crucial role in RT. The starting relation is the celebrated Clausius–Planck's inequality, found in any textbook of equilibrium thermodynamics, and stating that between two equilibrium sates A and B, one has

$$\Delta S \geq \int_A^B đQ/T. \tag{9.2}$$

Since the total quantity of heat $đQ$ results from the exchange with the exterior through the boundaries and the presence of internal sources, the above relation may be written as

$$\frac{\mathrm{d}}{\mathrm{d}t}\int_V \rho s \,\mathrm{d}V \geq -\int_\Sigma \frac{1}{T}\boldsymbol{q}\cdot\boldsymbol{n}\,\mathrm{d}\Sigma + \int_V \rho\frac{r}{T}\,\mathrm{d}V, \tag{9.3}$$

where s is the specific entropy, V is the total volume, and $\boldsymbol{n}$ is the outwards unit normal to the bounding surface Σ. In local form, (9.3) writes as

$$\rho\dot{s} + \nabla\cdot\frac{\boldsymbol{q}}{T} - \frac{\rho r}{T} \geq 0. \tag{9.4}$$

It is worth to note that the particular form (9.4) of the entropy inequality is restricted to the class of materials for which the entropy supply is given by $\rho r/T$ and the entropy flux by $\boldsymbol{q}/T$. For a more general expression of the entropy flux, see extended irreversible thermodynamics (Chap. 7). After elimination of r between the energy balance equation (9.1c) and inequality (9.4) and introduction of Helmholtz's free energy $f(=u-Ts)$, one comes out with

$$-\rho(\dot{f}+s\dot{T})+\boldsymbol{\sigma}:\nabla\boldsymbol{v}-\boldsymbol{q}\cdot\frac{\nabla T}{T}\geq 0, \tag{9.5}$$

which is referred to as the Clausius–Duhem's or the *fundamental inequality*. It is easily checked (see Problem 9.1) that the left-hand side of (9.5) represents the rate of dissipated energy $T\sigma^s$ per unit volume when the entropy flux is given by q/T.

9.2.2 Axiom of Memory

If it is admitted that the present is influenced not only by the present state but also by the past history, the constitutive relations will depend on the whole history of the independent variables. If $\varphi(t)$ designates an arbitrary function of time, say the temperature or the strain tensor, its history up to the time t is defined by $\varphi^t=\varphi(t-t')$ with $0\leq t'<\infty$.

The axiom of memory asserts that the behaviour of the system is completely determined by the history of the set of selected independent variables. This means that the free energy, the entropy, the heat flux and the stress tensor, for instance, will be expressed as functionals of the history of the independent variables. Considering the problem of heat conduction in a rigid isotropic material, an example of constitutive equation with memory is Fourier's generalized law

$$\boldsymbol{q}(t)=\int_{-\infty}^{t}\lambda(t-t')\nabla T(t')\mathrm{d}t', \tag{9.6}$$

where $\lambda(t-t')$ is the memory kernel. When this expression is substituted in the energy balance, one obtains an integro-differential equation for the temperature field, after use is made of $\dot{u}=c\dot{T}$ with c the specific heat capacity. If the memory kernel takes the form of an exponential like $(-\lambda/\tau)\exp[-(t-t')/\tau]$, it is left as an exercise (see Problem 9.2) to show that the time derivative of (9.6) is given by

$$\tau\dot{\boldsymbol{q}}=-\boldsymbol{q}-\lambda\nabla T, \tag{9.7}$$

which is the same Cattaneo equation as in EIT. It is important to realize that this result has been obtained by considering only the temperature as single state variable. This is a characteristic of RT where the state space is generally restricted to the classical variables, i.e. mass, velocity (or deformation), and temperature, while the fluxes are expressed in terms of integral constitutive equations containing the whole history of the independent variables. Instead of assuming that $\boldsymbol{q}$ depends on the whole history of temperature field, in practical applications it is assumed that $\boldsymbol{q}$ is a function of ∇T and its higher-order time derivatives. If the memory is very short in time, one may restrict this sequence to a limited number of terms. But even in this case, RT offers an interesting formalism which departs radically from that of classical irreversible thermodynamics.

In most situations, the notion of *fading memory* is also introduced. Accordingly, the distant history has little influence on the present state; although history is often described by an exponentially decreasing function of time, it could take more general forms as a sum of exponentials or of Gaussian memories. However, to avoid heavy mathematical developments, we shall suppose from now on that the materials forget their past experience quasi-instantaneously so that memory effects can be neglected.

In some versions of RT, the description in terms of histories is substituted by the state-process formalism (Noll 1974; Coleman and Owen 1974). Following these authors, to each thermodynamical system is associated a pair formed by an instantaneous state and a process describing the temporal evolution of the state space. The methods, tools and prospects of this theory present similar features with these described in this chapter; an exhaustive analysis of Coleman and Owen's approach can also be found in Silhavy (1997).

9.2.3 Axiom of Equipresence

This axiom states that if a variable is present in one constitutive relation, then there is no reason why it should not be present in all the other constitutive equations, until it is proved otherwise. The condition for the presence or absence of an independent variable is essentially determined by the Clausius–Duhem's inequality. It should be realized that there is no physical justification to such an axiom, which is merely a mathematical convenience in the determination of constitutive relations.

9.2.4 Axiom of Local Action

It is admitted that a material particle is only influenced by its immediate neighbourhood and that it is insensitive to what happens at distant points. Practically, it means that second and higher-order space derivatives are excluded from the constitutive relations. Higher spatial gradients have however been included in some developments of the theory on non-local actions.

9.2.5 Axiom of Material Frame-Indifference

Generally speaking, this axiom claims that the response of a system must be independent of the motion of the observer. As the most general motion of an observer, identified as a rigid coordinate system, is constituted by a translation and a rotation, the axiom implies that the constitutive equations must be invariant under the Euclidean transformation

$$\boldsymbol{x}^*(t) = \mathbf{Q}(t) \cdot \boldsymbol{x}(t) + \boldsymbol{c}(t). \tag{9.8}$$

The quantity $\mathbf{Q}(t)$ is an arbitrary, real, proper orthogonal, time-valued tensor satisfying

$$\mathbf{Q} \cdot \mathbf{Q}^{\mathrm{T}} = \mathbf{Q}^{\mathrm{T}} \cdot \mathbf{Q} = \mathbf{I}, \quad \det \mathbf{Q} = 1, \tag{9.9}$$

$\boldsymbol{c}(t)$ is an arbitrary time-dependent vector; $\boldsymbol{x}(t)$, the position vector of a material point at the present time and $\boldsymbol{x}^*(t)$ is the position occupied after having undergone a rotation (first term in the right-hand side of (9.8)) and a translation (second term in the right-hand side). In the particular case $\mathbf{Q} = \mathbf{I}$ and $\boldsymbol{c}(t) = \boldsymbol{v}_0 t$ with $\boldsymbol{v}_0$ a constant velocity, (9.8) reduces to a Galilean transformation.

When the Euclidean group (9.8) acts on a tensor of rank $n (n = 0, 1, 2)$, the latter is said to be *objective* if it transforms according to

$$a^* = a \quad \text{(objective scalar)}, \tag{9.10a}$$

$$\boldsymbol{a}^* = \mathbf{Q} \cdot \boldsymbol{a} \, \text{(objective vector)}, \tag{9.10b}$$

$$\boldsymbol{A}^* = \mathbf{Q} \cdot \mathbf{A} \cdot \mathbf{Q}^{\mathrm{T}} \quad \text{(objective tensor)}. \tag{9.10c}$$

We directly observe that the velocity $\boldsymbol{v}^* = \mathrm{d}\boldsymbol{x}^*/\mathrm{d}t$ is not an objective vector as it transforms as

$$\boldsymbol{v}^* = \mathbf{Q} \cdot \boldsymbol{v} + \dot{\boldsymbol{Q}} \cdot \boldsymbol{x} + \dot{\boldsymbol{c}},$$

which is not of the form (9.10b) because of the presence of the second and third terms in the right-hand side of the above relation; the same is true for the acceleration, which is not objective. It is easily checked (see Problems 9.3 and 9.4) that the velocity gradient, the angular velocity tensor and the material time derivative of objective vectors or tensors are non-objective; in contrast, quantities like the temperature gradient, the mass density gradient or the symmetric velocity gradient tensor are objective.

We are now in position to propose a more precise formulation of the *principle of material frame-indifference* which rests on the following requirements. First, the primitive variables as temperature, energy, entropy, free energy and energy supply are by essence objective scalars, the body force and the heat flux are objective vectors while the stress tensor is an objective tensor. Second, the constitutive relations are objective, i.e. form invariant with respect to the Euclidean transformation (9.8). Third, the constitutive equations, which reflect the material properties of a body, cannot depend on the angular velocity of the reference frame. To give an example, the Newton's equation of rational mechanics, when formulated in a non-inertial rotating frame, is Euclidean invariant but it depends explicitly on the angular velocity of the frame and therefore, it does not fulfil the axiom of material frame-indifference.

9.3 Application to Thermoelastic Materials

Consider a deformable, anisotropic elastic solid. Under the action of external forces and heating, the material changes its configuration from a non-deformed one with mass density ρ_0 to a deformed state with mass density ρ. The position of the material points is denoted by $\boldsymbol{X}$, in the non-deformed configuration and by $\boldsymbol{x} = \chi(\boldsymbol{X}, t)$, in the deformed one with $\boldsymbol{u} = \boldsymbol{x} - \boldsymbol{X}$ the displacement vector.

Loyal to our principle of simplicity, we shall restrict the analysis to *linear thermoelasticity*, i.e. small deformations and small temperature increments $T - T_0$ with respect to a reference temperature T_0. A more general description implying large deformations is treated in Box 9.1. Within the limit of small deformations, the density ρ remains constant and the balance equations for the displacement vector $\boldsymbol{u}$ and the temperature field read as (Eringen 1967; Truesdell and Toupin 1960)

$$\rho \ddot{\boldsymbol{u}} = \nabla \cdot \boldsymbol{\sigma} + \rho \boldsymbol{F}, \tag{9.11}$$

$$\rho \dot{u} = -\nabla \cdot \boldsymbol{q} + \boldsymbol{\sigma} : \dot{\boldsymbol{\varepsilon}} + \rho r, \tag{9.12}$$

where $\varepsilon = \frac{1}{2}[\nabla \boldsymbol{u} + (\nabla \boldsymbol{u})^{\mathrm{T}}]$ stands for the symmetric strain tensor. These relations contain unknown quantities as u (or equivalently f), $\boldsymbol{\sigma}$ and $\boldsymbol{q}$ which must be specified by constitutive equations, compatible with Clausius–Duhem's inequality. In (9.12), the scalar $\boldsymbol{u}$ (internal energy) should not be confused with the vector $\boldsymbol{u}$ (displacement vector) appearing in (9.11).

A thermoelastic material is defined by the following constitutive equations

$$f = f(T, \nabla T, \boldsymbol{\varepsilon}), \tag{9.13}$$

$$s = s(T, \nabla T, \boldsymbol{\varepsilon}), \tag{9.14}$$

$$\boldsymbol{\sigma} = \boldsymbol{\sigma}(T, \nabla T, \boldsymbol{\varepsilon}), \tag{9.15}$$

$$\boldsymbol{q} = \boldsymbol{q}(T, \nabla T, \boldsymbol{\varepsilon}), \tag{9.16}$$

where use has been made of the axiom of equipresence, s has been included among the constitutive relations as it figures explicitly in Clausius–Duhem's relation. By no means are the above constitutive equations the most general that one could propose but they appear as particularly useful to describe a large class of deformable elastic solids. As a consequence of (9.13) and using the chain differentiation rule, one can write the time derivative of f as

$$\dot{f} = \frac{\partial f}{\partial T}\dot{T} + \frac{\partial f}{\partial(\nabla T)} \cdot \dot{\nabla T} + \frac{\partial f}{\partial \varepsilon} : \dot{\varepsilon}. \tag{9.17}$$

It is left as an exercise (see Problem 9.5) to prove that the Clausius–Duhem's inequality (9.5) will take the form

$$-\rho(\dot{f} + s\dot{T}) + \boldsymbol{\sigma} : \dot{\varepsilon} - \frac{\boldsymbol{q} \cdot \nabla T}{T} \geq 0, \tag{9.18}$$

and, after substitution of (9.17),

$$-\rho\left(s + \frac{\partial f}{\partial T}\right)\dot{T} - \left(\rho\frac{\partial f}{\partial \varepsilon} - \boldsymbol{\sigma}\right) : \dot{\varepsilon} - \rho\frac{\partial f}{\partial \nabla T} \cdot \dot{\nabla T} - \boldsymbol{q} \cdot \frac{\nabla T}{T} \geq 0, \tag{9.19}$$

which is obviously linear in $\dot{T}, \dot{\varepsilon}$, and $\dot{\nabla T}$. Moreover, since there exist body forces and energy supplies that ensure that the balance equations of momentum and energy are identically satisfied, these laws do not impose constraints on $\dot{T}$, $\dot{\varepsilon}$, and $\dot{\nabla T}$, which can therefore take arbitrary prescribed values. In order that the entropy inequality (9.19) holds identically, it is then necessary and sufficient that the coefficient of each time derivative vanishes. As a consequence, it follows that

$$s = -\partial f / \partial T, \tag{9.20a}$$

$$\sigma = \rho \partial f / \partial \varepsilon, \tag{9.20b}$$

$$\partial f / \partial(\nabla T) = 0. \tag{9.20c}$$

It is concluded from (9.20c) that the free energy f does not depend on the temperature gradient and on account of (9.20a) and (9.20b), the same observation holds for the entropy s and the stress tensor $\boldsymbol{\sigma}$ so that (9.13)–(9.15) will take the form

$$f = f(T, \varepsilon), \quad s = s(T, \varepsilon), \quad \boldsymbol{\sigma} = \boldsymbol{\sigma}(T, \varepsilon). \tag{9.21}$$

From (9.20a), (9.20b), and (9.20c), we can write the differential expression

$$\mathrm{d}f = -s\mathrm{d}T + \rho^{-1}\boldsymbol{\sigma} : \mathrm{d}\varepsilon \tag{9.22}$$

or, equivalently,

$$T\,\mathrm{d}s = \mathrm{d}u - \rho^{-1}\boldsymbol{\sigma} : \mathrm{d}\varepsilon, \tag{9.23}$$

which is the Gibbs equation for thermoelastic bodies. Note that this relation has not been assumed as a starting point but has been derived within the formalism.

Expanding f up to the second order in $\boldsymbol{\sigma}$ and $T - T_0$, one obtains, in Cartesian coordinates, and using Einstein's summation convention,

$$f = \frac{1}{2\rho}C_{ijkl}\varepsilon_{ij}\varepsilon_{kl} - \frac{1}{2}\frac{c_\varepsilon}{T_0}(T - T_0)^2 - \frac{1}{\rho}\beta_{ij}\varepsilon_{ij}(T - T_0), \tag{9.24}$$

where C_{ijkl} is the fourth-order tensor of elastic moduli; c_ε, the heat capacity and β_{ij} is the second-order tensor of thermal moduli. In virtue of (9.20a) and (9.20b), the corresponding linear constitutive equations of s and σ_{ij} are given by

$$s = \frac{c_\varepsilon}{T_0}(T - T_0) + \frac{1}{\rho}\beta_{ij}\varepsilon_{ij}, \tag{9.25}$$

$$\sigma_{ij} = C_{ijkl}\varepsilon_{kl} - \beta_{ij}(T - T_0). \tag{9.26}$$

The result (9.26) is the well-known Neumann–Duhamel's relation of thermoelasticity, which simplifies to Hooke's law when the temperature is uniform ($T = T_0$).

Going back to Clausius–Duhem's expression (9.19), the latter reduces to the remaining inequality

$$\boldsymbol{q}(T, \nabla T, \boldsymbol{\varepsilon}) \cdot \nabla T \leq 0, \tag{9.27}$$

which reflects the property that heat flows spontaneously from high to low temperatures. Defining the heat conductivity tensor by

$$\mathbf{K}(T, \boldsymbol{\varepsilon}) = -\left(\frac{\partial \boldsymbol{q}(T, \nabla T, \boldsymbol{\varepsilon})}{\partial(\nabla T)}\right)_{\nabla T = 0} \tag{9.28}$$

and expanding $\boldsymbol{q}$ around $\nabla T = 0$ with T and $\boldsymbol{\varepsilon}$ fixed, one obtains in the neighbourhood of $\nabla T = 0$, i.e. by omitting non-linear terms,

$$\boldsymbol{q}(T, \nabla T, \boldsymbol{\varepsilon}) = \boldsymbol{q}(T, 0, \boldsymbol{\varepsilon}) - \mathbf{K}(T, \boldsymbol{\varepsilon}) \cdot \nabla T. \tag{9.29}$$

Substitution of (9.29) in (9.27) yields,

$$\boldsymbol{q}(T, 0, \boldsymbol{\varepsilon}) \cdot \nabla T - \nabla T \cdot \mathbf{K}(T, \boldsymbol{\varepsilon}) \cdot \nabla T \leq 0. \tag{9.30}$$

Since this relation must be satisfied for all ∇T, it is required that

$$\boldsymbol{q}(T, 0, \boldsymbol{\varepsilon}) = 0, \quad \nabla T \cdot \mathbf{K}(T, \boldsymbol{\varepsilon}) \cdot \nabla T \geq 0. \tag{9.31}$$

From this, it is immediately concluded that in an admissible thermoelastic process, the heat flux is zero when the temperature gradient vanishes and the heat conductivity tensor, which is independent of ∇T, is positive definite. In fact, only the symmetric part $\mathbf{K}^{\mathrm{sym}} = \mathbf{K} + \mathbf{K}^{\mathrm{T}}$ of $\mathbf{K}$ enters the statement (9.31b) to be positive definite, this is so because for the skew part $\mathbf{K}^{\mathrm{skew}} = \mathbf{K} - \mathbf{K}^{\mathrm{T}}$, one has $\nabla T \cdot \mathbf{K}^{\mathrm{skew}} \cdot \nabla T = 0$. It follows from the above considerations that expansion (9.29) becomes

$$\boldsymbol{q}(T, \nabla T, \boldsymbol{\varepsilon}) = -\mathbf{K}(T, \boldsymbol{\varepsilon}) \cdot \nabla T. \tag{9.32}$$

This is a generalization of Fourier's equation in which the thermal conductivity is a function not only of the temperature but also of the strain tensor.

We observe also from (9.32) that in a thermoelastic solid, it is not possible to produce a heat flux by a deformation only, at uniform temperature. This means that the presence of a piezoelectric effect is excluded in thermoelastic bodies.

The generalization of the above results to large deformations is straightforward and is presented in the Box 9.1.

Box 9.1 Finite Deformations in Thermoelastic Materials

It is usual in solid mechanics to introduce the following quantities (e.g. Eringen 1962):

$\mathbf{F} = \nabla\chi$ (deformation tensor, $F_{ij} = \partial x_i/\partial X_j$ in Cartesian coordinates, with x_i and X_j the position in the deformed and non-deformed reference configuration, respectively),
$\mathbf{E} = \frac{1}{2}(\mathbf{F}^{\mathrm{T}} \cdot \mathbf{F} - \mathbf{I})$ (Green symmetric strain tensor),
$\mathbf{L} = \dot{\mathbf{F}} \cdot \mathbf{F}^{-1} = \nabla \boldsymbol{v}$ (strain rate tensor, $(\nabla \boldsymbol{v})_{ij} = \partial v_j/\partial x_i$ in Cartesian coordinates),
$\mathbf{V} = \operatorname{sym}\mathbf{L}$ (symmetric part of $\mathbf{L}$),
$\mathbf{T} = (\rho_0/\rho)\boldsymbol{\sigma} \cdot \mathbf{F}^{-T}$ (first Piola–Kirchhoff stress tensor),
$\mathbf{S} = \mathbf{F}^{-1} \cdot \mathbf{T}$ (symmetric second Piola–Kirchhoff stress tensor),
$\boldsymbol{Q} = (\rho_0/\rho)\mathbf{F}^{-1} \cdot \boldsymbol{q}$ (heat flux measured in the reference non-deformed configuration).

The balance laws of mass, momentum and energy take the form

$$\rho_0/\rho = \det \mathbf{F},$$
$$\rho_0 \dot{\boldsymbol{v}} = \nabla \cdot \mathbf{T} + \rho_0 \boldsymbol{F},$$
$$\rho_0 \dot{u} = -\nabla \cdot \boldsymbol{q} + \mathbf{S} : \dot{\mathbf{E}} + \rho r.$$

The results derived for small deformations are still valid at the condition to replace everywhere $\boldsymbol{\varepsilon}$ by $\mathbf{E}, \boldsymbol{\sigma}$ by $\mathbf{S}$, $\boldsymbol{q}$ by $\boldsymbol{Q}$, and ∇T by $\boldsymbol{G} = \mathbf{F}^{\mathrm{T}} \cdot \nabla T$, the temperature gradient with respect to the reference configuration. For example, the Clausius–Duhem's inequality reads as

$$-\rho_0 \left(\frac{\partial f}{\partial T} + s\right) \dot{T} - \left(\rho_0 \frac{\partial f}{\partial \mathbf{E}} - \mathbf{S}\right) \cdot \dot{\mathbf{E}} - \rho_0 \frac{\partial f}{\partial \boldsymbol{G}} \cdot \dot{\boldsymbol{G}} - \boldsymbol{Q} \cdot \frac{\boldsymbol{G}}{T} \geq 0,$$

leading to the restrictions,

$$\partial f/\partial T = -s, \quad \partial f/\partial \mathbf{E} = \mathbf{S}/\rho_0, \quad \partial f/\partial \boldsymbol{G} = 0,$$

with a generalized Fourier law given by

$$\boldsymbol{Q}(T, \boldsymbol{G}, \mathbf{E}) = -\mathbf{K}(T, \mathbf{E}) \cdot \boldsymbol{G}.$$

Although most of the applications of RT have been devoted to solid mechanics, there is no problem to adapt the above considerations to hydrodynamics, as shown in Sect. 9.4.

9.4 Viscous Heat Conducting Fluids

To avoid lengthy mathematical developments with constitutive equations in functional form, we shall consider a particular class of fluids characterized by isotropy, absence of memory and described by the following set of constitutive equations:

$$f = f(v, T, \mathbf{V}, \nabla T), \tag{9.33}$$

$$s = s(v, T, \mathbf{V}, \nabla T), \tag{9.34}$$

$$\boldsymbol{\sigma} = \boldsymbol{\sigma}(v, T, \mathbf{V}, \nabla T), \tag{9.35}$$

$$\boldsymbol{q} = \boldsymbol{q}(v, T, \mathbf{V}, \nabla T), \tag{9.36}$$

wherein $v = \rho^{-1}$ is the specific volume, $\mathbf{V} = \frac{1}{2}[\nabla \boldsymbol{v} + (\nabla \boldsymbol{v})^{\mathrm{T}}]$ the symmetric velocity gradient tensor. Note that neither the velocity field $\boldsymbol{v}$ nor the velocity gradient $\nabla \boldsymbol{v}$ have been included in the set of independent variables, because they are not objective quantities and therefore do not satisfy the axiom of frame-indifference. Instead, the tensor $\mathbf{V}$ has been selected as it meets the property to be objective. Observe in passing that the above response functions are in full agreement with the axiom of equipresence. More restrictions on the constitutive relations are placed by Clausius–Duhem's inequality (9.5). Introducing (9.33)–(9.36) in inequality (9.5) and applying the chain differentiation rule to calculate $\dot{f}$, one obtains

$$-\rho\left(s + \frac{\partial f}{\partial T}\right)\dot{T} - \rho\frac{\partial f}{\partial \mathbf{V}} : \dot{\mathbf{V}} - \rho\frac{\partial f}{\partial(\nabla T)}\cdot\dot{(\nabla T)} - \frac{1}{T}\boldsymbol{q}\cdot\nabla T - \left(\frac{\partial f}{\partial v}\mathbf{I} - \boldsymbol{\sigma}\right) : \mathbf{V} \geq 0, \tag{9.37}$$

wherein use has been made of the continuity equation $\rho\dot{v} = \nabla \cdot \boldsymbol{v} = \mathbf{V} : \mathbf{I}$. On looking at inequality (9.37), we see that it is a linear expression in the time derivatives $\dot{T}$, $\dot{\mathbf{V}}$, and $\dot{(\nabla T)}$; if it is assumed that there are body forces and energy supply that ensure that the momentum and energy balance equations are identically satisfied, one can assign to these derivatives arbitrary and independent values. Clearly inequality (9.37) will not hold unless the coefficients of these derivatives are zero, which leads to the following results:

$$s = -\frac{\partial f}{\partial T}, \tag{9.38}$$

$$\frac{\partial f}{\partial \mathbf{V}} = 0, \tag{9.39}$$

$$\frac{\partial f}{\partial(\nabla T)} = 0. \tag{9.40}$$

Relation (9.38) is classical and from the next ones, it is deduced that the free energy f (and as a corollary the entropy s) is a function of v and T alone so that

$$f = f(v, T), \quad s = s(v, T). \tag{9.41}$$

Defining the equilibrium pressure p by $p = -\partial f/\partial v$, which is justified as p can only depend on the "equilibrium" variables v and T, and combining with (9.38), one finds the following Gibbs equation:

$$\mathrm{d}f = -s\,\mathrm{d}T - p\,\mathrm{d}v. \tag{9.42}$$

It is important to notice that, in contrast with classical irreversible thermodynamics where the Gibbs relation is *postulated* from the outset, in RT it is a *derived* result.

Furthermore, in virtue of the results (9.38)–(9.40), the Clausius–Duhem's inequality reduces to

$$\left(-\frac{\partial f}{\partial v}\mathbf{I} + \boldsymbol{\sigma}\right) : \mathbf{V} - \frac{1}{T}\boldsymbol{q} \cdot \nabla T \geq 0. \tag{9.43}$$

Introducing the viscous stress tensor $\boldsymbol{\sigma}^{(\mathrm{v})}$ defined by

$$\boldsymbol{\sigma}^{(\mathrm{v})}(v, T, \mathbf{V}, \nabla T) = \boldsymbol{\sigma}(v, T, \mathbf{V}, \nabla T) + p(v, T)\mathbf{I}, \tag{9.44}$$

(9.43) is written as

$$\boldsymbol{\sigma}^{(\mathrm{v})} : \mathbf{V} - \boldsymbol{q} \cdot \frac{\nabla T}{T} \geq 0, \tag{9.45}$$

and represents the rate of energy dissipated per unit volume of the fluid.

Explicit expressions for the constitutive equations for $\boldsymbol{q}$ and $\boldsymbol{\sigma}^{(\mathrm{v})}$ are directly obtained by using the representation theorems for isotropic tensors (Truesdell and Toupin 1960). In the linear approximation, when second and higher-order terms in ∇T and $\mathbf{V}$ are omitted, it is left as an exercise (Problem 9.7) to prove that

$$\boldsymbol{q} = -\lambda(v, T)\nabla T, \quad \boldsymbol{\sigma}^{(\mathrm{v})} = \gamma(v, T)(\nabla \cdot \boldsymbol{v})\mathbf{I} + 2\eta(v, T)\mathbf{V}. \tag{9.46}$$

One recognizes the Fourier law with λ the heat conductivity coefficient and the Newton's law of hydrodynamics, with η the dynamical shear viscosity and the coefficient γ related to the bulk viscosity; of course, the scalars λ, γ, and η are all three functions of v and T. The fluids described by the linear constitutive relations (9.46) are usually referred to as Fourier–Stokes–Newton fluids. After substitution of (9.46) in (9.45), it is observed that satisfaction of Clausius–Duhem's relation yields the following inequalities, which are well known in fluid mechanics:

$$\lambda > 0, \quad \gamma + \frac{2}{3}\eta > 0, \quad \eta > 0. \tag{9.47}$$

Summarizing, we can assert that, for the class of fluids described by constitutive equations of the form $\phi = \phi(v, T, \nabla T, \mathbf{V})$ with $\phi \equiv f, s, \boldsymbol{q}, \boldsymbol{\sigma}$, respectively:

1. The thermodynamic potentials f and s do not depend on ∇T and $\mathbf{V}$, so that $f = f(v, T)$ and $s = s(v, T)$.
2. The classical Gibbs equation (9.42) or equivalently $T\mathrm{d}s = \mathrm{d}u + p\,\mathrm{d}v$ is demonstrated to remain valid.
3. In the linear approximation, the constitutive equations for $\boldsymbol{q}$ and $\boldsymbol{\sigma}^{(\mathrm{v})}$ are the traditional equations of Fourier and Stokes–Newton, respectively.
4. As a side result of Clausius–Duhem's inequality, the heat conductivity and the viscosity coefficients are shown to be positive.

Despite the fact that rational thermodynamics radiates some taste of elegance and generality, it has been the subject of acrid criticisms, which are discussed in Sect. 9.5.

9.5 Comments and Critical Analysis

The axiomatic approach of RT has been the subject of severe critical observations (e.g. Lavenda 1979; Woods 1981; Rivlin 1984) for the lack of physical background and touch with experiments. In what follows, we shall shortly discuss the most frequently criticisms addressed against the formalism.

9.5.1 The Clausius–Duhem's Inequality

The basic idea is to use the Clausius–Duhem's inequality to place restrictions on the form of the constitutive equations. However, the original formulation of Clausius–Duhem's inequality is restricted to processes taking place between two *equilibrium states*. In RT, it is applied, without any justification, to arbitrary processes taking place between *non-equilibrium states*.

Furthermore, the local Clausius–Duhem's inequality amounts to admit that the positiveness of the entropy production σ^s is a necessary and sufficient condition allowing to restrict the range of acceptable constitutive relations. However, it is not proved that $\sigma^s \geq 0$ remains valid when truncated expressions of σ^s are used, as is generally the case in RT. By performing a series expansion as $\sigma^s = \sigma_1 + \sigma_2 + \sigma_3 + \cdots$, where $\sigma_j (j = 1, 2, 3, \ldots)$ is the entropy production at the j order, it is absolutely not ensured that the entropy production remains positive definite at any order of approximation.

9.5.2 Axiom of Phlogiston

The term *phlogiston* is borrowed from Woods (1981) and it designates a negative mass medium permeating all bodies and expelled by heat. The axiom concerns the property that in any process, time derivatives such as $\dot{T}$, $\nabla \dot{T}$, $\dot{\mathbf{V}}$ can be given arbitrary and independent values of $T, \nabla T$, and $\mathbf{V}$. In normal circumstances, in the balance equations of momentum and energy, the body force $\boldsymbol{F}$ and the source term r are known and take well-specified values: the deformation (or velocity) and temperature fields are then determined by solving the equations after that initial and boundary conditions are given. In RT, the procedure is reversed: for any value of the velocity $\boldsymbol{v}$ (or deformation $\boldsymbol{u}$) and the temperature T fields, it is admitted that one can select appropriately $\boldsymbol{F}$ and r so that the equations of momentum and energy are identically satisfied and do not impose any restriction on the set of variables. In a real experience, $\boldsymbol{F}$ and r are specified by physical rules, which are beyond our control and therefore they cannot be specified throughout the medium and in that respect, it can be said that the phlogiston axiom destroys the empirical content of the balance equations of momentum and energy. The presence of quantities $\boldsymbol{F}$ and r is required to guarantee that the state variables and their time derivatives can be varied independently. This limits seriously the domain of applicability of the theory as it fails to describe processes for which the variables are not independent of their time variations.

The problems linked with the use of the phlogiston axiom can be however circumvented thanks to an elegant technique proposed by Liu (1972) and outlined in Appendix 1.

9.5.3 The Meaning of Temperature and Entropy

It is important to realize that in a majority of works on RT, entropy and temperature are considered as primitive undefined objects and their physical meaning is not a subject of deep concern. Regarding the entropy, it is simply given by a constitutive relation expressing its functional dependence with respect to the selected variables while the temperature T remains an undefined variable which is specified by the laws it satisfies. Quoting Truesdell (1968), "As for physical meaning, I claim no physical applicability for anything I ever say... Whether a theory applies to a given piece of material at a given time is something very important, but something that the theorist cannot be expected to tell, in thermodynamics or any other theory." It must however be added that in the later developments (e.g. Day 1972; Coleman and Owen 1974; Serrin 1979; Coleman et al. 1981; Kratochvil and Silhavy 1982), more attention has been paid to prove the existence of an entropy as a state function as well as an absolute non-equilibrium temperature. The existence of entropy and absolute temperature is no longer postulated but deduced from statements involving cyclic processes; nevertheless, the fundamental problem of an unambiguous definition of temperature and entropy outside equilibrium remains called into question. For example, some people have questioned the measurability of the variable T introduced in the theory: a priori there should be no reason to identify this T with the temperature given by a gas thermometer or a thermocouple. Recent investigations (Muschik 1977; Casas-Vázquez and Jou 2003; Crisanti and Ritort 2003) have brought out that a precise definition of temperature and entropy is of prime importance in non-equilibrium thermodynamics. In addition, it was proved that the entropy used in rational thermodynamics is not unique. Meixner (1973a, 1973b) has shown that there exists an infinity of functionals, all deserving the name of entropy, that possess the property to satisfy the Clausius–Duhem's inequality and that there is no criterion that permits to favour one definition over the others. Meixner's arguments were reinforced by a more particular analysis by Day (1977) who demonstrated the non-uniqueness of entropy in the case of heat conduction in a rigid isotropic body with memory. Another example of a system having many different types of entropy was given by Coleman and Owen (1975) and concerns elastic–plastic materials. The failure of the entropy to be not unique is not eliminated in the state-process version of the theory.

9.5.4 Axiom of Frame-Indifference

This axiom, that requires the invariance of constitutive relations under time-dependent accelerations and rotations of the actual reference frame, has been the subject of intense debate and controversy. As first observed by Müller (1972) and Edelen and McLennan (1973), the axiom excludes physical processes which are Galilean invariant. Frame-indifference is not satisfied in a whole series of disciplines; in particular in rational mechanics. Newton's law of motion formulated in a non-inertial system is "objective" but not frame dependent as the non-inertial forces depend explicitly on the angular velocity of the reference frame; in the kinetic theory of gases, the Burnett equations generalizing the classical Fourier–Stokes–Newton's laws have also been shown to be frame dependent; in the theory of turbulence, it is experimentally observed that the turbulent viscosity takes different values according reference frame is inertial or non-inertial (Lumley 1983). Further violations of material indifference are found in rheology (Bird and de Gennes 1983) and molecular hydrodynamics (Hoover et al. 1981). Another illustration of non-respect of frame-indifference is provided by the classical theory of irreversible processes: referring to Chap. 2, we know that the phenomenological coefficients $L_{\alpha\beta}$ are depending on the angular velocity when measured in a rotating frame. It can be said that actually, there are serious evidences against the universality of the axiom of frame-indifference which has to be regarded as "a convenience rather than a principle", as concluded by Edelen and McLennan (1973). Notwithstanding, axiom of frame-indifference is widely adopted throughout rational thermodynamics and has disturbed people working in continuum mechanics. Fortunately, most results of continuum mechanics are established in inertial frames so that the effects of non-inertial forces can be ignored.

9.5.5 The Entropy Flux Axiom

It is also largely admitted in rational thermodynamics that the entropy flux is given by $\boldsymbol{J}^s = \boldsymbol{q}/T$ (plus eventually a term $-\mu \boldsymbol{J}/T$ in presence of matter diffusion) as one may expect from the Clausius–Planck's formula. Although this result is correct in the linear approximation, this is certainly not true in higher-order theories as confirmed by the kinetic theory of gases (Grad 1958) or extended irreversible thermodynamics. Referring to Chap. 7, for a wide class of processes, the entropy flux for pure substances is not the heat flux divided by the temperature but of the more general form

$$\boldsymbol{J}^s = \frac{\boldsymbol{q}}{T} - \frac{1}{3}\beta' \,(\mathrm{tr}\,\boldsymbol{\sigma}^{\mathrm{v}})\boldsymbol{q} - \beta'' \overset{0}{\boldsymbol{\sigma}}{}^{\mathrm{v}} \cdot \boldsymbol{q}, \tag{9.48}$$

wherein β' and β'' are phenomenological coefficients and $\overset{0}{\boldsymbol{\sigma}}{}^{\mathrm{v}}$ is the deviatoric part of the viscous stress tensor $\boldsymbol{\sigma}^{\mathrm{v}}$. This result is one of the most important differences between rational thermodynamics and extended irreversible thermodynamics.

9.5.6 The Axiom of Equipresence

This axiom states that all constitutive relations depend precisely on the same set of variables, unless it is proved otherwise. As a matter of fact, the conclusion always shows that the response functions do not depend generally on the whole set of variables. Of course, there is no physical argument for such an axiom, which is essentially a mathematical convenience while formulating constitutive equations. It is certainly overweening to elevate equipresence to the status of axiom or principle because it represents merely a technical commodity.

To conclude, RT is an axiomatic theory characterized by generality and mathematical elegance. Unfortunately, mathematical rigour has been obtained at the detriment of physical insight and this explains some lack of interest from some corporations of physicists and engineers. Nevertheless, it is our opinion that it is a theory that deserves a close attention. As mentioned earlier, some of the criticisms should be moderated and some of them are even avoidable, for instance the phlogiston can be circumvented by appealing to Liu's technique while other criticisms, like these addressed again the axiom of equipresence, are of minor consequence. It was the merit of RT to be the first non-equilibrium formalism to get rid of the local equilibrium hypothesis and to go beyond the linear regime to which is restricted the classical theory of non-equilibrium processes. RT has been applied to a huge number of problems mainly in the fields of non-linear elasticity, coupled mechanical, thermal and electro-magnetic phenomena, rheology, wave propagation, and shock waves. Most of them are published in the "Archive of Rational Mechanics and Analysis" which is the privileged tribune of "rationalists". Moreover, the tools of RT have been widely applied in other formalisms as rational extended thermodynamics (Müller and Ruggeri 1998), theories with internal variables (Maugin 1999) or GENERIC (Öttinger 2005).

Appendix 1: Liu's Lagrange Multipliers

An elegant alternative to the admissibility axiom of rational thermodynamics, i.e. the necessary and sufficient conditions to satisfy the Clausius–Duhem's inequality, was proposed by Liu (1972). He was able to show that the entropy inequality (9.5) becomes valid for completely arbitrary variations of the variables at the condition to consider the balance equations of mass, momentum and energy as mathematical constraints. To be more explicit, each balance equation is multiplied by an appropriate factor, named *Lagrange multiplier* by analogy with the extremization problem in mathematics, and the resulting vanishing quantity is added to the left-hand side of the entropy inequality (9.5). Let us illustrate the procedure by means of the simple example of heat conduction in a rigid isotropic body without source term. The behaviour of the temperature field is governed by one single field equation, namely the energy balance

$$\rho \dot{u} = -\nabla \cdot \boldsymbol{q}, \tag{9.49}$$

it being understood that there are restrictions placed by the entropy inequality

$$\rho \dot{s} + \nabla \cdot \left(\frac{\boldsymbol{q}}{T} \right) \geq 0. \tag{9.50}$$

Closure relations are provided by the following set of constitutive equations:

$$u = u(T, \nabla T), \quad s = s(T, \nabla T), \quad \boldsymbol{q} = \boldsymbol{q}(T, \nabla T). \tag{9.51}$$

The requirement that the temperature field satisfying the entropy inequality (9.50) must also be a solution of the energy balance equation (9.49) is interpreted by Liu as a constraint. It was proved by Liu that this constraint can be eliminated by the introduction of *Lagrange multipliers* and by writing the entropy inequality under the new form

$$\rho \dot{s} + \nabla \cdot \left(\frac{\boldsymbol{q}}{T} \right) - \Lambda(\rho \dot{u} + \nabla \cdot \boldsymbol{q}) \geq 0; \tag{9.52}$$

the quantity Λ designates the Lagrange multiplier, which depends generally on the whole set of variables and must be determined from the formalism. After that the constitutive relations (9.51) are inserted in inequality (9.52) and all differentiations are performed, one obtains a relation that is linear in the arbitrary derivatives $\dot{T}$, ∇T, $\nabla(\nabla T)$:

$$\rho\left(\frac{\partial s}{\partial T}-\Lambda\frac{\partial u}{\partial T}\right)\dot{T}+\rho\left(\frac{\partial s}{\partial(\nabla T)}-\Lambda\frac{\partial u}{\partial(\nabla T)}\right)\cdot(\dot{\nabla T})$$
$$+\left(\frac{1}{T}-\Lambda\right)\frac{\partial \boldsymbol{q}}{\partial(\nabla T)}:\nabla(\nabla T)+\left[\left(\frac{1}{T}-\Lambda\right)\frac{\partial \boldsymbol{q}}{\partial T}-\frac{1}{T^2}\boldsymbol{q}\right]\cdot\nabla T\geq 0. \quad (9.53)$$

Since this inequality could be violated unless the coefficients of these derivatives vanish, it is found that:

$$\frac{\partial s}{\partial T}-\Lambda\frac{\partial u}{\partial T}=0,\quad \frac{\partial s}{\partial(\nabla T)}-\Lambda\frac{\partial u}{\partial(\nabla T)}=0,\quad \Lambda=\frac{1}{T}. \tag{9.54}$$

An important result is the third relation from which it is concluded that the Lagrange multiplier can be identified as the inverse of the temperature. Moreover if it is recalled that the free energy is defined as $f=u-Ts$, it follows from the first and the second relation (9.54) that

$$\frac{\partial f}{\partial T}=-s,\quad \frac{\partial f}{\partial(\nabla T)}=0. \tag{9.55}$$

This implies that f is independent of ∇T and that f satisfies the classical Gibbs relation $\mathrm{d}f=-s\mathrm{d}T$. Another consequence of (9.55) is that the entropy s is also independent of the temperature gradient; this property is also shared by the internal energy u, in virtue of the definition of f.

There still remains from (9.53) the residual inequality

$$\boldsymbol{q}\cdot\nabla T\leq 0. \tag{9.56}$$

Clearly, the simplest way to guarantee that (9.56) is negative definite is to assume that

$$\boldsymbol{q}=-\lambda(T)\nabla T\quad\text{with}\quad\lambda\geq 0, \tag{9.57}$$

and we are back with Fourier's law. From now on, the procedure is classical. Replacing (9.57) in the energy balance equation (9.49) and writing for $u(T)$ a constitutive equation such that $\dot{u}=c\dot{T}$, with c the specific heat capacity, one is led to the diffusion equation $c\dot{T}=\nabla\cdot(\lambda\nabla T)$ which, after that initial and boundary conditions are specified, allows us to determine the temperature distribution in the body.

Appendix 2: Rational Extended Thermodynamics

The formulation of EIT, in Chap. 7, was inspired by the concepts and methods of CIT. But EIT may equivalently be described by making use of the tools and structure of rational thermodynamics: it is then referred to as rational extended thermodynamics.

As an illustration, consider the problem of hyperbolic heat conduction in a rigid isotropic body, in absence of an energy source term. The relevant variables are the internal energy u and the heat flux $\boldsymbol{q}$. The time evolution of u is governed by the balance law of energy (9.49) while the evolution equation of $\boldsymbol{q}$ will be cast in the general form

$$\rho\dot{\boldsymbol{q}} = -\nabla\cdot\mathbf{Q} + \boldsymbol{\sigma}^q, \tag{9.58}$$

where $\mathbf{Q}$ (a second-order tensor) denotes the flux of the heat flux and $\boldsymbol{\sigma}^q$ (a vector) is a source term. These quantities must be formulated by means of constitutive equations which, for simplicity, will be given by

$$\mathbf{Q} = \mathbf{Q}(u, \boldsymbol{q}) = a(u, q^2)\mathbf{I}, \quad \boldsymbol{\sigma}^q = \boldsymbol{\sigma}^q(u, \boldsymbol{q}) = b(u, q^2)\boldsymbol{q}, \tag{9.59}$$

wherein the scalars $a(u, q^2)$ and $b(u, q^2)$ are unknown functions of u and $\boldsymbol{q}^2$ to be determined.

Following Liu's technique, the entropy inequality will be formulated in such a way that the constraints imposed by the energy balance and (9.58) are explicitly introduced via the Lagrange multipliers $\Lambda_0(u, \boldsymbol{q})$ (a scalar) and $\boldsymbol{\Lambda}_1(u, \boldsymbol{q})$ (a vector), so that

$$\rho\dot{s} + \nabla\cdot\boldsymbol{J}^s - \Lambda_0(\rho\dot{u} + \nabla\cdot\boldsymbol{q}) - \Lambda_1\cdot(\rho\dot{\boldsymbol{q}} + \nabla\cdot\mathbf{Q} - \boldsymbol{\sigma}^q) \geq 0, \tag{9.60}$$

where $s(u, q^2)$ and $\boldsymbol{J}^s(u, \boldsymbol{q})$ are arbitrary functions of u and $\boldsymbol{q}$. By differentiating s and $\boldsymbol{J}^s$ with respect to u and $\boldsymbol{q}$, and rearranging the various terms, one may rewrite (9.60) as:

$$\begin{aligned}\rho\left(\frac{\partial s}{\partial u} - \Lambda_0\right)\dot{u} + \rho\left(2\frac{\partial s}{\partial q^2}\boldsymbol{q} - \Lambda_1\right)\cdot\dot{\boldsymbol{q}} + \frac{\partial \boldsymbol{J}^s}{\partial u}\cdot\nabla u + \frac{\partial \boldsymbol{J}^s}{\partial \boldsymbol{q}} : \nabla\boldsymbol{q}\\ -\Lambda_0\nabla\cdot\boldsymbol{q} - \Lambda_1\cdot\nabla u\frac{\partial a}{\partial u} - 2\boldsymbol{\Lambda}_1\boldsymbol{q} : \nabla\boldsymbol{q}\frac{\partial a}{\partial q^2} + b\boldsymbol{\Lambda}_1\cdot\boldsymbol{q} \geq 0.\end{aligned} \tag{9.61}$$

Since inequality (9.61) is linear in the arbitrary derivatives $\dot{u}$, $\dot{\boldsymbol{q}}$, ∇u, $\nabla\boldsymbol{q}$, positiveness of (9.61) requires that their respective factors vanish, from which it results that

$$\frac{\partial s}{\partial u} = \Lambda_0(\equiv\theta^{-1}), \quad 2\frac{\partial s}{\partial q^2}\boldsymbol{q} = \boldsymbol{\Lambda}_1(\equiv\gamma(u, q^2)\boldsymbol{q}) \tag{9.62}$$

and

$$\frac{\partial \boldsymbol{J}^s}{\partial u} = \frac{\partial a}{\partial u}\boldsymbol{\Lambda}_1, \quad \frac{\partial \boldsymbol{J}^s}{\partial \boldsymbol{q}} = \Lambda_0\mathbf{I} + 2\Lambda_1\boldsymbol{q}\frac{\partial a}{\partial q^2}, \tag{9.63}$$

wherein we have identified Λ_0 with θ^{-1}, the inverse of a non-equilibrium temperature (see Box 7.4) and, without loss of generality, $\boldsymbol{\Lambda}_1$ with $\gamma(u, q^2)\boldsymbol{q}$ where γ is an arbitrary function of u and q^2.

Taking into account of the results (9.62) and (9.63), the entropy inequality (9.61) reduces to the residual inequality

$$b\boldsymbol{\Lambda}_1\cdot\boldsymbol{q} \geq 0. \tag{9.64}$$

The fact that the entropy flux is an isotropic function implies that

$$\boldsymbol{J}^s = \varphi(u, q^2)\boldsymbol{q}, \tag{9.65}$$

which, substituted in (9.63), yields

$$\frac{\partial \varphi}{\partial u} = \gamma \frac{\partial a}{\partial u} \quad \text{(a)} \quad \text{and} \quad (\varphi - \theta^{-1})\mathbf{I} + 2\left(\frac{\partial \varphi}{\partial q^2} - \gamma \frac{\partial a}{\partial q^2}\right)\boldsymbol{qq} = 0 \quad \text{(b)}. \tag{9.66}$$

Since the dyadic product $\boldsymbol{qq}$ is generally not zero, it follows that

$$\frac{\partial \varphi}{\partial q^2} = \gamma \frac{\partial a}{\partial q^2} \quad \text{(a)} \quad \text{and} \quad \varphi = \theta^{-1} \quad \text{(b)}. \tag{9.67}$$

The second result (9.67) is important as it indicates that φ is equal to the inverse of the temperature and as a consequence that the entropy flux (9.65) is given by the usual expression

$$\boldsymbol{J}^s = \boldsymbol{q}/\theta. \tag{9.68}$$

Moreover, in virtue of (9.66a) and (9.66b), one has

$$\mathrm{d}\theta^{-1} \equiv \mathrm{d}\varphi = \frac{\partial \varphi}{\partial u}\mathrm{d}u + \frac{\partial \varphi}{\partial q^2}\mathrm{d}q^2 = \gamma \mathrm{d}a, \tag{9.69}$$

a result that will be exploited to obtain the final expression of the evolution equation of the heat flux. Indeed, after making use of the results $\boldsymbol{\sigma}^q = b\boldsymbol{q}$ and $\nabla \cdot \mathbf{Q} = \nabla a = \gamma^{-1}\nabla\theta^{-1}$ in (9.58), it is found that

$$\rho\dot{\boldsymbol{q}} = -\frac{1}{\gamma}\nabla\theta^{-1} + b\boldsymbol{q}. \tag{9.70}$$

Dividing both members of (9.70) by b and setting $\rho/b = -\tau$, $(\gamma b)^{-1} = L$, one recovers a *Cattaneo-type relation*

$$\tau\dot{\boldsymbol{q}} = L\nabla\theta^{-1} - \boldsymbol{q}, \tag{9.71}$$

wherein τ can be identified as a relaxation time and L as a generalized heat conductivity. In terms of these coefficients, the Lagrange multiplier Λ_1 is given by

$$\boldsymbol{\Lambda}_1 = \gamma\boldsymbol{q} = -(\tau/\rho L)\boldsymbol{q}. \tag{9.72}$$

Let us finally derive the *Gibbs equation*, it follows directly from (9.62) that

$$\mathrm{d}s = \frac{\partial s}{\partial u}\mathrm{d}u + \frac{\partial s}{\partial q^2}\mathrm{d}q^2 = \frac{\partial s}{\partial u}\mathrm{d}u + \boldsymbol{\Lambda}_1 \cdot \mathrm{d}\boldsymbol{q} = \theta^{-1}\mathrm{d}u - \left(\frac{\tau}{\rho L}\right)\boldsymbol{q}\cdot\mathrm{d}\boldsymbol{q}, \tag{9.73}$$

where the identification (9.72) has been introduced. It is worth to stress that this relation is similar to (7.24). Equality of the second-order derivatives of s in (9.73) leads to the Maxwell relation $\partial\theta^{-1}/\partial q^2 = \frac{1}{2}(\partial\gamma/\partial u)$ and, after integration,

$$\theta^{-1}(u, q^2) = T^{-1}(u) + \frac{1}{2}\int \frac{\partial\gamma(u, q^2)}{\partial u}\mathrm{d}q^2, \tag{9.74}$$

where $T(u)$ is the local equilibrium temperature. Clearly, the non-equilibrium temperature reduces to its equilibrium value when the factor γ (or equivalently the Lagrange multiplier Λ_1) is independent of u.

We close this analysis with some considerations about the sign of the various coefficients. Stability of (local) equilibrium requires that the second variation of entropy with respect to the state variables u and $\boldsymbol{q}$ is negative definite, so that in particular,

$$\frac{\partial^2 s}{\partial q^2} = \gamma \leq 0, \tag{9.75}$$

but, referring to the entropy inequality (9.64) which can be cast in the form $b\gamma q^2 \geq 0$, it follows that $b < 0$ whence $L \equiv (b\gamma)^{-1} \geq 0$ and $\tau = -(\rho/b) \geq 0$. These results confirm the positive definite property of the heat conductivity coefficient and the relaxation time.

As compared with the results of Appendix 1, it is seen that in the present description, the non-equilibrium entropy (9.73) depends on the heat flux. This is due to the fact that $\boldsymbol{q}$ is assumed to have an evolution equation (9.58) of its own, leading to the introduction of a supplementary Lagrange multiplier in equation (9.60); instead, in Appendix 1 the temperature gradient was not assumed to be described by a proper evolution equation so that no extra Lagrange multiplier was needed.

9.6 Problems

9.1. *Dissipated energy.* Verify that the left-hand side of the Clausius–Duhem's inequality (9.5) can be identified with the rate of dissipated energy $T\sigma^s$ at the condition that the corresponding entropy flux is given by $\boldsymbol{J}^s = \boldsymbol{q}/T$.

9.2. *Generalized Fourier's law.* Show that the generalized Fourier law as the time derivative of the heat flux $\boldsymbol{q}(t) = \int_{-\infty}^{t} \lambda(t-t')\nabla T(t')\mathrm{d}t'$ is equal to Cattaneo's relation $\dot{\boldsymbol{q}} = -\boldsymbol{q}/\tau - (\lambda/\tau)\nabla T$ when the memory kernel is given by the expression $\lambda(t-t') = -(\lambda/\tau) \times \exp[-(t-t')/\tau]$, where λ is the heat conductivity and τ is a constant relaxation time.

9.3. *Objectivity.* Show that the skew-symmetric (antisymmetric) part of the velocity gradient tensor $\mathbf{W} = \frac{1}{2}[(\nabla \boldsymbol{v}) - (\nabla \boldsymbol{v})^{\mathrm{T}}]$ transforms as $\mathbf{W}^* = \mathbf{Q}\cdot\mathbf{W}\cdot\mathbf{Q}^{\mathrm{T}} + \dot{\mathbf{Q}}\cdot\mathbf{Q}^{\mathrm{T}}$ under the Euclidean transformation (9.8).

9.4. *Objectivity.* Prove that (a) ∇T and the displacement vector $\boldsymbol{u}$ of elasticity are objective vectors; (b) the symmetric part of the velocity gradient $\mathbf{V} = \frac{1}{2}[(\nabla \boldsymbol{v}) + (\nabla \boldsymbol{v})^{\mathrm{T}}]$ and the gradient of an objective vector are objective tensors; and (c) the deformation gradient $\mathbf{F}$ and the material time derivative of an objective vector, say the heat flux $\boldsymbol{q}$, are not objective.

9.5. *Clausius–Duhem's inequality.* Establish the Clausius–Duhem's inequality, respectively, in the case of small elastic deformations (see (9.18)) and large deformations (refer to Box 9.1).

9.6. *Large elastic deformations.* Consider a material body defined by the set of variables T, ∇T, $\mathbf{F}$, and $\dot{\mathbf{F}}$ where T is the temperature and $\mathbf{F}$ the deformation tensor. Determine the restrictions placed on the constitutive equations of f (free energy), s (entropy), $\mathbf{S}$ (Piola stress tensor), and $\boldsymbol{Q}$ (heat flux vector).

9.7. *Isotropic tensors.* Referring to the theorems of representation of isotropic tensors (e.g. Truesdell and Toupin 1960), the constitutive equation of the viscous stress tensor of a heat conducting viscous fluid reads as

$$\boldsymbol{\sigma}^{\mathrm{v}} = \alpha(\rho, T)\mathbf{I} + \gamma(\rho, T)(\nabla \cdot \boldsymbol{v})\mathbf{I} + 2\eta(\rho, T)\mathbf{V},$$

when the second- and higher-order terms in $\mathbf{V}$ are omitted. (a) Show that α is zero, as a result of the property that the viscous stress tensor has to be zero at equilibrium. (b) Decomposing the symmetric velocity gradient tensor in a bulk and a deviatoric part $\mathbf{V} = \frac{1}{3}(\nabla \cdot \boldsymbol{v})\mathbf{I} + \overset{0}{\mathbf{V}}$, with $\overset{0}{\mathbf{V}}$ the traceless deviator, show that

$$\boldsymbol{\sigma}^{\mathrm{v}} = \zeta(\rho, T)(\nabla \cdot \boldsymbol{v})\mathbf{I} + 2\eta(\rho, T)\overset{0}{\mathbf{V}},$$

with $\zeta = \gamma + \frac{2}{3}\eta$ designating the bulk viscosity. (c) Verify that $\zeta > 0$, $\eta > 0$.

9.8. *Parabolic heat conduction.* Consider a one-dimensional rigid heat conductor defined by the following constitutive equations:

$$f = f(T, \partial T/\partial x), \quad s = s(T, \partial T/\partial x), \quad \boldsymbol{q} = \boldsymbol{q}(T, \partial T/\partial x).$$

and the energy balance equation

$$\rho \partial u/\partial t = \partial q/\partial x + \rho r.$$

Determine the resulting heat conduction equation for the temperature field, given by

$$\rho T \frac{\partial^2 f}{\partial T^2}\frac{\partial T}{\partial t} = -\frac{\partial q}{\partial T}\frac{\partial T}{\partial x} - \frac{\partial q}{\partial(\partial T/\partial x)}\frac{\partial^2 T}{\partial x^2} + \rho r.$$

Show that this equation is of the parabolic type if $\partial^2 f/\partial T^2 > 0$ and $\partial q/\partial T < 0$ so that no propagation of wave can occur.

9.9. *Hyperbolic heat conduction.* Reconsider Problem 9.8 but with $T, \partial T/\partial t$, $\partial T/\partial x$ as variables. Show that in the present case, the temperature equation is of the hyperbolic type.

9.10. *Viscous heat conducting fluid.* A viscous heat conducting fluid is characterized by the following constitutive equation $\phi = \phi(v, T, \nabla v, \nabla T, \mathbf{V})$ with $\phi \equiv f, s, \boldsymbol{q}, \boldsymbol{\sigma}$. Determine the restrictions placed by Clausius–Duhem's inequality on the constitutive equations.

9.11. *Thermodiffusion.* Study the problem of thermodiffusion in a binary mixture of non-viscous fluids within the framework of rational thermodynamics.

9.12. *Liu's technique.* Reformulate the problem of a heat conducting viscous fluid (see Sect. 9.4) by using Liu's technique.

Chapter 10
Hamiltonian Formalisms

A Mathematical Structure of Reversible and Irreversible Dynamics

In the theories discussed so far, the limitations on the form of the constitutive equations arise essentially from the application of the second law of thermodynamics. However, no restrictions were placed on the reversible terms of the time evolution equations, as they do not contribute to the entropy production. Such terms may be either gyroscopic forces, as Coriolis force, or convected time derivatives, as Maxwell or corotational derivative, which are of frequent use in rheology. It is shown here that requirement of a Hamiltonian structure leads to restrictions on the reversible part of the evolution equations.

Hamiltonian formulations have always played a central role not only in mechanics but also in thermodynamics. They have been identified at different levels of description: from the microscopic one (classical mechanics, kinetic theory) to the macroscopic one (theory of elasticity, frictionless fluid flows, equilibrium thermodynamics). Despite their theoretical appeal, and except rare attempts, Hamiltonian methods have not been fully exploited in presence of irreversible effects. The reasons that militate in favour of a Hamiltonian description are numerous: the first one is conciseness, as the whole set of balance equations are expressed in terms of a limited number of potentials, generally, one single potential is sufficient. From a practical point of view, there exist, in parallel, powerful numerical methods, which have been developed for Hamiltonian systems to obtain approximate solutions, as the task of scientists is not only to derive equations but also to find solutions. Moreover, besides their power of synthesis and their practical interest, Hamiltonian descriptions are also helpful for the physical interpretation of the processes: indeed, the generating potentials may generally be identified with well-defined physical quantities as the mechanical energy, the energy of deformation, the thermodynamic potentials, etc. Hamiltonian techniques have also a wide domain of applicability, as they are not restricted to the linear regime. Finally, they place several restrictions on the possible forms of the constitutive equations, complementing those provided by the second law of thermodynamics.

The main part of this chapter will be devoted to the presentation of a general equation for the non-equilibrium reversible–irreversible coupling, abbreviated as GENERIC, which consists in a generalization of the Poisson bracket formalism originally proposed in the framework of classical mechanics (Grmela and Öttinger 1997; Öttinger and Grmela 1997; Öttinger 2005). The principal motivation for developing GENERIC is the modelling of the flow properties of rheological fluids; this is achieved by formulating general time evolution equations taking the same universal form whatever the nature of the state variables. Another particularity of GENERIC is that the evolution equations are expressed in terms of two appropriate thermodynamic potentials, called generators, taking the form of the total energy and a dissipation potential. GENERIC can be viewed as an extension of both Hamilton's equations and Landau's potential (Landau 1965). Before examining the main tenets of GENERIC and illustrating its application by a number of simple examples, let us preliminarily recall the Hamiltonian description of classical mechanics.

10.1 Classical Mechanics

Consider a collection of N particles characterized by the set of variables $\boldsymbol{x} = (\boldsymbol{r}, \boldsymbol{p})$ with $\boldsymbol{r} = (\boldsymbol{r}_1, \ldots, \boldsymbol{r}_N)$ and $\boldsymbol{p} = (\boldsymbol{p}_1, \ldots, \boldsymbol{p}_N)$ denoting the position and momentum vectors of the particles. Denoting by $E = K + V$ the total energy (kinetic and potential energies) or the Hamiltonian of the system, it is well known that the time evolution of the set of variables $\boldsymbol{x}$ is given by the Hamilton equations

$$\frac{\mathrm{d}\boldsymbol{x}}{\mathrm{d}t} = \hat{L}\frac{\partial E}{\partial \boldsymbol{x}}, \tag{10.1}$$

with

$$\hat{L} = \begin{pmatrix} 0 & 1 \\ -1 & 0 \end{pmatrix},$$

or, more explicitly,

$$\frac{\mathrm{d}\boldsymbol{r}}{\mathrm{d}t} = \frac{\partial E}{\partial \boldsymbol{p}}, \quad \frac{\mathrm{d}\boldsymbol{p}}{\mathrm{d}t} = -\frac{\partial E}{\partial \boldsymbol{r}}. \tag{10.2}$$

The Poisson matrix $\hat{L}$ expresses the *reversible* kinematics of $\boldsymbol{x}$; reversibility means invariance with respect to the change $t \to -t$, it being understood that concomitantly $\boldsymbol{r}, \boldsymbol{p} \to \boldsymbol{r}, -\boldsymbol{p}$ and $E(\boldsymbol{r}, \boldsymbol{p}) \to E(\boldsymbol{r}, -\boldsymbol{p})$. Instead of working with the matrix $\hat{L}$, an equivalent way to formulate the dynamical equations (10.2) is to introduce the Poisson bracket

$$\{A, B\} = \frac{\partial A}{\partial \boldsymbol{r}}\frac{\partial B}{\partial \boldsymbol{p}} - \frac{\partial A}{\partial \boldsymbol{p}}\frac{\partial B}{\partial \boldsymbol{r}}, \tag{10.3}$$

which is related to $\hat{L}$ by

$$\{A, B\} = \left(\frac{\partial A}{\partial \boldsymbol{x}}, \hat{L}\frac{\partial B}{\partial \boldsymbol{x}}\right) \tag{10.4}$$

with $(\cdot,\cdot)$ denoting the scalar product, whereas A and B are regular functions of $\boldsymbol{x}$. The evolution of $\boldsymbol{x}$ is now governed by

$$\frac{\mathrm{d}A}{\mathrm{d}t} = \{A, E\}, \tag{10.5}$$

where $A(\boldsymbol{x})$ is an arbitrary function of $\boldsymbol{x}$, not dependent explicitly on time. This is directly seen by writing explicitly both members of (10.5), i.e.

$$\frac{\partial A}{\partial \boldsymbol{r}}\frac{\mathrm{d}\boldsymbol{r}}{\mathrm{d}t} + \frac{\partial A}{\partial \boldsymbol{p}}\frac{\mathrm{d}\boldsymbol{p}}{\mathrm{d}t} = \frac{\partial A}{\partial \boldsymbol{r}}\frac{\partial E}{\partial \boldsymbol{p}} - \frac{\partial A}{\partial \boldsymbol{p}}\frac{\partial E}{\partial \boldsymbol{r}}, \tag{10.6}$$

and after identifying the coefficients of $\partial A/\partial \boldsymbol{r}$ and $\partial A/\partial \boldsymbol{p}$ in both sides of (10.6), one recovers indeed the evolution equations (10.2). An illustrative example is presented in Box 10.1.

Box 10.1 A Classical Mechanics Illustration: The Harmonic Oscillator

Consider a particle of constant mass m fixed at the end of a spring of stiffness H. The problem is assumed to be one dimensional with x denoting the position of the particle and p its momentum. The total energy is

$$E(x,p) = \frac{1}{2}\frac{p^2}{m} + \frac{1}{2}Hx^2. \tag{10.1.1}$$

In virtue of (10.2), the evolution equations of x and p are given by

$$\frac{\mathrm{d}x}{\mathrm{d}t} = \frac{p}{m}, \quad \frac{\mathrm{d}p}{\mathrm{d}t} = -Hx. \tag{10.1.2}$$

After elimination of p, one recovers Newton's law

$$m\frac{\mathrm{d}^2x}{\mathrm{d}t^2} = -Hx. \tag{10.1.3}$$

Making use of the definition (10.4) of the Poisson bracket $\{A, B\}$ one has

$$\{A, B\} = \begin{pmatrix}\frac{\partial A}{\partial x} & \frac{\partial A}{\partial p}\end{pmatrix}\begin{pmatrix}0 & 1\\ -1 & 0\end{pmatrix}\begin{pmatrix}\frac{\partial B}{\partial x}\\ \frac{\partial B}{\partial p}\end{pmatrix}, \tag{10.1.4}$$

which is in agreement with the result (10.3). Accordingly, the Poisson bracket $\{A, E\}$ describing the harmonic oscillator is given by

$$\{A, E\} = \frac{\partial A}{\partial x}\frac{p}{m} - \frac{\partial A}{\partial p}Hx \tag{10.1.5}$$

and the corresponding evolution equation (10.5) takes the form

$$\frac{\partial A}{\partial x}\frac{\mathrm{d}x}{\mathrm{d}t}+\frac{\partial A}{\partial p}\frac{\mathrm{d}p}{\mathrm{d}t}=\frac{\partial A}{\partial x}\frac{p}{m}-\frac{\partial A}{\partial p}Hx. \tag{10.1.6}$$

Comparison of the coefficients of $\partial A/\partial x$ and $\partial A/\partial p$ in both sides of (10.1.6) gives back the equations of motion (10.1.2).

Expression (10.4) is a Poisson bracket if the two following conditions are fulfilled:

(1)

$$\{A,B\}=-\{B,A\}\ \text{(antisymmetry)}, \tag{10.7}$$

(2)

$$\{A,\{B,C\}\}+\{B,\{C,A\}+\{C,\{A,B\}\}=0\ (\text{ Jacobi's identity}). \tag{10.8}$$

The antisymmetry property of the Poisson bracket implies the antisymmetry of operator $\hat{L}$; the Jacobi's identity imposes additional severe restrictions and expresses the time structure invariance of the Poisson bracket $\{A,B\}$. It is precisely this identity what gives information on the reversible contributions to the dynamics (for instance, non-linear convective contributions to the time derivatives), which is not available from the positiveness of the entropy production.

Evolution equations as (10.1) or its equivalent (10.5) are reversible time evolution equations as they are invariant with respect to time reversal $t\to -t$, a result well known in classical mechanics. In more general situations as these encountered in continuum physics, $\mathrm{d}\boldsymbol{x}/\mathrm{d}t$ is the sum of a reversible and a non-reversible contribution and we will see in Sect. 10.2 how to integrate irreversible dynamics in the framework of GENERIC. In mathematical terminology, the manifold of the phase space, a Hamiltonian, and an antisymmetric matrix relating the time derivative of the variables with the partial derivatives of the Hamiltonian with respect to its variables are known as a symplectic manifold.

10.2 Formulation of GENERIC

GENERIC may be considered as an extension of Landau's idea accordingly the time evolution of a state variable $\boldsymbol{x}$, like mass density or energy, towards its equilibrium value $\boldsymbol{x}_{\mathrm{eq}}$ is governed by the relation

$$\frac{\mathrm{d}\boldsymbol{x}}{\mathrm{d}t}=-\frac{\delta\Psi}{\delta\boldsymbol{x}}, \tag{10.9}$$

where Ψ is a given potential which is minimum in the equilibrium or in the steady state:

$$\frac{\delta \Psi}{\delta \boldsymbol{x}} = 0 \text{ at } \boldsymbol{x} = \boldsymbol{x}_{\text{eq}}, \tag{10.10}$$

with $\delta/\delta\boldsymbol{x}$ denoting the functional or Volterra derivative with respect to the variable $\boldsymbol{x}$; if $\Psi = \int \psi \,\mathrm{d}V$ is a simple scalar function of the variables, then $\delta\Psi/\delta\boldsymbol{x}$ reduces to the usual partial derivative $\delta\Psi/\delta\boldsymbol{x} = \partial\psi/\partial\boldsymbol{x}$; the situation is more complicated when Ψ depends in addition on the gradient of $\boldsymbol{x}$, if $\boldsymbol{x}$ is assumed to be a scalar, then

$$\frac{\delta \Psi}{\delta x} = \frac{\partial \psi}{\partial x} - \nabla \cdot \left[\frac{\partial \Psi}{\partial (\nabla x)} \right]. \tag{10.11}$$

However, a relaxation-type equation as (10.9) describing the irreversible approach to equilibrium is too restrictive and cannot pretend to cope with general processes of continuum physics.

When both reversible and irreversible processes are present, one formulates, in the framework of GENERIC, a general time evolution equation in which the evolution of a variable $\boldsymbol{x}$ is expressed in terms of two potentials, the total energy E of the overall system and a dissipation function Ψ: explicitly, one has

$$\frac{\mathrm{d}\boldsymbol{x}}{\mathrm{d}t} = \hat{L}\frac{\delta E}{\delta \boldsymbol{x}} + \frac{\delta \Psi}{\delta(\delta S/\delta \boldsymbol{x})}. \tag{10.12}$$

The quantity S has the physical meaning of the entropy of the overall system. The dissipation potential Ψ, which is a real-valued function of the derivatives $\delta S/\delta x$, possesses the following properties: $\Psi(0) = 0$, it is minimum at $\boldsymbol{x}=0$ and is convex in the neighbourhood of 0. Note that in the GENERIC framework, it is generally assumed that the overall system is isolated from its environment so that

$$\frac{\mathrm{d}E}{\mathrm{d}t} = 0, \quad \frac{\mathrm{d}S}{\mathrm{d}t} > 0. \tag{10.13}$$

Notice also that in the particular case that Ψ is a quadratic function of $\delta \boldsymbol{S}/\delta\boldsymbol{x}$, of the form $\frac{1}{2}(\partial S/\partial \boldsymbol{x}) \cdot \hat{M} \cdot (\partial S/\partial \boldsymbol{x})$, (10.12) takes the more familiar form

$$\frac{\mathrm{d}\boldsymbol{x}}{\mathrm{d}t} = \hat{L}\frac{\delta E}{\delta \boldsymbol{x}} + \hat{M}\frac{\delta S}{\delta \boldsymbol{x}}. \tag{10.14}$$

The matrices $\hat{L}$ and $\hat{M}$, operating on the functional derivatives of E and S, produce the reversible and irreversible contributions to the evolution of $\boldsymbol{x}$, respectively. These matrices must satisfy some conditions: $\hat{L}$ must be antisymmetric and verify Jacobi's identity, and $\hat{M}$ must be symmetric and positive-definite to ensure the positiveness of the dissipation rate. However, the restrictions on $\hat{L}$ and $\hat{M}$ mentioned above are not sufficient to guarantee a thermodynamically consistent description of the dynamics of the system, and two supplementary restrictions, called degeneracy conditions, are introduced, namely

$$\hat{L}\frac{\delta S}{\delta \boldsymbol{x}} = 0, \tag{10.15}$$

$$\hat{M}\frac{\delta E}{\delta \boldsymbol{x}} = 0. \tag{10.16}$$

The first condition expresses that entropy cannot contribute to the reversible nature of $\hat{L}$ and therefore is not modified by the reversible part of the dynamics; the second condition implies the conservation of total energy by the dissipative contribution to the dynamics. Relations (10.12) or (10.14) represent the GENERIC extension of the Landau equation (10.9) and express the universal structure of non-equilibrium thermodynamics, which is completely specified by the knowledge of the four quantities E, S (or Ψ), $\hat{L}$, and $\hat{M}$.

By analogy with classical mechanics where the Hamilton equations (10.1) can be replaced by the Poisson brackets (10.5), it is equivalent to write (10.14) in the form

$$\frac{\mathrm{d}A}{\mathrm{d}t} = \{A, E\} + [A, S], \tag{10.17}$$

where

$$\{A, E\} = \left(\frac{\delta A}{\delta \boldsymbol{x}}, \hat{L}\frac{\delta E}{\delta \boldsymbol{x}}\right) \tag{10.18}$$

is a Poisson bracket with the antisymmetry property $\{A, E\} = -\{E, A\}$, whereas

$$[A, S] = \left(\frac{\delta A}{\delta \boldsymbol{x}}, \hat{M}\frac{\delta S}{\delta \boldsymbol{x}}\right) \tag{10.19}$$

is the so-called Landau symmetric bracket, i.e. $[A, S] = [S, A]$ satisfying in addition the positiveness property $[S, S] > 0$. In terms of the dissipation potential Ψ, the above bracket will be given the form

$$[A, S] = \left(\frac{\delta A}{\delta \boldsymbol{x}}, \frac{\delta \Psi}{\delta(\delta S/\delta \boldsymbol{x})}\right), \tag{10.20}$$

which is a generalization of (10.19). In the foregoing, we will illustrate the use of GENERIC with two examples: isothermal hydrodynamics and matter diffusion in a binary moving mixture. Several applications to polymer solutions have been worked out, but we refer to Öttinger and Grmela (1997) and Öttinger (2005) for an overview of this topic.

10.2.1 Classical Navier–Stokes' Hydrodynamics

Let us consider the motion of a compressible one-component viscous fluid assumed to take place under *isothermal* conditions. The following four steps govern the construction of the GENERIC formalism:

Step 1. *Selection of the state variables*
Just like in other thermodynamic theories, the selection of state variables is subordinated by the nature of the process under consideration and the degree of accuracy that one wishes to achieve. For the present problem, the set $\boldsymbol{x}$ of variables is conveniently selected as:

$$\boldsymbol{x} : \rho(\text{mass density}), \boldsymbol{u}(\text{momentum}). \tag{10.21}$$

In GENERIC, it is more convenient to work with the momentum rather than with the velocity field $\boldsymbol{v}(=\boldsymbol{u}/\rho)$. The pressure p is not included among the variables because it will be expressed in terms of the independent fields by means of a constitutive equation.

Step 2. *Thermodynamic potential*
Since the temperature T is fixed and uniform, the Helmholtz free energy is the potential that will play the central role, it is expressed by

$$\Phi(\rho, \boldsymbol{u}) = \int \phi(\rho, \boldsymbol{u})\mathrm{d}V = E(\rho, \boldsymbol{u}) - TS(\rho), \tag{10.22}$$

with the total energy and the total entropy given by

$$E(\rho, \boldsymbol{u}) = \int [(\boldsymbol{u}\cdot\boldsymbol{u})(2\rho)^{-1} + \varepsilon(\rho)]\mathrm{d}V, \tag{10.23a}$$

$$S(\rho) = \int s(\rho)\mathrm{d}V, \tag{10.23b}$$

where ε and s denote, respectively, the internal energy and the entropy referred per unit volume; just like ε, the entropy s cannot depend on the momentum $\boldsymbol{u}$ as it is a specific thermodynamic quantity. Here we use the notation ε instead of u to designate the internal energy, thus avoiding the confusion with the momentum $\boldsymbol{u}$.

For further purpose, it is interesting to observe that $\phi_{u_\alpha} = u_\alpha/\rho$ is the velocity field; the notation ϕ_{u_α} standing for $\partial\phi/\partial u_\alpha$ where $\partial/\partial u_\alpha$ is the partial derivative with respect to the component u_α of the momentum.

Step 3. *Hamiltonian reversible dynamics*
Our purpose is to specify the reversible contribution to the evolution equations of hydrodynamics. Let $A = \int a\,\mathrm{d}V$ denote a regular function of ρ and $\boldsymbol{u}$ whose time evolution is given by

$$\frac{\mathrm{d}A}{\mathrm{d}t} = \int (a_\rho \partial_t \rho + a_{u_\alpha}\partial_t u_\alpha)\mathrm{d}V, \tag{10.24}$$

where $a_\rho = \partial a/\partial\rho$ and $a_{u_\alpha} = \partial a/\partial u_\alpha$.

Following the lines of thought of GENERIC, the time evolution of A can be cast in the form

$$\frac{\mathrm{d}A}{\mathrm{d}t} = \{A, \Phi\}, \tag{10.25}$$

where $\{A, \Phi\}$ is a Poisson bracket which, in the present problem, is defined as

$$\{A, \Phi\} = \int \{\rho[(\partial_\alpha a_\rho)\phi_{u_\alpha} - (\partial_\alpha \phi_\rho)a_{u_\alpha}] + u_\gamma[(\partial_\alpha a_{u_\gamma})\phi_{u_\gamma} - (\partial_\alpha \phi_{u_\gamma})a_{u_\alpha}]\}\mathrm{d}V \tag{10.26}$$

with ∂_α denoting the spatial derivative with respect to the x_α coordinate, we have also used the summation convention on repeated indices. A systematic construction of the expression of the Poisson bracket (10.26) has been developed on general arguments based on the group theory by Grmela and Öttinger (1997) and Öttinger (2005). Substituting (10.24) and (10.26) in (10.25), one obtains

$$\int [a_\rho(\partial_t \rho) + a_{u_\alpha}(\partial_t u_\alpha)]\mathrm{d}V$$
$$= \int \{a_\rho[-\partial_\gamma(\rho\phi_{u_\gamma})] - a_{u_\alpha}[\partial_\gamma(u_\alpha\phi_{u_\gamma}) + \rho\partial_\alpha\phi_\rho + u_\gamma\partial_\alpha u_\gamma]\}\mathrm{d}V, \tag{10.27}$$

after that the right-hand side of (10.26) has been integrated by parts and the boundary conditions have been selected to make all the integrals over the boundary equal to zero.

By identification of the coefficients of a_ρ and a_{u_α}, it is found that

$$\partial_t \rho = -\partial_\gamma(\rho\phi_{u_\gamma}), \tag{10.28}$$
$$\partial_t u_\alpha = -\partial_\gamma(u_\alpha\phi_{u_\gamma}) - \rho\partial_\alpha\phi_\rho - u_\gamma\partial_\alpha\phi_{u_\gamma}. \tag{10.29}$$

To recover the Euler equations of hydrodynamics, one has to identify the two last terms of the right-hand side of (10.29) with the pressure gradient, i.e.

$$\partial_\alpha p = \rho\partial_\alpha\phi_\rho + u_\gamma\partial_\alpha\phi_{u_\gamma}. \tag{10.30}$$

Moreover, from the chain differentiation rule, one obtains

$$\partial_\alpha \phi = \phi_\rho\partial_\alpha\rho + \phi_{u_\gamma}\partial_\alpha u_\gamma,$$

which, coupled to (10.30) leads to

$$p = -\phi + \rho\phi_\rho + u_\gamma\phi_{u_\gamma} = -\varepsilon + Ts + \rho\mu. \tag{10.31}$$

This is the well-known Euler equation of equilibrium thermodynamics after that $\phi_\rho = \mu$ has been identified with the chemical potential. As shown earlier, one has $\phi_{u_\alpha} = v_\alpha$ so that (10.28) and (10.29) take the form

$$\partial_t \rho = -\partial_\gamma(\rho v_\gamma), \tag{10.32}$$
$$\partial_t u_\alpha = -\partial_\gamma(p\delta_{\alpha\gamma} + u_\alpha u_\gamma \rho^{-1}), \tag{10.33}$$

which are the equations describing the behaviour of an Euler fluid in hydrodynamics.

When the time evolution of the system is described by the Euler equations (10.32) and (10.33), it is checked that

$$\frac{\mathrm{d}\Phi}{\mathrm{d}t} = 0, \tag{10.34}$$

which follows directly from (10.25) as a consequence of the antisymmetry property of the Poisson bracket. It is also verified that the relations (10.32) and (10.33) are invariant with respect to the time reversal $t \rightarrow -t$ and this justifies the denomination "reversible" or "non-dissipative".

Step 4. *Irreversible dynamics*
In the presence of dissipation, the free energy Φ will diminish in the course of time

$$\frac{\mathrm{d}\Phi}{\mathrm{d}t} < 0 \tag{10.35}$$

and the general time evolution equation is

$$\frac{\mathrm{d}A}{\mathrm{d}t} = \{A, \Phi\} - [A, \Phi], \tag{10.36}$$

where, referring to (10.20), the dissipative bracket is defined as

$$[A, \Phi] = [\delta A/\delta u_\alpha, \delta \Psi/\delta \phi_{u_\alpha}] = \int a_{u_\alpha} \left[(\partial \psi/\partial \phi_{u_\alpha}) - \partial_\gamma \left(\partial \psi/\partial(\partial_\gamma \phi_{u_\alpha})\right)\right] \mathrm{d}V \tag{10.37}$$

with $\Psi = \int \psi \,\mathrm{d}V$ the dissipation potential. Two remarks are in form about the formulation of (10.36) and (10.37). First, when compared to (10.17), the change of sign in front of the dissipative bracket is a consequence of the use of the free energy Φ instead of the entropy S as basic function; second, there is no contribution of the mass density to (10.37) because ρ is a conserved quantity.

To reproduce the classical Navier–Stokes' relation, select ψ as given by the quadratic form

$$\psi = \frac{1}{4}(\partial_\gamma \phi_{u_\alpha} + \partial_\alpha \phi_{u_\gamma})\eta(\partial_\gamma \phi_{u_\alpha} + \partial_\alpha \phi_{u_\gamma}) > 0, \tag{10.38}$$

wherein η is a phenomenological coefficient, the dynamic viscosity, which is a positive quantity to guarantee the positive definite property of the dissipation potential. It is directly checked that the integrant of (10.37) is

$$-a_{u_\alpha} \partial_\gamma \left[\eta(\partial_\alpha \phi_{u_\gamma} + \partial_\gamma \phi_{u_\alpha})\right]. \tag{10.39}$$

Following the same procedure as in Step 3, one obtains the next evolution equations for ρ and u_α:

$$\partial_t \rho = -\partial_\gamma(\rho \phi_{u_\gamma}), \tag{10.40}$$

$$\partial_t u_\alpha = -\partial_\gamma(u_\alpha \phi_{u_\alpha}) - \partial_\alpha p - \partial_\gamma P_{\alpha\gamma}, \tag{10.41}$$

where the symmetric pressure tensor $P_{\alpha\gamma}$ is expressed by

$$P_{\alpha\gamma} = -\frac{1}{2}\eta(\partial_\alpha \phi_{u_\gamma} + \partial_\gamma \phi_{u_\alpha}) = -\frac{1}{2}\eta(\partial_\alpha v_\gamma + \partial_\gamma v_\alpha). \tag{10.42}$$

The last term in (10.41) represents the irreversible contribution to the momentum equation as it is not invariant to the time reversal. From the identification of (10.42), it is seen that the dissipation potential (10.38) can be cast in the form $-\mathbf{P}^{\mathrm{v}} : \mathbf{V}$, which is (2.68) for the viscous dissipation found in the classical formulation of hydrodynamics. By incorporating in the relation (10.38) of the dissipation potential an extra term proportional to the square of $\partial_\gamma \phi_{u_\gamma}$, one would recover Stokes' law between the bulk viscous pressure and the divergence of the velocity. External forces, like gravity, can also be incorporated by simply adding the corresponding potential, namely $-\int \rho \boldsymbol{g} \cdot \boldsymbol{x}\, \mathrm{d}V$ to the total energy given by (10.23a).

The above results are important because they exhibit the GENERIC structure of the familiar Navier–Stokes' equations. Indeed, accordingly to the general rules of GENERIC, the evolution equations of hydrodynamics can be formulated in terms of two potentials: the Helmholtz free energy and the dissipation potential; moreover, there exists an acute separation between reversible and irreversible contributions. It is also important to mention some properties of the solutions of (10.40) and (10.41): first, when the system is left during a sufficiently long time outside the influence of external constraints, it will tend to an equilibrium state characterized by $\phi_{u_\alpha} = 0$, $\phi_\rho = 0$ for $t \to \infty$. Second, if the potential Φ is convex ($\Phi \geq 0$), and because $\mathrm{d}\Phi/\mathrm{d}t < 0$, Φ plays the role of a Lyapounov function ensuring the stability of equilibrium.

It was stated in Sect. 10.1 that an equivalent description of GENERIC consists in expressing the evolution equations in terms of the two potentials E and S and the operators $\hat{L}$ and $\hat{M}$. However, this kind of approach requires some expertise, technique and feeling, which is outside the scope of the present introductory monograph and therefore we refer the reader to specialized works (Grmela and Öttinger 1997; Öttinger and Grmela 1997; Öttinger 2005).

In the present section, we have considered the flow of a one-component fluid. In Sect. 10.2.2, we will examine the more general situation of the motion of a binary mixture to exhibit the coupling between matter diffusion and momentum transport.

10.2.2 Fickian Diffusion in Binary Mixtures

We consider an isotropic mixture of two Newtonian viscous fluids perfectly miscible and chemically inert, the whole system is in motion in absence of external forces and its temperature is assumed to be uniform. Let ρ_1 and ρ_2 be the mass densities of the two components and $\boldsymbol{u}_1$ and $\boldsymbol{u}_2$ their barycentric momenta, the average barycentric momentum is given by $\boldsymbol{u} = \boldsymbol{u}_1 + \boldsymbol{u}_2$. The presentation of Sect. 10.2.1 will be traced back and first applied to Fickian diffusion.

Step 1. *State variables*
The space of basic variables for the present problem is formed by the set $\boldsymbol{x} : \boldsymbol{u}, \rho_1, \rho_2$. However, for practical reasons, it is more convenient to work with the following equivalent set

$$\boldsymbol{x} : \boldsymbol{u}, \rho, c, \tag{10.43}$$

where $\rho = \rho_1 + \rho_2$ is the total mass density and $c = \rho_1/\rho$ is the mass fraction of one of the constituents, say component 1 (note that $c + c_2 = 1$, with $c_2 = \rho_2/\rho$).

Step 2. *Thermodynamic potential*
We will use the same Helmholtz potential as in Sect. 10.2.1, i.e.

$$\Phi(\rho, c, \boldsymbol{u}) = E(\rho, c, \boldsymbol{u}) - TS(\rho, c) \tag{10.44}$$

with $E = \int [\boldsymbol{u} \cdot \boldsymbol{u}/(2\rho) + \varepsilon(\rho, c)]\mathrm{d}V$ where ε is the total internal energy per unit volume of the binary mixture. At this level of description, we have neglected the contribution from the diffusion kinetic energy, which is justified in diluted systems.

Step 3. *Reversible dynamics*
The reversible dynamics is written in the form $\mathrm{d}A/\mathrm{d}t = \{A, \Phi\}$ where the Poisson bracket is now given by (Grmela et al. 1998):

$$\begin{aligned}\{A, \Phi\} = \int \{&\rho[(\partial_\alpha a_\rho)\phi_{u_\alpha} - (\partial_\alpha \phi_\rho)a_{u_\alpha}] + u_\gamma[(\partial_\alpha a_{u_\gamma})\phi_{u_\alpha} \\ &-(\partial_\alpha \phi_{u_\gamma})a_{u_\alpha}] - (\partial_\alpha c)(a_c\phi_{u_\alpha} - \phi_c a_{u_\alpha})\}\mathrm{d}V.\end{aligned} \tag{10.45}$$

This is the same relation as (10.26) to which we have added a third-term expressing the contribution from the mass fraction c. The resulting evolution equations are obtained as before, i.e. by writing explicitly $\mathrm{d}A/\mathrm{d}t$ in terms of the variables, by integrating (10.45) by parts and by identifying the coefficients of a_ρ, a_{u_α}, and a_c, respectively. These operations lead to the following reversible part of the evolution equations

$$\partial_t \rho = -\partial_\gamma(\rho\phi_{u_\gamma}), \tag{10.46}$$

$$\partial_t u_\alpha = -\partial_\gamma(u_\alpha \phi_{u_\gamma}) - \partial_\alpha p, \tag{10.47}$$

$$\partial_t c = -(\partial_\gamma c)\phi_{u_\gamma}, \tag{10.48}$$

where the thermostatic pressure is still given by $p = -\phi + \rho\phi_\rho + u_\gamma \phi_{u_\gamma}$. It is easily checked that the above equations are invariant with respect to time reversal.

Step 4. *Irreversible dynamics*
To recover Fick's law of diffusion, it is necessary to introduce the dissipation potential. The latter is a generalization of the potential used in (10.38), written here as ψ (10.38), with a supplementary term,

$$\psi = \psi(10.38) + \frac{1}{2}(\partial_\gamma \phi_c) D^* (\partial_\gamma \phi_c). \tag{10.49}$$

The contribution coming from diffusion is the last term at the right-hand side of (10.49), D^* is a positive phenomenological coefficient allowed to depend on ρ and c. After that dissipation has been included, the evolution equations (10.46)–(10.48) take the final form

$$\partial_t \rho = -\partial_\gamma(\rho v_\gamma), \tag{10.50}$$

$$\partial_t u_\alpha = -\partial_\gamma(u_\alpha v_\gamma) - \partial_\alpha p - \partial_\gamma P_{\alpha\gamma}, \tag{10.51}$$

$$\partial_t c = -v_\gamma \partial_\gamma c + \partial_\gamma(D^* \partial_\gamma \phi_c), \tag{10.52}$$

wherein the two first relations are the classical continuity and momentum equations whereas the last one is the diffusion equation; the irreversible contributions are the last term in (10.51), with $P_{\alpha\gamma}$ still given by (10.42), and the last term of (10.52). Denoting by J_γ, the γ component of the flux of matter, it follows from (10.52) that

$$J_\gamma = -D^* \partial_\gamma \phi_c. \tag{10.53}$$

Assuming that ϕ is a quadratic function of c with $\phi_c = D_1 c$, where D_1 is a phenomenological constant, one recovers Fick's law

$$J_\gamma = -\rho D \partial_\gamma c, \tag{10.54}$$

at the condition to define the coefficient of diffusion D as $D_1 D^*/\rho$; observe that in contrast to D_1, the factor D is not necessarily a constant.

One of the merits of GENERIC is to provide a systematic way to derive the evolution equations whatever the nature and the number of basic variables. Of course, to obtain explicit relations, we need constitutive relations for the potentials ϕ and ψ, which will be determined by the physico-chemical properties of the system under study.

Although the above analysis gives an overall description of diffusion in a two-constituent mixture, it cannot pretend to describe the behaviour of each individual component; in particular, one has no information about the time evolution of the particular velocities $\boldsymbol{u}_1$ and $\boldsymbol{u}_2$. This would require the introduction of $\boldsymbol{u}_1$ and $\boldsymbol{u}_2$ as independent variables, instead of the single barycentric momentum $\boldsymbol{u}$. Such a task will be achieved in Sect. 10.2.3.

10.2.3 Non-Fickian Diffusion in Binary Mixtures

Our objective is twofold: to derive the evolution equations of the two individual components, and to generalize the Fick's law to non-steady situations. For the sake of simplicity, we assume that the barycentric momentum $\boldsymbol{u}$ is equal to zero; this assumption implies that the total variation of the total mass density ρ is negligible in the course of time. In addition, we assume that both fluids are non-viscous. A more general analysis, wherein the overall velocity does not vanish and where viscosity is not neglected, can be found in Grmela et al. (2003).

1. *State variables*
   Here we make the one-to-one transformation
$$\rho_1, \rho_2, \boldsymbol{u}_1, \boldsymbol{u}_2 \rightarrow \rho, c, \boldsymbol{u}, \boldsymbol{w}, \tag{10.55}$$
   where as above ρ designates the total mass density, c is the mass fraction of component 1, $\boldsymbol{u}$ is the barycentric momentum while the vector $\boldsymbol{w}$ stands for
$$\boldsymbol{w} = \rho^{-1}(\rho_2 \boldsymbol{u}_1 - \rho_1 \boldsymbol{u}_2). \tag{10.56}$$
   As stated above, from now on, one supposes that $\boldsymbol{u} = 0$ (no convection). Once c, ρ, and $\boldsymbol{w}$ are determined, one obtains directly ρ_1, ρ_2, $\boldsymbol{u}_1$ and $\boldsymbol{u}_2$ from the inverse transformation $\rho_1 = c\rho, \rho_2 = (1-c)\rho, \boldsymbol{u}_1 = -\boldsymbol{u}_2 = \boldsymbol{w}$.
2. *Thermodynamic potential*
   The latter is similar to (10.44) but the specific kinetic energy $u_1^2(2\rho_1)^{-1} + u_2^2(2\rho_2)^{-1}$ is now equal to the diffusion kinetic energy $w^2/2\rho c(1-c)$ instead of the barycentric kinetic energy $u^2/2\rho$, as in Sect. 10.2.2.
3. *Hamiltonian reversible dynamics*
   The Hamiltonian setting is equivalent to that proposed in the previous sections; inspired by (10.26) and (10.45), we will write
$$\begin{aligned}\{A, \Phi\} = \int \{&\rho c(1-c)[\partial_\alpha(a_c/\rho)\phi_{w_\alpha} - \partial_\alpha(\phi_c/\rho)a_{w_\alpha}] \\ &-(1-c)w_\gamma[(\partial_\alpha a_{w_\gamma})\phi_{w_\alpha} - (\partial_\alpha \phi_{w_\gamma})a_{w_\alpha}] \\ &-w_\gamma[\partial_\alpha(c a_{w_\gamma})\phi_{w_\alpha} - \partial_\alpha(c\phi_{w_\gamma})a_{w_\alpha}]\}\, \mathrm{d}V. \end{aligned}\tag{10.57}$$
   It follows that the non-dissipative contribution to the evolution equations will be
$$\partial_t \rho = 0, \tag{10.58}$$
$$\partial_t c = -\partial_\gamma[\rho c(1-c)\phi_{w_\gamma}]/\rho, \tag{10.59}$$
$$\begin{aligned}\partial_t w_\alpha = &-\rho c(1-c)\partial_\alpha(\phi_c/\rho) + c\partial_\gamma(w_\alpha \phi_{w_\gamma}) \\ &-\partial_\gamma[(1-c)w_\alpha\phi_{w_\gamma}] - (1-c)w_\gamma(\partial_\alpha\phi_{w_\gamma}) + w_\gamma\partial_\alpha(c\phi_{w_\gamma}). \end{aligned}\tag{10.60}$$
   Expression (10.59) is the classical diffusion equation wherein the flux of matter is given by $J_\gamma = \rho c(1-c)\phi_{w_\gamma}$. Relation (10.60) represents the

most original contribution of GENERIC formalism to this problem, as it is a time evolution equation for the relative velocity $\boldsymbol{w}$, which has no counterpart in the classical theory of irreversible processes. More about its physical content will be discussed below.

4. *Dissipation*
As viscous contributions are neglected, the dissipation potential is simply given by

$$\psi = \frac{1}{2}\phi_{w_\alpha}\Lambda\phi_{w_\alpha}, \tag{10.61}$$

where Λ is a phenomenological coefficient which is positive to guarantee the positiveness of the dissipation potential. The complete set of evolution equations is still given by (10.58)–(10.60) at the condition to complement the evolution equation (10.60) for w_α by the dissipative term $-\Lambda\phi_{w_\alpha}$.

It is interesting to observe that by inserting this dissipative term in (10.60), one obtains an equation of the Cattaneo type. By omitting in (10.60) the non-linear contributions in the fluxes and making use of the result $\phi_{w_\alpha} = w_\alpha/\rho c(1-c)$, one arrives at

$$\partial_t w_\alpha + [\Lambda/\rho c(1-c)]w_\alpha = -\rho c(1-c)\partial_\alpha(\phi_c/\rho), \tag{10.62}$$

which is clearly of the form

$$\tau\partial_t w_\alpha + w_\alpha = -D^*\partial_\alpha(\phi_c/\rho), \tag{10.63}$$

where $\tau = 2\rho c(1-c)\Lambda^{-1}$ corresponds to the relaxation time of the diffusive momentum and $D^* = [\rho c(1-c)]^2\Lambda^{-1}$ to a diffusion-like coefficient. Expression (10.63) is an example of non-Fickian law. Non-Fickian diffusion occurs for instance when a solvent diffuses into a medium characterized by a timescale which is the same or larger than the timescale on which the penetration takes place. This situation arises when solvents penetrate into glassy polymers. Another example in which diffusion exhibits non-classical features is the mixture of ^{3}He–^{4}He isotopes below the lambda point (Lhuillier et al. 2003; Lebon et al. 2003).

If we assume that w_α evolves much faster than the other state variables ($\partial_t w_\alpha = 0$), then (10.63) reduces to the standard description of Fick's diffusion. Indeed, in virtue of the definition of the diffusion flux $J_\alpha = \rho c(1-c)\phi_{w_\alpha} \equiv w_\alpha$, (10.63) becomes simply

$$J_\alpha = -D^*\partial_\alpha(\phi_c/\rho), \tag{10.64}$$

which is the classical Fick's law.

10.3 Final Comments

In this section, GENERIC has been presented as a general formalism to describe systems driven out of equilibrium. The main feature of this theory is that it uses two generating potentials: the total energy for reversible dynamics and entropy (or more generally the dissipation potential) for irreversible dynamics. The principal objective is to derive the relevant evolution equations of the process under study. The main result of the study is that it offers a structure which is applicable whatever the level of description, either macroscopic, mesoscopic, or microscopic. The basic ideas underlying GENERIC have been discussed thanks to two simple examples: classical hydrodynamics and diffusion of two perfectly miscible fluids.

It is instructive to recapitulate how we obtained the time evolution equations. First, one has to define the space of the state variables $\boldsymbol{x}$. This choice depends on our goals and the nature of the system under investigation; it is made on the basis of some intuition guided by our insight of the physics of the problem. In classical hydrodynamics, the solution is unambiguous: the state variables are the classical hydrodynamic fields, i.e. mass, momentum, and energy or temperature. However, this classical approach fails when dealing with complex fluids, like non-Newtonian or polymer solutions. Due to the active role of the microscopic structure of the complex fluids in the macroscopic time evolution, the choice of the state variables is not universal. For example, to describe complex fluids, the hydrodynamic variables will be complemented by structural variables, sometimes called internal variables as in Chap. 8. This is in opposition with the use of functionals as in rational thermodynamics. We emphasize that the choice of state variables represents always a first step to make and that this is always an assumption. It should also be stressed that GENERIC does not offer any help in the selection of the basic variables.

The next step is to formulate the corresponding set of evolution equations. In classical hydrodynamics, they are the usual balance equations of mass, momentum, and energy complemented by appropriate constitutive equations allowing us to "close" the set of balance equations. In GENERIC, one makes a clear distinction between the reversible and the irreversible contributions to the kinematics of the field variables. The non-dissipative part is obtained explicitly as a straightforward consequence of the dynamics expressed by the Poisson brackets or equivalently by the skew-symmetric operator $\hat{L}$. The reversible part of the evolution equation is completely specified by the kinematics, which means that explicit relations are obtained for the evolution equations including expressions for the scalar thermostatic pressure and eventually other extra reversible contributions like reversible stress tensors. After the evolution equations have been derived, we have no freedom to use other physical requirements to cancel, add or modify the governing evolution relations. The only thing that we can change is the Poisson bracket, indeed we cannot exclude the possibility that other Poisson brackets, different from these used in the preceding sections, provide a better physical description of the physical process.

How do we find the Poisson brackets? A first type of arguments consists in proceeding by trials and errors and, guided by our physical intuition, by comparing the results as arising from the Poisson kinematics with results derived from other approaches. A second more sophisticated method consists in using the Lie group of transformations as addressed in Öttinger's book (2005) to which we refer for more details. The dissipative part of the evolution equations, as specified by bracket (10.20) is derived after an explicit expression for the dissipation potential ψ has been determined, the latter being compelled to obey the properties to be zero at equilibrium and to be a convex function of the variables outside equilibrium. To establish a valuable expression for the dissipation potential is, generally speaking, not an easy task as it should involve material information related to non-equilibrium parameters as viscosity, diffusion coefficient, thermal conductivity, hydrodynamic interaction tensors, etc.

In short, the formulation of the evolution equations of GENERIC requires the knowledge of the following entities: the Poisson bracket, the energy E (or an equivalent thermodynamic potential as Helmholtz free energy), the dissipation potential ψ, and entropy S. The three functions E, S and ψ contain the individual features of the medium under study while to be sure that we are in presence of a true Poisson bracket, we have to check the Jacobi's identity $\{A,\{B,C\}\} + \{B,\{C,A\} + \{C,\{A,B\}\}$; this represents a heavy and tedious task but is of importance because it expresses that the reversible dynamics has a Hamiltonian structure. In recent publications about GENERIC, there is a tendency in favour of the use of the following building blocks: the two generators E (energy) and S (entropy) and the two operators $\hat{L}$ and $\hat{M}$.

As shown in Sect. 10.3, the splitting of the total pressure tensor as $p\delta_{\alpha\beta} + P_{\alpha\beta}$ into a reversible part $p\delta_{\alpha\beta}$ and an extra viscous tensor $P_{\alpha\beta}$ arises as a result of the GENERIC approach. This is not so in the context of other formalisms as classical irreversible thermodynamics, extended irreversible thermodynamics or rational thermodynamics, where this splitting is imposed a priori. Moreover, in most non-equilibrium thermodynamic theories, it is the entropy production that is put into the focus; the part that does not produce entropy, the so-called gyroscopic forces of the form $\omega \times \boldsymbol{x}$ where ω is the angular velocity, or the convected parts of the generalized time derivatives remain largely undetermined. It is a success of GENERIC to incorporate naturally such gyroscopic effects thanks to the Poisson structure of the evolution equations and to propose explicit relations for these non-dissipative contributions. In that respect, it is worth noting that Jacobi's identity imposes that the convected derivatives fulfil the principle of material frame-indifference; this exhibits the strong link between Jacobi's identity and frame-indifference and sheds a new lighting on this controversial principle (see Sect. 9.5).

Note also that in GENERIC, even the well-known and well-tested balance equations of mass, momentum, and energy are re-derived while such relations are usually taken for granted in other thermodynamic theories. Moreover, the governing equations in GENERIC are generally non-linear and this is the reason why they find a natural domain of application in rheology. Another difference between GENERIC and some other thermodynamic approaches is that in the latter, the Onsager reciprocal relations are only applied in the linear regime; in GENERIC, this property is taken for granted even in the non-linear domain as attested by the requirement that the symmetry of operator $\hat{M}$ is to be universally satisfied.

It should also be realized that in GENERIC, it is generally understood that the system under consideration is isolated from its environment as it is assumed that the total energy is conserved $\mathrm{d}E/\mathrm{d}t = 0$ and the total entropy is increasing with time $\mathrm{d}S/\mathrm{d}t > 0$. This means that the theory is not well adapted for describing open and/or externally forced systems as for instance Bénard's convection in a fluid submitted to an external temperature gradient. In the case of non-isolated systems, one should, as in classical thermodynamics, introduce some kind of hypothetical reservoirs but such an operation may be delicate and obscure the formalism (Grmela 2001). In a recent paper, Öttinger (2006) suggests to split the Poisson and dissipative brackets into bulk and boundary contributions.

Moreover, in presence of coupled heat and mass transport, the procedure dictated by GENERIC is rather intricate and requires the introduction of unusual state variables like some particular combinations of the entropy flux and the mass flux (Grmela et al. 2003). Finally, similarly to other approaches, GENERIC formalism remains silent about the physical meaning of some basic quantities such as non-equilibrium temperature.

GENERIC has been the subject of several applications in statistical mechanics, molecular dynamics, quantum mechanics and macroscopic physics, mainly in rheology (Öttinger 2005) where one of the most important problems is to derive constitutive equations consistent with the dynamics of the basic variables. A number of new results produced by GENERIC should also be mentioned, for instance the formulation of generalized reptation models (Öttinger 1999a), of equations in relativistic hydrodynamics and cosmology (Ilg and Öttinger 1999; Öttinger 1999b), new models for heat transfer in nano-systems (Grmela et al. 2005). GENERIC may also be considered as an extension of the bracket formulation as proposed by Beris and Edwards (1994) and Beris (2003).

The essential merit of GENERIC is its unifying universality as it is applicable whatever the level of description of the material system, either macroscopic, mesoscopic, or microscopic. It should also be realized that GENERIC, as well as extended thermodynamics, internal variables theories or rational thermodynamics, are not exclusive of each other: they complement harmoniously and offer, in most circumstances, results that are in agreement with each other. Finally, yet importantly, it is expected that a better comprehension of all these theories will provide a step forward the formulation of a desirable unified theory of irreversible processes.

10.4 Problems

10.1. *Poisson bracket.* Show that the scalar product $(\partial A/\partial \boldsymbol{x}, \hat{L}\partial B/\partial \boldsymbol{x})$ with $\hat{L} = \begin{pmatrix} 0 & 1 \\ -1 & 1 \end{pmatrix}$ is equivalent to the Poisson bracket $\{A, B\}$.

10.2. *Jacobi's identity.* Verify that by taking $\hat{L} = \begin{pmatrix} 0 & 1 \\ -1 & 0 \end{pmatrix}$, the bracket (10.4) satisfies Jacobi's identity (10.8).

10.3. *Degeneracy conditions.* Prove that the results $\mathrm{d}E/\mathrm{d}t = 0$, $\mathrm{d}S/\mathrm{d}t > 0$ may be considered as a direct consequence of the degeneracy conditions $\hat{L}\,\delta S/\delta \boldsymbol{x} = 0$ and $\hat{M}\,\delta E/\delta \boldsymbol{x} = 0$.

10.4. *Degeneracy conditions.* Check that the operators $\hat{L}$ and $\hat{M}$ given by (10.46) verify the degeneracy properties $\hat{L}\,\delta S/\delta \boldsymbol{x} = 0, \hat{M}\,\delta E/\delta \boldsymbol{x} = 0$.

10.5. *Isothermal fluid.* Show that in the case of a one-component isothermal fluid characterized by the variables $\boldsymbol{x} = (\rho, u_\alpha)$, the functional derivatives $\delta E/\delta \boldsymbol{x}$ and $\delta S/\delta \boldsymbol{x}$ are given by

$$\frac{\delta E}{\delta \boldsymbol{x}} = \begin{pmatrix} -u^2/2\rho + \partial\varepsilon/\partial\rho \\ u_\alpha/\rho \end{pmatrix}, \quad \frac{\delta S}{\delta \boldsymbol{x}} = \begin{pmatrix} -\mu/T \\ 0 \end{pmatrix},$$

wherein μ designates the chemical potential and T the uniform temperature.

10.6. *Non-isothermal fluid.* Derive the GENERIC building blocks E, S, L and M for a one-component fluid in presence of temperature gradients (Öttinger and Grmela 1997).

10.7. *Matter diffusion.* Establish the relevant equations of matter diffusion in binary mixtures in the simplified case of non-viscous fluids and uniform temperature.

10.8. *Non-Fickian diffusion.* Repeat the analysis of Sect. 10.2.2 when it is assumed that the barycentric velocity $\boldsymbol{u}$ does not vanish.

10.9. *Thermodiffusion.* Consider the more complex problem of coupling of heat and mass transport (thermodiffusion) in a two-component system. Determine the evolution equations for the total mass density, the mass fraction, the momentum, and the energy in the framework of the above Hamiltonian formalism. *Hint*: See Grmela et al. (2003).

10.10. *Extended irreversible thermodynamics.* Show that the evolution equations governing the behaviour of the dissipative fluxes of extended irreversible thermodynamics possess a Hamiltonian structure (Grmela and Lebon 1990).

Chapter 11
Mesoscopic Thermodynamic Descriptions
Thermodynamics and Fluctuations

The previous chapters were essentially concerned with the average behaviour of thermodynamic variables without reference to their fluctuations. The latter are not only a subject of intense study in statistical mechanics, but also play a significant role in non-equilibrium thermodynamics, like near critical points, non-equilibrium instabilities and, more generally, in processes taking part at small time and space scales. They played an important role in the original derivation of Onsager–Casimir's reciprocal relations, and in modern formulations of the fluctuation–dissipation relations expressing the transport coefficients in terms of the correlation functions of the fluctuations of the fluxes. Furthermore, when the fluctuations become important, they must be included amongst the set of independent variables. The status of fluctuations being intermediate between the macroscopic and the microscopic, it is justified to refer to them as mesoscopic variables.

In the first part of this chapter is presented an introductory overview of the theory of fluctuations. After recalling Einstein's theory of fluctuations around equilibrium, we derive Onsager–Casimir's reciprocal relation, and discuss the fluctuation–dissipation theorem, which is important in non-equilibrium statistical mechanics. Afterwards, we analyse two formalisms which have influenced the recent developments of irreversible thermodynamics, namely Keizer's theory, where a non-equilibrium entropy is defined in terms of the second moments of fluctuations, and the so-called mesoscopic thermodynamics, in which the distribution function of the fluctuations is itself the central variable.

11.1 Einstein's Formula: Second Moments of Equilibrium Fluctuations

In 1902, Einstein proposed a relation between the probability of the fluctuation of a thermodynamic variable with respect to its equilibrium value and the entropy change in isolated systems. This was obtained by inverting the well-known Boltzmann–Planck's formula

$$S = k_{\mathrm{B}} \ln W, \tag{11.1}$$

where S is the total entropy of the system, W is the probability of the macrostate, and k_{B} is the Boltzmann constant. With Einstein, let us write

$$W \approx \exp(\Delta S / k_{\mathrm{B}}), \tag{11.2}$$

where ΔS is the change of entropy associated to the fluctuations, namely

$$\Delta S = S - S_{\mathrm{eq}}. \tag{11.3}$$

As an example, consider the fluctuations of energy in an isolated system composed of two subsystems in thermal contact. In equilibrium, both of them have the same temperature, and the total internal energy is distributed in such a way that the total entropy is maximum. However, despite the total internal energy remains constant, both subsystems may exhibit fluctuations in their respective internal energies.

If the fluctuations are small, ΔS may be expanded around equilibrium, so that

$$\Delta S \approx (\delta S)_{\mathrm{eq}} + \tfrac{1}{2}(\delta^2 S)_{\mathrm{eq}}, \tag{11.4}$$

δS and $\delta^2 S$ being the first and the second differentials of the entropy. For an isolated system, one has $(\delta S)_{\mathrm{eq}} = 0$ and $(\delta^2 S)_{\mathrm{eq}} \leq 0$, because the total entropy is maximum at equilibrium. Introduction of these results into (11.2) yields the well-known Einstein's formula for the probability of fluctuations,

$$W \approx \exp\left(\frac{1}{2}\frac{\delta^2 S}{k_{\mathrm{B}}}\right). \tag{11.5}$$

Note that this relation is restricted to second-order developments and is only applicable to isolated systems.

To extend (11.5) to non-isolated systems, Einstein suggested to express the probability of fluctuations as

$$W \approx \exp\left(-\frac{\Delta \mathcal{A}}{k_{\mathrm{B}} T_0}\right), \tag{11.6}$$

where $\mathcal{A}$, the availability, is a new quantity defined as

$$\mathcal{A} = U - T_0 S + p_0 V, \tag{11.7}$$

in which T_0 and p_0 are the temperature and pressure of the environment. Thus, $\mathcal{A}$ is not an intrinsic property of the system alone but of the ensemble formed by it and its surroundings. It can be shown that $\Delta \mathcal{A}$ measures the maximum amount of useful work which can be extracted from the system during exchanges with the outside world. If the system and the surroundings are in thermal and mechanical equilibrium, $T = T_0$ and $p = p_0$, the change in $\mathcal{A}$ is given by

$$\Delta\mathcal{A} = \Delta U - T\Delta S + p\Delta V = \Delta G, \tag{11.8}$$

where G is the Gibbs' function, so that (11.6) takes the form

$$W \approx \exp(-\Delta G/k_{\mathrm{B}}T). \tag{11.9}$$

Similarly, for a system at fixed V and T, $\Delta\mathcal{A} = \Delta F$, with F the Helmholtz free energy, and the probability of fluctuations becomes

$$W \approx \exp(-\Delta F/k_{\mathrm{B}}T). \tag{11.10}$$

For small fluctuations, $\Delta\mathcal{A}$ can be expanded as

$$\Delta\mathcal{A} \approx \delta\mathcal{A}_{\mathrm{eq}} + \frac{1}{2}(\delta^2\mathcal{A})_{\mathrm{eq}} + \cdots,$$

where the first-order term is zero because at equilibrium G is minimum at fixed T and p (or, respectively, F at fixed V and T). As a consequence,

$$W \approx \exp\left(-\frac{1}{2}\frac{\delta^2 G}{k_{\mathrm{B}}T}\right), \tag{11.11a}$$

$$W \approx \exp\left(-\frac{1}{2}\frac{\delta^2 F}{k_{\mathrm{B}}T}\right). \tag{11.11b}$$

These equations can still be expressed in terms of $\delta^2 S$ because $\delta^2 F$ at constant T and V (or $\delta^2 G$ at fixed p and T) is proportional to $\delta^2 S$. Indeed, remember that $F = U - TS$; when one studies the fluctuations of F at constant T and V, U is the independent variable and S is a function of U so that $\delta^2 F = -T\delta^2 S$. From (11.11a), one recovers (11.5), but with $\delta^2 S$ computed at constant T and V. Similarly, it is easy to see that $\delta^2 G = -T\delta^2 S$ when combined with (11.11b), again leads to (11.5) but with $\delta^2 S$ computed at constant T and p. These observations indicate that (11.5) is applicable even outside isolated systems, at the condition to specify explicitly which variables are kept fixed.

In an isolated system, information about the fluctuations around equilibrium states implies the knowledge of $\delta^2 S$ in terms of δU and δV. The second variation of S is directly obtained from the Gibbs' equation $\delta S = T^{-1}\delta U + pT^{-1}\delta V$ which, introduced in (11.5) yields

$$W \propto \exp\left\{-\frac{1}{2k_{\mathrm{B}}}\left[(T^{-1})_U(\delta U)^2 + 2(T^{-1})_V\delta U\delta V + (T^{-1}p)_V(\delta V)^2\right]\right\}, \tag{11.12}$$

where subscripts U and V stand for partial differentiation with respect to these variables.

Recall the mathematical result that for a Gaussian distribution function of the form

$$W \approx \exp(-\tfrac{1}{2}E_{ij}\delta x_i\delta x_j), \tag{11.13}$$

the second moments of the fluctuations are given by

$$\langle \delta x_i \delta x_j \rangle = (E^{-1})_{ij}, \tag{11.14}$$

where δx_i designates the fluctuation of x_i and the brackets $\langle \ldots \rangle$ denote the average over the probability distribution (11.13). The second moments of the fluctuations of U and V may be obtained from (11.12) and (11.14) by taking into account that the matrix corresponding to the second derivatives of the entropy with respect to these variables is

$$E_{UV} = -\frac{1}{k_\mathrm{B}} \begin{pmatrix} (T^{-1})_U & (T^{-1})_V \\ (T^{-1}p)_U & (T^{-1}p)_V \end{pmatrix}. \tag{11.15}$$

By inverting the matrix and introducing the result in (11.14), it is left as an exercise to prove that (see Problem 11.1)

$$\begin{aligned} \langle \delta U \delta U \rangle &= k_\mathrm{B} C_p T^2 - 2k_\mathrm{B} T^2 pV\alpha + k_\mathrm{B} T p^2 V \kappa_T, \\ \langle \delta U \delta V \rangle &= k_\mathrm{B} T^2 V\alpha - k_\mathrm{B} TpV\kappa_T, \\ \langle \delta V \delta V \rangle &= k_\mathrm{B} TV\kappa_T, \end{aligned} \tag{11.16}$$

where C_p is the heat capacity at constant pressure; α is the coefficient of thermal expansion, and κ_T is the isothermal compressibility. Near critical points, when some of the coefficients in (11.16) diverge, the second moments of the fluctuations become very large and they have a clear macroscopic influence on the behaviour of the system. The analysis of this behaviour has been a main topic of research in statistical mechanics from the 1960s to 1980s (e.g. Callen 1985; Landau and Lifshitz 1980; Reichl 1998).

There is a natural distinction in the fluctuation theory between statics and dynamics. The former is related to properties occurring at equal times, as discussed in the present section, the latter is concerned with their evolution during finite time intervals. This aspect is discussed in the next sections in relation with the derivation of Onsager–Casimir's relations and the dissipation–fluctuation theorem.

11.2 Derivation of the Onsager–Casimir's Reciprocal Relations

The analysis of the evolution of fluctuations in the course of time led Onsager to establish his famous reciprocal relations, which were introduced as a phenomenological postulate in Chap. 2, and whose microscopic derivation is outlined here.

Assume that the entropy of an isolated system is a function of several variables $A_1, \ldots, A_n$ and denote by α_β the fluctuations from their average equilibrium value $\langle A_\beta \rangle$, i.e. $\alpha_\beta = A_\beta - \langle A_\beta \rangle$. For small fluctuations around equilibrium, the entropy may be written as

$$S = S_{\text{eq}} - \frac{1}{2} G_{\beta\gamma} \alpha_\beta \alpha_\gamma, \tag{11.17}$$

with

$$G_{\beta\gamma} = \left(\frac{\partial^2 S}{\partial A_\beta \partial A_\gamma} \right)_{\text{eq}}, \quad (G_{\beta\gamma} = G_{\gamma\beta}). \tag{11.18}$$

Summation with respect to repeated subindices is understood. It follows from Einstein's relation (11.5) that the second moments of the fluctuations are given by

$$\langle \alpha_\beta \alpha_\gamma \rangle = k_{\text{B}} (G^{-1})_{\beta\gamma}. \tag{11.19}$$

In his original demonstration, Onsager assumed that the decay of spontaneous fluctuations of the system obeys the same phenomenological laws that describe the evolution of the perturbations produced by external causes. Furthermore, he supposed that these laws are linear, so that the evolution of α_β is governed by

$$\frac{\mathrm{d}\alpha_\beta}{\mathrm{d}t} = -M_{\beta\gamma} \alpha_\gamma \tag{11.20}$$

with $M_{\beta\gamma}$ a matrix independent of α_γ.

Moreover, Onsager defines the thermodynamic fluxes J_β as the time derivatives of the fluctuations α_β and the thermodynamic forces X_β as the derivatives of the entropy with respect to α_β, i.e.

$$J_\beta = \frac{\mathrm{d}\alpha_\beta}{\mathrm{d}t}, \tag{11.21a}$$

$$X_\beta = \left(\frac{\partial \Delta S}{\partial \alpha_\beta} \right) = -G_{\beta\gamma} \alpha_\gamma. \tag{11.21b}$$

With these definitions of J_β and X_β, the time derivative of the entropy is bilinear in the fluxes and the forces

$$\frac{\mathrm{d}S}{\mathrm{d}t} = -G_{\beta\gamma} \alpha_\gamma \frac{\mathrm{d}\alpha_\beta}{\mathrm{d}t} = J_\beta X_\beta, \tag{11.22}$$

while the evolution equations (11.20) can be cast in the form of linear relations between fluxes and forces:

$$J_\beta = L_{\beta\gamma} \boldsymbol{X}_\gamma. \tag{11.23}$$

Indeed, it suffices to introduce the definitions (11.21) into (11.23) to obtain

$$\frac{\mathrm{d}\alpha_\beta}{\mathrm{d}t} = -L_{\beta\gamma} G_{\gamma\eta} \alpha_\eta. \tag{11.24}$$

Comparison with (11.20) leads to the identification

$$\mathbf{L} = \mathbf{M} \cdot \mathbf{G}^{-1}. \tag{11.25}$$

In a steady state, which by definition is invariant with respect to a translation in time, one has for any couple of variables α_β and α_γ

$$\langle \alpha_\beta(0)\alpha_\gamma(t)\rangle = \langle \alpha_\beta(0+\tau)\alpha_\gamma(t+\tau)\rangle. \tag{11.26}$$

Since this equality is valid for any values of t and τ; it is in particular true for $\tau = -t$, from which

$$\langle \alpha_\beta(0)\alpha_\gamma(t)\rangle = \langle \alpha_\beta(-t)\alpha_\gamma(0)\rangle. \tag{11.27}$$

Assume that the variable A_β has a well-defined time-reversal parity under the transformations $t \to -t$, by definition, even variables will transform as $\alpha_\beta(-t) = \alpha_\beta(t)$, while odd variables will behave as $\alpha_\beta(t) = -\alpha_\beta(-t)$. Mass and energy are examples of even variables, and momentum and heat flux are odd variables. This allows us to write (11.27) in whole generality as

$$\langle \alpha_\beta(0)\alpha_\gamma(t)\rangle = \varepsilon_\beta\varepsilon_\gamma\langle \alpha_\beta(t)\alpha_\gamma(0)\rangle, \tag{11.28}$$

where $\varepsilon_\beta = +1$ or -1 according the variable is even or odd. Note that (11.28) implies that equal-time correlations between even and odd variables vanish in equilibrium. Relation (11.28) expresses the microscopic reversibility of the system and is of fundamental importance in deriving the Onsager–Casimir's relations.

Moreover, direct integration of the evolution equation (11.20) yields

$$\alpha_\beta(t) = \exp[-\mathbf{M}t]_{\beta\gamma}\alpha_\gamma(0). \tag{11.29}$$

For small values of t, one may expand (11.29) in Taylor's series, by keeping only first-order terms in t, and (11.28) takes then the form

$$\langle \alpha_\beta(0)(\delta_{\gamma\eta} - M_{\gamma\eta}t)\alpha_\eta(0)\rangle = \varepsilon_\beta\varepsilon_\gamma\langle(\delta_{\beta\eta} - M_{\beta\eta}t)\alpha_\eta(0)\alpha_\gamma(0)\rangle, \tag{11.30}$$

from which

$$M_{\gamma\eta}\langle \alpha_\beta(0)\alpha_\eta(0)\rangle = \varepsilon_\beta\varepsilon_\gamma M_{\beta\eta}\langle \alpha_\eta(0)\alpha_\gamma(0)\rangle. \tag{11.31}$$

since $\delta_{\gamma\eta} = 1$ if $\gamma = \eta$, $\delta_{\gamma\eta} = 0$ otherwise. Expressing the second moments in terms of the matrix $\mathbf{G}$ of the second derivatives of the entropy, (11.31) can be written as

$$M_{\gamma\eta}(G^{-1})_{\beta\eta} = \varepsilon_\beta\varepsilon_\gamma M_{\beta\eta}(G^{-1})_{\eta\gamma}. \tag{11.32}$$

From relation (11.25) between $\mathbf{L}$ and $\mathbf{M}$, and the property that $\mathbf{G}$ is a symmetric matrix by construction, it follows that

$$L_{\gamma\beta} = \varepsilon_\beta\varepsilon_\gamma L_{\beta\gamma}, \tag{11.33}$$

which are the well-known Onsager–Casimir's reciprocal relations. Historically, Onsager derived the reciprocity relations in 1931, but only in the case of even variables. The extension to odd variables was achieved in 1945 by Casimir.

The above analysis exhibits clearly the main assumptions underlying the derivation of the Onsager–Casimir's relations, namely linear evolution equations and the hypothesis that the behaviour of fluctuations is described on the average by relations mimicking the macroscopic transport laws. In Chap. 2 these relations were presented as a macroscopic postulate, whereas here it follows from dynamical equations and statistical considerations.

In presence of a magnetic field $\boldsymbol{\mathcal{H}}$, the operation of time reversal will only change the internal magnetic field arising from the motion of the particles, which is automatically reversed under time reversal. However, because of the Lorentz force, which is proportional to the vector product $\boldsymbol{v} \times \boldsymbol{\mathcal{H}}$ of the velocities and the magnetic induction, the external magnetic field $\boldsymbol{\mathcal{H}}$ must be inverted in order that Lorentz's force remains unchanged; this guarantees that the reversal motion follows the same path as the direct one. The same comment applies when the system rotates as a whole with angular velocity $\boldsymbol{\omega}$; the sign of the angular velocity must be changed because the Coriolis force is proportional to the vector product $\boldsymbol{\omega} \times \boldsymbol{v}$ of the velocities of the particles and the angular velocity of the system. Therefore, in presence of an external magnetic induction $\boldsymbol{\mathcal{H}}$ and angular rotation $\boldsymbol{\omega}$, the Onsager–Casimir's relations read as

$$L_{\gamma\beta}(\boldsymbol{\mathcal{H}}, \boldsymbol{\omega}) = \varepsilon_\beta \varepsilon_\gamma L_{\beta\gamma}(-\boldsymbol{\mathcal{H}}, -\boldsymbol{\omega}). \qquad (11.34)$$

The importance and the limitations of the Onsager–Casimir's relations have been underlined in Chap. 2. Here, the role of the time-reversal parity of the several variables becomes clear, as well as the reason why $\boldsymbol{\mathcal{H}}$ and $\boldsymbol{\omega}$ must be inverted to keep the validity of the reciprocal relations.

11.3 Fluctuation–Dissipation Theorem

In his paper on Brownian motion published in 1905, Einstein established that molecular collisions between the fluid and the Brownian particles are the source of friction or dissipation and moreover responsible for the occurrence of fluctuations in the position of the particles. Einstein's idea was further explored by many authors, and was the basis for the derivation of the so-called fluctuation–dissipation theorem, expressing that fluctuations and dissipative coefficients are related to each other. Langevin's stochastic analysis of the motion of a Brownian particle, and Nyquist study of the fluctuations of the electric current in a conductor contributed further to a better understanding of the relation between fluctuations and dissipation. Since the 1950s, these analyses were set on a more general basis by Callen and Welton (1951),

Callen and Greene (1952), and Tisza and Manning (1957). It is not here our purpose to enter into the details of their derivations, but rather to underline the general ideas, because of their importance in statistical thermodynamics.

One of the main objectives of non-equilibrium statistical mechanics is to provide, at the microscopic level, an expression for the memory function relating a flux $\boldsymbol{J}(t)$, at time t, to its conjugate force $\boldsymbol{X}(t')$, at previous times $t' \leq t$:

$$\boldsymbol{J}(t) = \int_{-\infty}^{0} \mathbf{K}(t-t') \cdot \boldsymbol{X}(t')\mathrm{d}t'; \tag{11.35}$$

$\mathbf{K}(t-t')$ is the so-called memory function or memory kernel. As an example, consider the Maxwell–Cattaneo evolution equation written formally as,

$$\tau \frac{\partial \boldsymbol{J}}{\partial t} = -(\boldsymbol{J} - \mathbf{L} \cdot \boldsymbol{X}), \tag{11.36}$$

with $\mathbf{L}$ the corresponding transport coefficient. Integration of (11.36) yields, for a zero initial flux at time $t' = -\infty$,

$$\boldsymbol{J}(t) = \int_{-\infty}^{0} \frac{\mathbf{L}}{\tau} \exp\left[-(t-t')/\tau\right] \cdot \boldsymbol{X}(t')\mathrm{d}t'. \tag{11.37}$$

By comparing (11.37) and (11.35), one may identify the memory function as

$$\mathbf{K}(t-t') = \frac{\mathbf{L}}{\tau} \exp\left(-\frac{t-t'}{\tau}\right), \tag{11.38}$$

indicating that in the linearized version of extended irreversible thermodynamics (EIT), the memory function decays exponentially. More general versions of transport theory can be found in modern statistical mechanics, based on N-particle distribution functions instead of the more restrictive one-particle distribution function, and on the Liouville equation for the description of its evolution.

The most significant feature of the fluctuation–dissipation theorem is that the memory kernel is related to the correlation function of the fluctuations of the corresponding fluxes through (McQuarrie 1976; Reichl 1998)

$$K(t-t') = \frac{1}{k_{\mathrm{B}}T} \langle \delta J(t) \delta J(t') \rangle, \tag{11.39}$$

angular brackets stand for the average over an equilibrium distribution function.

As particular examples of (11.39), the memory functions generalizing the thermal conductivity and shear viscosity are written as

$$\lambda_{ij}(t-t') = \frac{V}{k_{\mathrm{B}}T^2}\langle \delta q_i(t)\delta q_j(t')\rangle. \tag{11.40}$$

$$\eta_{ijkl}(t-t') = \frac{V}{k_{\mathrm{B}}T}\langle \delta P^v_{ij}(t)\delta P^v_{kl}(t')\rangle. \tag{11.41}$$

The knowledge of the quantities inside the brackets of (11.40) and (11.41) in microscopic terms opens the way to the determination of the memory functions in the most general situations, from dilute ideal gases to dense fluids.

In terms of the memory functions (11.40) and (11.41), the Fourier's and Newton–Stokes' equations are generalized in the form

$$q_i(t) = -\int_{-\infty}^{0} \lambda_{ij}(t-t')\frac{\partial T(t')}{\partial x_j}\mathrm{d}t', \tag{11.42a}$$

$$P^v_{ij}(t) = -\int_{-\infty}^{0} \eta_{ijkl}(t-t')V_{kl}(t')\mathrm{d}t'. \tag{11.42b}$$

When the relaxation times of the fluctuations of the fluxes are very short, in such a way that the memory functions are negligible except for very small values of $t-t'$, the bounds of the integrals may be extended to infinity and the thermodynamic forces may be taken out of the integral with the value corresponding to $t=t'$. Doing so, one recovers from (11.40) and (11.41), the Green–Kubo's formula relating the dissipative coefficients of heat conductivity and shear viscosity to the fluctuations of the heat flux and the momentum flux, respectively:

$$\lambda_{ij} = (k_{\mathrm{B}}T^2)^{-1}\int_0^{\infty} \langle \delta q_i(0)\delta q_j(t)\rangle \mathrm{d}t, \tag{11.43a}$$

$$\eta_{ijkl} = (k_{\mathrm{B}}T)^{-1}\int_0^{\infty} \langle \delta P^v_{ij}(0)\delta P^v_{kl}(t)\rangle \mathrm{d}t. \tag{11.43b}$$

The Green–Kubo's relations are useful for calculating the dissipative transport coefficients from first principles, independently of the density of the system, in contrast with the classical calculations starting from the Boltzmann equation, whose validity is limited to a low-density range (Reichl 1998).

Despite their elegance and generality, the simplicity of the above results is misleading because of the complexity of the microscopic Liouville operator describing the evolution of the fluctuations of the fluxes. This is why some approximations must be introduced, usually in the form of mathematical models for the memory functions, as a prerequisite for solving practical problems. Therefore, formalisms such as EIT may be of valuable assistance to model the evolution of the fluctuations of the fluxes.

11.4 Keizer's Theory: Fluctuations in Non-Equilibrium Steady States

11.4.1 Dynamics of Fluctuations

Keizer (1978, 1983, 1987) proposed to regard the fluctuations as the foundations for a definition of entropy outside equilibrium. More recently (Reguera 2000, Rubi 2004), attempts have been made to upgrade the fluctuations to the status of independent variables besides their usual average values. This topic will be analysed in the next section.

Keizer combines the molecular picture of Boltzmann and the stochastic picture of Onsager, and lays the foundations of a statistical thermodynamics of equilibrium and non-equilibrium steady states. Steady states are, like equilibrium states, time-independent but they involve constant inputs and outputs per unit time, i.e. they require more parameters, like the fluxes, to specify the state.

Consider a large molecular system undergoing a series of elementary processes characterized by a set of extensive variables $\boldsymbol{x}(= U(\text{energy}), V, N \dots N$ (volume), $N_1, \dots, N_n$, with N_i the number of particles of component i). According to the Boltzmann–Planck's postulate, the probability of a fluctuation $\delta\boldsymbol{x}$ around local equilibrium is given by

$$W(\delta\boldsymbol{x}) \approx \exp(\Delta S/k_\mathrm{B}) = \{(2\pi)^{-(n+2)} \det[(-S_\mathrm{eq}/k_\mathrm{B})]\}^{1/2} \exp(\delta^2 S_\mathrm{eq}/2k_\mathrm{B}), \tag{11.44}$$

with

$$(S_\mathrm{eq})_{ij} = (\partial^2 S/\partial x_i \partial x_j)_\mathrm{eq}, \quad \delta^2 S_\mathrm{eq} = (\partial^2 S/\partial x_i \partial x_j)_\mathrm{eq} \delta x_i \delta x_j, \tag{11.45}$$

after performing a Taylor expansion around equilibrium and using the maximum entropy principle. Like in Sect. 11.1, $\delta\boldsymbol{x}$ stands for the fluctuations of the variables with respect to their equilibrium values (namely $\delta\boldsymbol{x}(\boldsymbol{r},t) \equiv \boldsymbol{x}(\boldsymbol{r},t) - \boldsymbol{x}_\mathrm{eq} \equiv (\delta U, \delta V, \delta N_1, \dots, \delta N_n)$. From now on, we shall omit for simplicity to write the time and position dependence of the variables.

A main feature of Keizer's theory is that the elementary molecular processes can be described by "canonical" kinetic evolution equations of the form

$$\frac{\mathrm{d}x_i}{\mathrm{d}t} = \sum_k \omega_{ki}(V_k^+ - V_k^-) + J_i(\boldsymbol{x},t) \equiv R_i(\boldsymbol{x},t). \tag{11.46}$$

The first term in the right-hand side represents the dissipative part of the evolution and J_i the non-dissipative part, R_i is a compact notation to express the sum of both terms. The quantity ω_{ki} stands for $\omega_{ki} = x_{ki}^+ - x_{ki}^-$, with the + or − upper indices denoting the forward or backward variation of the values of the extensive variable i in the elementary process k, finally $V_k^{\pm}$ is given by

$$V_k^{\pm} = \Omega_k^{\pm} \exp\left[-\frac{1}{k_{\mathrm{B}}}\sum_j x_{kj}^{\pm}\left(\frac{\partial S}{\partial x_j}\right)\right], \tag{11.47}$$

with $\Omega_k^{\pm}$ the intrinsic forward and backward rates of the elementary k process. It was shown by Keizer that the Fourier's, Fick's, and Newton–Stokes' transport laws as well as the chemical mass action law or the Boltzmann collision operator may be written in the canonical form (11.46).

It follows from (11.46) that the dynamics of fluctuations around a steady state may be described by

$$\frac{\mathrm{d}\delta x_i}{\mathrm{d}t} = \left(\frac{\partial R_i}{\partial x_j}\right)\delta x_j + f_i \equiv H_{ij}(\boldsymbol{x}^{\mathrm{ss}})\delta x_j + f_i, \tag{11.48}$$

where superscript ss stands for steady state, and f_i for a stochastic noise. By definition, one has $\langle f_i \rangle = 0$ and

$$\langle f_i(t) f_j(t')\rangle = \sum_j \omega_{ki}\left[V_k^{+}(\boldsymbol{x}^{\mathrm{ss}}) + V_k^{-}(\boldsymbol{x}^{\mathrm{ss}})\right]\omega_{kj}\delta(t-t') \equiv \gamma_{ij}(\boldsymbol{x}^{\mathrm{ss}})\delta(t-t'). \tag{11.49}$$

It follows from these results that the probability of fluctuations in a non-equilibrium steady state may be written in the Gaussian form

$$W(\delta\boldsymbol{x}) = [(2\pi)^{-(n+2)} \det \boldsymbol{\sigma}^{-1}] \exp\left[-\frac{1}{2}\delta\boldsymbol{x}^{\mathrm{T}}\cdot\boldsymbol{\sigma}^{-1}\cdot\delta\boldsymbol{x}\right], \tag{11.50}$$

with $\boldsymbol{\sigma}$ the covariance matrix,

$$\boldsymbol{\sigma} \equiv \langle \delta\boldsymbol{x}\delta\boldsymbol{x}\rangle^{\mathrm{ss}}, \tag{11.51}$$

given by the generalized fluctuation–dissipation expression

$$\mathbf{H}(\boldsymbol{x}^{\mathrm{ss}})\cdot\boldsymbol{\sigma} + \boldsymbol{\sigma}\cdot\mathbf{H}^{\mathrm{T}}(\boldsymbol{x}^{\mathrm{ss}}) = -\boldsymbol{\gamma}(\boldsymbol{x}^{\mathrm{ss}}). \tag{11.52}$$

Since the matrices $\mathbf{H}$ and $\boldsymbol{\gamma}$ are known from (11.48) and (11.49), the non-equilibrium covariance matrix (11.51) may directly be obtained from (11.52).

11.4.2 A Non-Equilibrium Entropy

An important pillar of Keizer's theory is the definition of a non-equilibrium entropy function S_{K} by interpreting (11.50) in an analogous way to (11.44), namely as

$$\frac{\partial^2 S_{\mathrm{K}}}{\partial x_i \partial x_j} \equiv -k_{\mathrm{B}}(\boldsymbol{\sigma}^{-1})_{ij}. \tag{11.53}$$

This is a differential equation for S_{K}, which may be solved provided that $\partial\sigma_{ij}/\partial x_k = \partial\sigma_{ik}/\partial x_j$, a condition that was explicitly studied by Keizer. At equilibrium, $\boldsymbol{\sigma}$ reduces to the equilibrium covariance matrix and S_{K} becomes identical to the equilibrium entropy. The physics underlying the above definition is that the entropy S_{K} is related to the fluctuations in the extensive variables in a way analogous to Boltzmann–Planck's formula (11.1).

When expressed in terms of S_{K}, the statistical distribution (11.53) is of the form $W \approx \exp(\delta^2 S_{\mathrm{K}}/2k_{\mathrm{B}})$ with

$$\delta^2 S_{\mathrm{K}} = \sum_{ij} (\partial^2 S_{\mathrm{K}}/\partial x_i \partial x_j)^{\mathrm{ss}} (x_i - x_i^{\mathrm{ss}})(x_i - x_i^{\mathrm{ss}}). \tag{11.54}$$

Since stability requirements impose that $\boldsymbol{\sigma}$ must be a positive definite matrix, it follows from (11.53) that in stable steady states $\delta^2 S_{\mathrm{K}} \leq 0$. Moreover, it was shown by Keizer that the second variation of S_{K} is a Lyapounov function, from which

$$\delta^2 S_{\mathrm{K}} \leq 0, \qquad \frac{\mathrm{d}}{\mathrm{d}t}\delta^2 S_{\mathrm{K}} \geq 0. \tag{11.55}$$

Since non-equilibrium states can only be sustained by the presence of fluxes (of energy, mass, etc.) between the system and reservoirs (of energy, mass, etc.), it is assumed by Keizer that S_{K} will depend not only on the intensive variables $\boldsymbol{x}$ but also on the fluxes $\boldsymbol{f}$ and, in addition, some intensive variables $\boldsymbol{\Gamma}$ characterizing the reservoirs, so that

$$S_{\mathrm{K}} = S_{\mathrm{K}}(\boldsymbol{x}; \boldsymbol{f}, \boldsymbol{\Gamma}). \tag{11.56}$$

In analogy with classical thermodynamics, Keizer defines intensive variables conjugate to the extensive variables x_i by

$$\phi_i = \frac{\partial S_{\mathrm{K}}}{\partial x_i}. \tag{11.57}$$

In particular, the intensive variables conjugate to the energy, volume and number of moles are related to generalized temperature T, pressure p, and chemical potentials μ_i, in such a way that the generalized Gibbs' equation for S_{K} will read as

$$\mathrm{d}S_{\mathrm{K}} = T^{-1}\mathrm{d}U + pT^{-1}\mathrm{d}V - \sum_i \mu_i T^{-1}\mathrm{d}N_i + \sum_j \frac{\partial S_{\mathrm{K}}}{\partial f_j}\mathrm{d}f_j + \sum_l \frac{\partial S_{\mathrm{K}}}{\partial \Gamma_l}\mathrm{d}\Gamma_l. \tag{11.58}$$

Note that (11.58) contains a contribution in the fluxes, which is analogous to EIT – the latter lacking the last term, related to the properties of the reservoirs.

Expanding $S_{\mathrm{K}}(\boldsymbol{x}; \boldsymbol{f}, \boldsymbol{\Gamma})$ around equilibrium results in

$$S_{\mathrm{K}}(\boldsymbol{x}; \boldsymbol{f}, \boldsymbol{\Gamma}) = S_{\mathrm{eq}}(\boldsymbol{x}) - \sum_i f_i v_i(\boldsymbol{x}; \boldsymbol{f}, \boldsymbol{\Gamma}), \tag{11.59}$$

where ν_i is the "force" conjugated to f_i. As a consequence, any intensive variable, like for instance the temperature T conjugated to the internal energy U through $1/T = \partial S_{\mathrm{K}}/\partial U$, will be given by

$$\frac{1}{T} = \frac{1}{T_{\mathrm{eq}}} - \sum_i f_i \frac{\partial v_i}{\partial U}. \tag{11.60}$$

This result shares some features with EIT, as it exhibits the property that the temperature is not equal to the (local) equilibrium temperature T_{eq}, but contains additional terms depending on the fluxes (see Box 7.3).

An alternative expression of the generalized entropy, more explicit in the fluctuations and based on (11.53), is

$$S_{\mathrm{K}}(\boldsymbol{x}; \boldsymbol{f}, \boldsymbol{\Gamma}) = S_{\mathrm{eq}}(\boldsymbol{x}) - \frac{1}{2} k_{\mathrm{B}} \sum_{ij} x_i x_j (\sigma^{-1} - \sigma_{\mathrm{eq}}^{-1})_{ij}, \tag{11.61}$$

where $x_0 = U$ corresponds to the internal energy. It is directly inferred from (11.61) that the temperature can be cast in the form

$$\frac{1}{T} = \frac{1}{T_{\mathrm{eq}}} - k_{\mathrm{B}} \sum_i x_i (\sigma_{i0}^{-1} - \sigma_{i0,\mathrm{eq}}^{-1}). \tag{11.62}$$

This result is important as it points out that non-equilibrium temperature (the same reasoning remains valid for the pressure or the chemical potential) is not identical to the local equilibrium one but is related to it through the difference of the non-equilibrium and equilibrium correlation function of the fluctuations.

A spontaneous variation of the extensive variables x_i will lead to a rate of change of entropy given by

$$\frac{\mathrm{d}S_{\mathrm{K}}}{\mathrm{d}t} = \sum_i \phi_i \frac{\mathrm{d}x_i}{\mathrm{d}t}. \tag{11.63}$$

Using the property that for small deviations with respect to the steady state

$$\phi_i - \phi_i^{\mathrm{ss}} = \sum_j \left(\frac{\partial^2 S}{\partial x_i \partial x_j} \right)^{\mathrm{ss}} (x_j - x_j^{\mathrm{ss}}), \tag{11.64}$$

(11.54) can be written as

$$\delta^2 S_{\mathrm{K}} = \sum_i (\phi_i - \phi_i^{\mathrm{ss}})(x_i - x_i^{\mathrm{ss}}) \tag{11.65}$$

and, after differentiation with respect to time,

$$\frac{\mathrm{d}}{\mathrm{d}t} \delta^2 S_{\mathrm{K}} = 2 \sum_i (\phi_i - \phi_i^{\mathrm{ss}}) \frac{\mathrm{d}x_i}{\mathrm{d}t} \geq 0. \tag{11.66}$$

Rearranging (11.66), one obtains

$$\sum_i \phi_i \frac{\mathrm{d}x_i}{\mathrm{d}t} \geq \sum_i \phi_i^{\mathrm{ss}} \frac{\mathrm{d}x_i}{\mathrm{d}t}. \tag{11.67}$$

The left-hand side of inequality (11.67) refers to the instantaneous values of the intensive variables whereas the right-hand side involves their average values in the steady state, but in virtue of (11.63), the left-hand side is also the rate of change of entropy so that finally,

$$\frac{\mathrm{d}S_{\mathrm{K}}}{\mathrm{d}t} \geq \sum_i \phi_i^{\mathrm{ss}} \frac{\mathrm{d}x_i}{\mathrm{d}t}, \tag{11.68}$$

which was called by Keizer a generalized Clausius inequality because it generalizes Clausius inequality $T_{\mathrm{R}}\mathrm{d}S/\mathrm{d}t > \mathrm{d}Q/\mathrm{d}t$ established for a system in equilibrium with a reservoir at temperature T_{R}.

Keizer's theory has been the subject of numerous applications as ion transport through biological membranes, isomerization reactions, fluctuations caused by electro-chemical reactions, light scattering under thermal gradients, laser heated dimerization, etc. (Keizer 1987).

11.5 Mesoscopic Non-Equilibrium Thermodynamics

At short time and small length scales, the molecular nature of the systems cannot be ignored. Classical irreversible thermodynamics(CIT) is no longer satisfactory; indeed, molecular degrees of freedom that have not yet relaxed to their equilibrium value will influence the global dynamics of the system, and must be incorporated into the description, as done for example in extended thermodynamics or in internal variables theories. Unlike these approaches, which use as variables the average values of these quantities and in contrast also with Keizer's theory that adds, as additional variables, the non-equilibrium part of the second moments of their fluctuations, mesoscopic non-equilibrium thermodynamics describes the system through a probability distribution function $P(\boldsymbol{x}, t)$, where $\boldsymbol{x}$ represents the set of all relevant degrees of freedom remaining active at the time and space scales of interest (Reguera 2004; Rubí 2004).

The main idea underlying mesoscopic non-equilibrium thermodynamics is to use the methods of CIT to obtain the evolution equation for $P(\boldsymbol{x}, t)$. The selected variables do not refer to the microscopic properties of the molecules, as for instance in the kinetic theory, but are obtained from an averaging procedure, as in macroscopic formulations. It is in this sense that the theory is mesoscopic, i.e. intermediate between macroscopic and microscopic descriptions. It shares some characteristics with extended thermodynamics or internal variables theories, like an enlarged choice of variables and statistical theories like the statistical concept of distribution function.

11.5.1 Brownian Motion with Inertia

As illustration, we will consider the role of inertial effects in the problem of diffusion of N non-interacting Brownian particles of mass m immersed in a fluid of volume V. Inertial effects are relevant when changes in spatial density occur at timescales comparable to the time required by the velocity distribution of particles to relax to its equilibrium Maxwellian value. At short timescales, the particles do not have time to reach the equilibrium velocity distribution, therefore the local equilibrium hypothesis cannot remain valid and fluctuations become relevant. They are modelled by introducing as extra variables the probability density $P(\boldsymbol{r}, \boldsymbol{v}, t)$ to find the system with position between $\boldsymbol{r}$ and $\boldsymbol{r} + \mathrm{d}\boldsymbol{r}$ and velocity between $\boldsymbol{v}$ and $\boldsymbol{v} + \mathrm{d}\boldsymbol{v}$ at time t. The problem to be solved is to obtain the evolution equation of $P(\boldsymbol{r}, \boldsymbol{v}, t)$.

The connection between entropy and probability of a state is given by the Gibbs' entropy postulate (e.g. de Groot and Mazur 1962), namely

$$s = s_{\mathrm{eq}} - \frac{k_{\mathrm{B}}}{m} \int P(\boldsymbol{r}, \boldsymbol{v}, t) \ln \frac{P(\boldsymbol{r}, \boldsymbol{v}, t)}{P_{\mathrm{eq}}(\boldsymbol{r}, \boldsymbol{v})} \mathrm{d}\boldsymbol{r}\, \mathrm{d}\boldsymbol{v}, \tag{11.69}$$

where s is the entropy per unit mass and $P_{\mathrm{eq}}(\boldsymbol{r}, \boldsymbol{v})$ the equilibrium distribution. By analogy with CIT, we formulate a Gibbs' equation of the form

$$T\,\mathrm{d}s = -\int \mu(\boldsymbol{r}, \boldsymbol{v}, t)\mathrm{d}P(\boldsymbol{r}, \boldsymbol{v}, t)\mathrm{d}\boldsymbol{r}\, \mathrm{d}\boldsymbol{v}, \tag{11.70}$$

where T is the temperature of a heat reservoir and μ the chemical potential per unit mass, which can be given the general form

$$\mu(\boldsymbol{r}, \boldsymbol{v}, t) = \mu_{\mathrm{eq}} + \frac{k_{\mathrm{B}}T}{m} \ln \frac{P(\boldsymbol{r}, \boldsymbol{v}, t)}{P_{\mathrm{eq}}(\boldsymbol{r}, \boldsymbol{v})} + K(\boldsymbol{r}, \boldsymbol{v}), \tag{11.71}$$

wherein $K(\boldsymbol{r}, \boldsymbol{v})$ is an extra potential which does not depend on $P(\boldsymbol{r}, \boldsymbol{v}, t)$. Since no confusion is possible between chemical potentials measured per unit mass or per mole, we have omitted the horizontal bar surmounting "μ". For an ideal system of non-interacting particles in absence of external fields, statistical mechanics considerations suggest to identify $K(\boldsymbol{r}, \boldsymbol{v})$ with the kinetic energy per unit mass

$$K(\boldsymbol{r}, \boldsymbol{v}) = \frac{1}{2}v^2. \tag{11.72}$$

The evolution equation for the distribution function will be given by the continuity equation in the position and velocity space, namely

$$\frac{\partial P}{\partial t} = -\frac{\partial \boldsymbol{J}_r}{\partial \boldsymbol{r}} - \frac{\partial \boldsymbol{J}_v}{\partial \boldsymbol{v}}, \tag{11.73}$$

where $\boldsymbol{J}_r$ and $\boldsymbol{J}_v$ are the probability fluxes in the $\boldsymbol{r}, \boldsymbol{v}$ space. The diffusion flux of particles in the physical $\boldsymbol{r}$ coordinate space is directly obtained by integration over the velocity space

$$\bar{\boldsymbol{J}}_r(\boldsymbol{r},t) = \int \boldsymbol{v} P(\boldsymbol{r},\boldsymbol{v},t)\mathrm{d}\boldsymbol{v}. \tag{11.74}$$

The fluxes $\boldsymbol{J}_r$ and $\boldsymbol{J}_v$ in (11.73) are not known a priori, and will be derived by following the methodology of non-equilibrium thermodynamic methods, i.e. by imposing the restrictions placed by the second law. By differentiating (11.70) with respect to time and using (11.73), it is found that the rate of entropy production can be written as

$$T\frac{\mathrm{d}s}{\mathrm{d}t} = \int \left(-\boldsymbol{J}_r \cdot \frac{\partial \mu}{\partial \boldsymbol{r}} - \boldsymbol{J}_v \cdot \frac{\partial \mu}{\partial \boldsymbol{v}}\right)\mathrm{d}\boldsymbol{r}\,\mathrm{d}\boldsymbol{v}, \tag{11.75}$$

after performing partial integrations and supposing that the fluxes vanish at the boundaries. As in classical irreversible thermodynamics, one assumes linear relations between the fluxes $\boldsymbol{J}$ and the forces, so that

$$\boldsymbol{J}_r = -L_{rr}\frac{\partial \mu}{\partial \boldsymbol{r}} - L_{rv}\frac{\partial \mu}{\partial \boldsymbol{v}}, \tag{11.76}$$

$$\boldsymbol{J}_v = -L_{vr}\frac{\partial \mu}{\partial \boldsymbol{r}} - L_{vv}\frac{\partial \mu}{\partial \boldsymbol{v}}, \tag{11.77}$$

where L_{ij} are phenomenological coefficients, to be interpreted later on. To ensure the positiveness of the entropy production (11.75), the matrix of these coefficients must be positive definite. Furthermore, if we take for granted the Onsager–Casimir's reciprocity relations, one has $L_{rv} = -L_{vr}$, with the minus sign because $\boldsymbol{r}$ and $\boldsymbol{v}$ have opposite time-reversal parity.

To identify the phenomenological coefficients, substitute (11.71) in (11.76) and impose the condition that the particle diffusion flux in the $\boldsymbol{r}$-space should be recovered from the flux in the $\boldsymbol{r},\boldsymbol{v}$ space, namely

$$\begin{aligned}\bar{\boldsymbol{J}}_r(\boldsymbol{r},t) &= \int \boldsymbol{v} P(\boldsymbol{r},\boldsymbol{v},t)\mathrm{d}\boldsymbol{v} = \int \boldsymbol{J}_r(\boldsymbol{r},\boldsymbol{v},t)\mathrm{d}\boldsymbol{v} \\ &= -\int \left(L_{rr}\frac{k_\mathrm{B}T}{m}\frac{1}{P}\frac{\partial P}{\partial \boldsymbol{r}} + L_{rv}\frac{k_\mathrm{B}T}{m}\frac{1}{P}\frac{\partial P}{\partial \boldsymbol{v}} + L_{rv}\boldsymbol{v}\right)\mathrm{d}\boldsymbol{v}. \end{aligned}\tag{11.78}$$

Since $P(\boldsymbol{r},\boldsymbol{v},t)$ is arbitrary, (11.78) may only be identically satisfied if $L_{rr}=0$, $L_{rv} = -P$, and the only left undetermined coefficient is L_{vv}. If it is taken as $L_{vv} = P/\tau$, where τ is a velocity relaxation time related to the inertia of the particles, (11.76) and (11.77) become, respectively,

$$\boldsymbol{J}_r = \left(\boldsymbol{v} + \frac{D}{\tau}\frac{\partial}{\partial \boldsymbol{v}}\right)P, \tag{11.79}$$

$$\boldsymbol{J}_v = -\left(\frac{D}{\tau}\frac{\partial}{\partial \boldsymbol{r}} + \frac{\boldsymbol{v}}{\tau} + \frac{D}{\tau^2}\frac{\partial}{\partial \boldsymbol{v}}\right)P, \tag{11.80}$$

where $D \equiv (k_\mathrm{B}T/m)\tau$ is identified as the diffusion coefficient. When these expressions are introduced into the continuity equation (11.73), it is found that

$$\frac{\partial P}{\partial t} = -\frac{\partial}{\partial \boldsymbol{r}} \cdot (\boldsymbol{v}P) + \frac{\partial}{\partial \boldsymbol{v}} \cdot \left(\frac{\boldsymbol{v}}{\tau} + \frac{D}{\tau^2}\frac{\partial}{\partial \boldsymbol{v}}\right) P. \tag{11.81}$$

This is the well-known Fokker–Planck's equation for non-interacting Brownian particles in presence of inertia and it is worth to stress that this equation has been derived only by thermodynamic methods, without explicit reference to statistical mechanics. Equilibrium situations corresponds to the vanishing of the fluxes, i.e. $\boldsymbol{J}_r = 0, \boldsymbol{J}_v = 0$ and a Gaussian probability distribution. In CIT, it is assumed that the probability distribution is the same as in equilibrium, i.e. Gaussian, centred at a non-zero average, and with its variance related to temperature in the same way as in equilibrium. Moreover, diffusion in the coordinate $\boldsymbol{r}$-space is much slower than in the velocity $\boldsymbol{v}$-space, so that it is justified to put $\boldsymbol{J}_v = 0$. From (11.80), it is then seen that $(\boldsymbol{v} + (D/\tau)\partial/\partial\boldsymbol{v})P = -D\partial P/\partial\boldsymbol{r}$ which, substituted in (11.79), yields Fick's law

$$\boldsymbol{J}_r = -D\frac{\partial \mu}{\partial \boldsymbol{r}}. \tag{11.82}$$

Far from equilibrium, neither the space nor the velocity distributions correspond to equilibrium and one has $\boldsymbol{J}_r \neq 0, \boldsymbol{J}_v \neq 0$. In this case, the velocity distribution may be very different from a Gaussian form, and it is not clear how to define temperature (see, however, Box 11.1 where an attempt to define a non-equilibrium temperature is presented).

Box 11.1 Non-Equilibrium Temperature

In the present formalism, it is not evident how to define a temperature outside equilibrium. To circumvent the problem, a so-called effective temperature has been introduced. It is defined as the temperature at which the system is in equilibrium, i.e. the one corresponding to the probability distribution at which the rate of entropy production vanishes. Substituting (11.79) and (11.80) in (11.75), it is easily proven that the entropy production can be written as

$$\frac{\mathrm{d}s}{\mathrm{d}t} = \int \frac{P}{T\tau}\left(\boldsymbol{v} + k_\mathrm{B}T\frac{\partial}{\partial \boldsymbol{v}}\ln P\right)^2 \mathrm{d}\boldsymbol{r}\mathrm{d}\boldsymbol{v}, \tag{11.1.1}$$

from that the effective temperature will be given by

$$\frac{1}{T_\mathrm{eff}} = -\frac{k_\mathrm{B}}{\boldsymbol{v}}\frac{\partial}{\partial \boldsymbol{v}}\ln P. \tag{11.1.2}$$

This expression can be rewritten in a form recalling that of the equilibrium temperature, i.e.

$$\frac{1}{T_\mathrm{eff}} = \frac{\partial s_\mathrm{eff}}{\partial e}, \tag{11.1.3}$$

at the condition to define an "effective" entropy by $s_\mathrm{eff} = -k_\mathrm{B}\ln P$ and an energy density by $e = (1/2)v^2$. Other definitions of effective temperature are possible. Indeed, by taking $e = (1/2)(\boldsymbol{v} - \langle \boldsymbol{v} \rangle)^2$, where $\langle \boldsymbol{v}^2 \rangle$ is the average value of v^2, the temperature would be the local equilibrium one, but this would not correspond to a zero entropy production. Moreover, as mentioned by Vilar and Rubí (2001), the effective temperature is generally a function of $\boldsymbol{r}, \boldsymbol{v}, t$; this means that, at a given position in space, there is no temperature at which the system can be at equilibrium, because $T(\boldsymbol{r}, \boldsymbol{v}, t) \neq T(\boldsymbol{r}, t)$. If it is wished to define a temperature at a position $\boldsymbol{x}$, it would depend on the way the additional degrees of freedom are eliminated.

One more remark is in form. It is rather natural to expect that the mesoscopic theory discussed so far will cope with evolution equations of the Maxwell–Cattaneo type, as the latter involve characteristic times comparable to the relaxation time for the decay of the velocity distributions towards its equilibrium value. Indeed by multiplying the Fokker–Planck's equation (11.81) by $\boldsymbol{v}$ and integrating over $\boldsymbol{v}$, one obtains

$$\tau \frac{\partial \bar{\boldsymbol{J}}}{\partial t} = -\bar{\boldsymbol{J}} - D\nabla n, \tag{11.83}$$

where $\bar{\boldsymbol{J}}$ is the diffusion flux in space, given by (11.64), $n(\boldsymbol{r}, t) = \int P(\boldsymbol{r}, \boldsymbol{v}, t)\mathrm{d}\boldsymbol{v}$ the particle number density, and $D = \langle v^2 \rangle \tau = (k_\mathrm{B}T/m)\,\tau$. This indicates that mesoscopic non-equilibrium thermodynamics is well suited to describe processes governed by relaxation equations of the Maxwell–Cattaneo type, just like EIT. But now, it is the probability distribution function which is elevated to the status of variable as basic variable rather than the average value of the diffusion flux. It should be added that working in the frame of EIT allows also obtaining the second moments of the fluctuations of $\boldsymbol{J}$ by combining the expression of extended entropy with Einstein's relation (11.5). The above considerations could leave to picture that mesoscopic thermodynamics is more general than EIT, however, when the second moments of fluctuations are taken as variables, besides their average value, EIT provides an interesting alternative more easier to deal with in practical situations.

11.5.2 Other Applications

The above results are directly generalized when more degrees of freedom than $\boldsymbol{r}$ and $\boldsymbol{v}$ are present. The probability density will then be a function of the whole set of degrees of freedom, denoted $\boldsymbol{x}$, so that $P = P(\boldsymbol{x}, t)$. The analysis performed so far can be repeated by replacing in all the mathematical expressions the couple $\boldsymbol{r}, \boldsymbol{v}$ by $\boldsymbol{x}$. In particular, the continuity equation will take the form

$$\frac{\partial P}{\partial t} = -\frac{\partial \boldsymbol{J}}{\partial \boldsymbol{x}}, \tag{11.84}$$

where $\boldsymbol{J}$ is given by the constitutive relation

$$\boldsymbol{J} = -\frac{L}{T}\frac{\partial \mu}{\partial \boldsymbol{x}}. \tag{11.85}$$

After substitution of (11.85) in (11.84) and use of (11.71) for the chemical potential, one obtains the kinetic equation

$$\frac{\partial P}{\partial t} = \frac{\partial}{\partial \boldsymbol{x}}\left(D\frac{\partial P}{\partial \boldsymbol{x}} + \frac{D}{k_{\mathrm{B}}T}\frac{\partial \Phi}{\partial \boldsymbol{x}}P\right), \tag{11.86}$$

which is a generalization of the Fokker–Planck's equation (11.81), where now Φ is not the kinetic energy as in (11.72) but it includes the potential energy related to the internal degrees of freedom.

Extension to non-linear situations, as in chemical reactions, does not raise much difficulty. If chemical processes are occurring at short timescales, they will generally take place from an initial to a final state through intermediate molecular configurations. Let the variable $\boldsymbol{x}$ characterize these intermediate states. The chemical potential is still given by (11.71) in which K is, for instance, a bistable potential whose wells correspond to the initial and final states while the maximum represents the intermediate barrier. Such a description is applicable to several problems as active processes, transport through membranes, thermionic emission, adsorption, nucleation processes (Reguera et al. 2005). Let us show, in particular, that mesoscopic thermodynamics leads to a kinetic equation where the reaction rate satisfies the mass action law. The linear constitutive law (11.85) is generalized in the form

$$\boldsymbol{J} = -k_{\mathrm{B}}L\frac{1}{z}\frac{\partial z}{\partial \boldsymbol{x}}, \tag{11.87}$$

where $z \equiv \exp(\mu/k_{\mathrm{B}}T)$ is the fugacity; an equivalent expression is

$$\boldsymbol{J} = -D\frac{\partial z}{\partial \boldsymbol{x}}, \tag{11.88}$$

where $D \equiv k_{\mathrm{B}}L/z$ represents the diffusion coefficient. Assuming D constant and integrating (11.88) from the initial state 1 to the final state 2 yields

$$\bar{J} \equiv \int_1^2 \boldsymbol{J}\,\mathrm{d}\boldsymbol{x} = -D(z_2 - z_1) = -D\left(\exp\frac{\mu_2}{k_{\mathrm{B}}T} - \exp\frac{\mu_1}{k_{\mathrm{B}}T}\right), \tag{11.89}$$

where $\bar{J}$ is the integrated rate. Expression (11.89) can alternatively be cast in the more familiar form of a kinetic law

$$\bar{J} = K[1 - \exp(-\mathcal{A}/k_{\mathrm{B}}T)], \tag{11.90}$$

where K stands for $D\exp(\mu_1/k_{\mathrm{B}}T)$ and $\mathcal{A} = \mu_2 - \mu_1$ is the affinity of the reaction. When $\mu_i/k_{\mathrm{B}}T \ll 1$, one recovers from (11.90), the classical linear phenomenological law of CIT (see Chap. 4)

$$\bar{J} = \frac{D}{k_{\mathrm{B}}T}\mathcal{A}. \tag{11.91}$$

Up to now, the theory has been applied to homogeneous systems characterized by the absence of gradients of thermal and mechanical quantities. The constraint of uniform temperature is now relaxed and the question is how to incorporate thermal gradients in the formalism. For the problem of Brownian motion, this is achieved by writing the Gibbs' equation as follows:

$$\mathrm{d}s = \frac{1}{T}\mathrm{d}u - \frac{1}{T}\int \mu \,\mathrm{d}P \,\mathrm{d}\boldsymbol{x}\,\mathrm{d}\boldsymbol{v}, \tag{11.92}$$

where u is the energy density. The corresponding constitutive equations are now given by (Reguera et al. 2005)

$$\boldsymbol{J}_q = -L_{TT}\nabla T/T^2 - \int k_{\mathrm{B}} L_{Tv} \frac{\partial}{\partial \boldsymbol{v}} \ln \frac{P}{P_{\mathrm{leq}}} \mathrm{d}\boldsymbol{v}, \tag{11.93}$$

$$\boldsymbol{J}_v = -L_{vT}\nabla T/T^2 - \int k_{\mathrm{B}} L_{vv} \frac{\partial}{\partial \boldsymbol{v}} \ln \frac{P}{P_{\mathrm{leq}}} \mathrm{d}\boldsymbol{v}, \tag{11.94}$$

where $\boldsymbol{J}_q$ is the heat flux, $\boldsymbol{J}_v$ the probability current, and P_{leq} the local equilibrium distribution function; L_{ij} are phenomenological coefficients forming a positive definite matrix and obeying the Onsager relation $L_{Tv} = -L_{vT}$. Equations (11.93) and (11.94) exhibit clearly the coupling between the two irreversible processes occurring in the system: diffusion probability and heat conduction. The corresponding evolution equation of the probability density is now

$$\frac{\partial P}{\partial t} = -\boldsymbol{v}\cdot\nabla P + \beta\frac{\partial}{\partial \boldsymbol{v}}\cdot\left(P\boldsymbol{v} + k_{\mathrm{B}}T\frac{\partial P}{\partial \boldsymbol{v}}\right) + \frac{\gamma}{T}\frac{\partial}{\partial \boldsymbol{v}}\cdot(P\nabla T), \tag{11.95}$$

where β is the friction coefficient of the particles and γ a coefficient related to L_{vT}.

To summarize, mesoscopic non-equilibrium thermodynamics has shown its applicability in a wide variety of situations where local equilibrium is never reached, as for instance relaxation of polymers or glasses, dynamics of colloids or flows of granular media. Such systems are characterized by internal variables exhibiting short length scales and slow relaxation times. The basic idea of this new theory is to incorporate in the description those variables, which have not yet relaxed to their local equilibrium values. As illustrative examples, we have discussed the problem of inertial effects in diffusion of Brownian particles where change of density take place at a timescale of the order of the time needed by the particles to relax to equilibrium, chemical reactions and thermodiffusion. The tools are borrowed from CIT (see Chap. 2) but the original contribution is to introduce, amongst the set of variables, the density probability distribution whose time evolution is shown to be governed by a Fokker–Planck's equation.

11.6 Problems

11.1. *Second moments.* (a) Show, from (11.12) and (11.15) that the second moments of the energy and the volume fluctuations are given by

$$\langle \delta U \delta U \rangle = -(k_{\mathrm{B}}/M)(\partial U/\partial T^{-1})_{T^{-1}p},$$
$$\langle \delta U \delta V \rangle = -(k_{\mathrm{B}}/M)(\partial V/\partial T^{-1})_{T^{-1}p},$$
$$\langle \delta V \delta V \rangle = -(k_{\mathrm{B}}/M)(\partial V/\partial T^{-1}p)_{T^{-1}}.$$

(b) Write explicitly the second-order derivatives appearing in (11.12) and (11.15), and derive expressions (11.16).

11.2. *Density fluctuations.* (a) From the second moments of the volume fluctuations, write the expression for the density fluctuations of a one-component ideal gas at 0°C and 1 atm, when the root-mean-square deviation in density is 1% of the average density of the system? (b) Show that near a critical point, where $(\partial p/\partial V)_T = 0$, these fluctuations diverge.

11.3. *Dielectric constant.* The dielectric constant ε of a fluid varies with the mass density according to the Clausius–Mossoti's relation

$$\frac{\varepsilon - 1}{\varepsilon + 2} = C\rho,$$

with C is a constant related to the polarizability of the molecules and ρ is the mass density. Show that the second moments of the fluctuations of ε are given by

$$\langle (\delta\varepsilon)^2 \rangle = \frac{k_{\mathrm{B}} T \kappa_T}{9V} \frac{(\varepsilon - 1)^2}{(\varepsilon + 2)^2},$$

with κ_T being the isothermal compressibility.

11.4. *Density fluctuations and non-locality.* The correlations in density fluctuations at different positions are usually written as $\langle \delta n(\boldsymbol{r}_1) \delta n(\boldsymbol{r}_2) \rangle \equiv \bar{n}\delta(\boldsymbol{r}_1 - \boldsymbol{r}_2) + \bar{n}\nu(r)$ with $\bar{n}$ the average value of the density, $r \equiv |\boldsymbol{r}_2 - \boldsymbol{r}_1|$, and $\nu(r)$ the correlation function. To describe such correlations, Ginzburg and Landau propose to include in the free energy a non-local term of the form

$$F(T, n, \nabla n) = F_{\text{eq}}(T, n) - \frac{1}{2} b (\nabla n)^2 = \frac{1}{2} a (n - \bar{n})^2 - \frac{1}{2} b (\nabla n)^2,$$

with $a(T)$ a function of temperature which vanishes at the critical point and $b > 0$ a positive constant. (a) If the density of fluctuations are expressed as

$$n - \bar{n} = \sum_k n_k \mathrm{e}^{\mathrm{i}\boldsymbol{k}\cdot\boldsymbol{r}},$$

show that

$$\langle |\delta n_k|^2 \rangle = \frac{k_B T}{V(a+bk^2)}.$$

(b) Taking into account that

$$\int \nu(r) e^{-i\boldsymbol{k}\cdot\boldsymbol{r}} dV = \frac{V}{n} \langle |\delta n_k|^2 \rangle - 1,$$

prove that

$$\nu(r) = \frac{k_B T a}{4\pi \bar{n} b} \frac{1}{r} \exp\left(-\frac{r}{\xi}\right),$$

where ξ is a correlation length, given by $\xi = (b/a)^{1/2}$. (Note that in the limit $\xi \to 0$, one obtains $\nu(r) = \delta(r)$, with $\delta(r)$ the Dirac's function, and that near a critical point the correlation length diverges.)

11.5. *Transport coefficients.* (a) Apply (11.43) to obtain the classical results $\eta = nk_B T\tau$ and $\lambda = (5k_B T^2/2m)\tau$. In terms of the peculiar molecular velocities c the microscopic operators for the fluxes are given by $\hat{P}_{12}^v = mc_1c_2$ and $\hat{q}_1 = \left(\frac{1}{2}mc^2 - \frac{5}{2}k_B T\right) c_1$. The equilibrium average should be performed over the Maxwell–Boltzmann distribution function. (b) Make a similar analysis for the memory kernel corresponding to the diffusion coefficient, namely $D(t-t') = (V/k_B T) \langle \delta J_i(t) \delta J_i(t') \rangle$.

11.6. *Second moments and EIT.* Apply Einstein's relation (11.5) to the entropy (7.61) of EIT, and find the second moments of the fluctuations of the fluxes around equilibrium. Note that the results coincide with (11.43), obtained from Green–Kubo's relations by assuming an exponential relaxation of the fluctuations of the fluxes.

11.7. *Fluctuations around steady states.* Assume with Keizer that Einstein's relation remains valid around a non-equilibrium steady state. Determine the second moments of the fluctuations of u and $\boldsymbol{q}$ around a non-equilibrium steady state characterized by a non-vanishing average heat flux $\boldsymbol{q}_0$. Compare the results with those obtained in equilibrium in Problem 11.6.

11.8. *Brownian motion.* Langevin proposed to model the Brownian motion of the particles by adding to the hydrodynamic friction force $-\zeta \boldsymbol{v}$ (ζ is the friction coefficient, $\boldsymbol{v}$ is the speed of the particle), a stochastic force $\boldsymbol{f}$ describing the erratic forces due to the collisions of the microscopic particles of the solvent, in such a way that

$$m\frac{d\boldsymbol{v}}{dt} = -\zeta \boldsymbol{v} + \boldsymbol{f}.$$

He assumed that $\boldsymbol{f}$ is white (without memory) and Gaussian, in such a way that $\langle \boldsymbol{f} \rangle = 0$, $\langle \boldsymbol{f}(t)\boldsymbol{f}(t+t') \rangle = B\delta(t')$. Using the result that in the long-time limit the equipartition condition $\langle \frac{1}{2}m\boldsymbol{v}^2 \rangle = \frac{3}{2}k_B T$ must be satisfied, show that $\langle \boldsymbol{f}(t)\boldsymbol{f}(t+t')\rangle = (\zeta/m)k_{\mathrm{B}}T\delta(t')$, i.e. $B = (\zeta/m)k_{\mathrm{B}}T$. This is similar to a fluctuation–dissipation relation, as it shows that the second moments of the fluctuating force are proportional to the friction coefficient. This is called a fluctuation–dissipation relation of the second kind.

11.9. *Generalized fluctuation–dissipation relation.* The result of the previous problem may be written in a more general way for any set $\boldsymbol{a}(t)$ of random variables satisfying a linear equation of the form $\mathrm{d}\boldsymbol{a}/\mathrm{d}t + \mathbf{H}\cdot\boldsymbol{a} = \boldsymbol{f}$ with $\mathbf{H}$ a friction matrix and $\boldsymbol{f}$ a white noise such that $\langle \boldsymbol{f}\rangle = 0, \langle \boldsymbol{f}(t)\boldsymbol{f}^{\mathrm{T}}(t+t')\rangle = \mathbf{B}\delta(t')$. Show that the matrix $\mathbf{B}$ must satisfy the fluctuation–dissipation relation

$$\mathbf{H}\cdot\mathbf{G} + \mathbf{G}\cdot\mathbf{H}^{\mathrm{T}} = \mathbf{B},$$

with $\mathbf{G} \equiv \langle \delta\boldsymbol{a}\,\delta\boldsymbol{a}^{\mathrm{T}}\rangle$ the matrix of the second moments of fluctuations of $\boldsymbol{a}(t)$.

11.10. *Non-equilibrium temperature.* To underline the connection between the non-equilibrium temperature in Keizer's formalism and in EIT, note that, when the heat flux $\boldsymbol{q}$ is the only relevant flux, EIT predicts that $\langle \delta u\,\delta u\rangle = \langle \delta u\,\delta u\rangle_{\mathrm{eq}} + \alpha' q^2$, where subscript eq mean local equilibrium. In this expression δu is the fluctuations of the internal energy with respect to its steady state average and α' a coefficient, whose explicit form is given in Jou et al. (2001). The above result can still be cast in the form

$$q^2 = [\langle \delta u\,\delta u\rangle - \langle \delta u\,\delta u\rangle_{\mathrm{eq}}]\,(\alpha')^{-1}.$$

Recalling that in EIT the non-equilibrium temperature T is given by

$$\frac{1}{T} = \frac{1}{T_{\mathrm{eq}}} - \frac{1}{2}\frac{\partial\alpha}{\partial u}q^2,$$

write this expression in terms of the second moments of energy fluctuations and compare with Keizer's expression (11.60).

Epilogue

By writing this book, our objective was threefold:

1. First, to go beyond equilibrium thermodynamics. Although it is widely recognized that equilibrium thermodynamics is a universal and well-founded discipline with many applications mainly in chemistry and engineering, it should be realized that its domain of application is limited to equilibrium states and idealized reversible processes, excluding dissipation. This is sufficient to predict the final equilibrium state, knowing the initial state, but it is silent about the duration and the nature of the actual process between the initial and the final equilibrium states, whence the need to go beyond equilibrium thermodynamics. Another reason is that, to foster the contact between micro- and macroscopic approaches, it is imperative to go beyond equilibrium as most of the microscopic theories deal with situations far from equilibrium. But the problem we are faced with is that the avenue of equilibrium thermodynamics bifurcates in many routes.
2. Our second objective was to propose a survey, as complete as possible, of the many faces of non-equilibrium thermodynamics. For pedagogical reasons, we have restricted the analysis to the simplest situations, emphasizing physical rather than mathematical aspects. The presentation of each theory is closed by a critical discussion from which can be concluded that none of the various approaches is fully satisfactory. It appears that each school has its own virtues but that, in practical situations, one of them may be preferable to another. It is our purpose that after gone through the present book, the reader will be able to make up his personal opinion. We do not pretend to have been everywhere fully objective and exhaustive. We have deliberately been silent about some valuable descriptions, as the entropy-free theory of Meixner (1973a, b), the Lagrangian formalism of Biot (1970), the variational analysis of Sieniutycz (1994), the thermodynamics of chaos (Beck and Schlögl 1993; Berdichevsky 1997; Gaspard 1998; Ruelle 1991), the statistical approach of Luzzi et al. (2001, 2002), and

Tsallis' non-extensive entropy formalism (2004). For reasons of place and unity, and despite their intrinsic interest, we have also deliberately omitted microscopic formulations, such as statistical mechanics (e.g. Grandy 1987), kinetic theory (e.g. Chapman and Cowling 1970), information theory (Jaynes 1963), molecular dynamics (Evans and Morriss 1990), and other kinds of computer simulations (e.g. Hoover 1999; Hutter and Jöhnk 2004). We apologize for these omissions; the main reason was our option to confine the volume of the book to a reasonable size rather than to write an extensive encyclopaedia. We have also bypassed some approaches either because of their more limited impact in the scientific community, or because they lack of sufficiently new fundamental ideas or techniques. Concerning this multiplicity, it may be asked why so many thermodynamics? A tentative answer may be found in the diversity of thought of individuals, depending on their roots, environment, and prior formation as physicists, mathematicians, chemists, engineers, or biologists. The various thermodynamic theories are based on different foundations: macroscopic equilibrium thermodynamics, kinetic theory, statistical mechanics, or information theory. Other causes of diversity may be found in the selection of the most relevant variables and the difficulty to propose an undisputed definition of temperature, entropy, and the second law outside equilibrium. At the exception of classical irreversible thermodynamics (CIT), it is generally admitted that non-equilibrium entropy depends, besides classical quantities as mass, energy, charge density, etc., on extra variables taking the form of dissipative fluxes in extended irreversible thermodynamics (EIT), internal structural variables in internal variables theories (IVT) and in GENERIC and probality distribution function in mesoscopic theories. Out of equilibrium, the constitutive relations can either be cast in the form of linear algebraic phenomenological relations as in CIT, integrals involving the memory as in rational thermodynamics (RT), or time evolution differential equations as in EIT, IVT, and GENERIC. The next natural question is then: what is the best approach? Although the present authors have their own (subjective) opinion, we believe that the final answer should be left to each individual reader but there is no doubt that trying to reach unanimity remains a tremendous challenging task.

3. We failed to meet a third objective, namely to bring a complete unity into non-equilibrium thermodynamics. Being aware about the role of non-uniformity and the importance of diversity, we realize that the achievement of such a unity may appear as illusory. However, we do not think that it is a completely desperate task; indeed, it is more than a dream to believe that in a near future it would be possible to summon up all the pieces of the puzzle and to build up a well-shaped, unique, and universal non-equilibrium thermodynamics. In that respect, we would like to stress that there exists a wide overlapping between the different schools. More specifically, all the theories contain as a special case the classical irreversible thermodynamics. The Cattaneo model of heat conduction is not typical of

EIT but can also be obtained in the framework of IVT, RT, GENERIC, and the mesoscopic description. It is our hope that the present book will contribute to promote reconciliation among the several approaches and foster further developments towards deeper and more unified formulations of thermodynamics beyond equilibrium.

By the way, it was also our purpose to convince the reader that "thermodynamics is the science of everything". Clearly, thermodynamics represents more than converting heat into work or calculating engine efficiencies. It is a multi-disciplinary science covering a wide variety of fields ranging from thermal engineering, fluid and solid mechanics, rheology, material science, chemistry, biology, electromagnetism, cosmology to economical, and even social sciences.

Among the several open and challenging problems, let us mention three of them. The first one is related to the limits of applicability of thermodynamics to small systems, like found in nano-technology and molecular biophysics. At the opposite, in presence of long-range interactions, such as gravitation, it may be asked how the second law should be formulated when these effects are dominant, like in cosmology. Finally, does thermodynamics conflict with quantum mechanics? how to reconcile the reversible laws of quantum theory and the subtleties of quantum entanglement of distant systems with the irreversible nature of thermodynamics?

By writing this book, we were guided by the intellectual ambition to better understand the frontiers and perspectives of the multi-faced and continuously changing domain of our knowledge in non-equilibrium thermodynamics. This remains clearly an unfinished task and we would like to think to have convinced the reader that, therefore, this fascinating story is far from

The End

References

Agrawal D.C. and Menon V.J., Performance of a Carnot refrigerator at maximum cooling power, J. Phys. A: Math. Gen. **23** (1990) 5319–5326

Alvarez X. and Jou D., Memory and nonlocal effects in heat transport: from diffusive to ballistic regimes, Appl. Phys. Lett. **90** (2007) 083109 (3 pp)

Andresen B., Finite-time thermodynamics in simulated annealing, in *Entropy and Entropy Generation* (Shiner J.S., ed.), Kluwer, Dordrecht, 1996

Angulo-Brown F., An ecological optimization criterion for finite-time heat engines, J. Appl. Phys. **69** (1991) 7465–7469

Anile A.M. and Muscato O., An improved thermodynamic model for carrier transport in semiconductors, Phys. Rev. B **51** (1995) 16728–16740

Anile A.M. and Pennisi S., Extended thermodynamics of the Blotekjaer hydrodynamical model for semiconductors, Continuum Mech. Thermodyn. **4** (1992) 187–197

Anile A.M., Allegretto W. and Ringhofer C., *Mathematical Problems in Semi-Conductor Physics*, Lecture Notes in Mathematics, vol. 1823, Springer, Berlin Heidelberg New York, 2003

Astumian R.D., Thermodynamics and kinetics of a Brownian motor, Science **276** (1997) 917–922

Astumian R.D. and Bier M., Fluctuation driven ratchets. Molecular motors, Phys. Rev. Lett. **72** (1994) 1766–1769

Bampi F. and Morro A., Nonequilibrium thermodynamics: a hidden variable approach, in *Recent Developments in Non-Equilibrium Thermodynamics* (Casas-Vázquez J., Jou D. and Lebon G., eds.), Lecture Notes in Physics, vol. 199, Springer, Berlin Heidelberg New York, 1984

Beck C. and Schlögl F., *Thermodynamics of Chaotic Systems*, Cambridge University Press, Cambridge, 1993

Bedeaux D., Mazur P. and Pasmanter R.A., The ballast resistor; an electro-thermal instability in a conducting wire I; the nature of the stationary states, Physica A **86** (1977) 355–382

Bejan A., *Entropy Generation Minimization*, CRC, Boca Raton, FL, 1996

Bénard H., Les tourbillons cellulaires dans une nappe liquide transportant la chaleur par conduction en régime permanent, Rev. Gén. Sci. Pures et Appliquées **11** (1900) 1261–1271, 1309–1318

Berdichevsky V.L., *Thermodynamics of Chaos and Order*, Longman, Essex, 1997

Bergé P., Pomeau Y. and Vidal C., *L'ordre dans le chaos*, Hermann, Paris, 1984

Beris A.N., Simple Non-equilibrium thermodynamics application to polymer rheology, in *Rheology Reviews* (Binding D.N. and Walters K., eds.), pp 33–75, British Society of Rheology, Aberystwyth, 2003

Beris A.N. and Edwards B.J., *Thermodynamics of Flowing Systems with Internal Microstructure*, Oxford Science Pubisher, Oxford, 1994

Biot M.A., *Variational Principles in Heat Transfer*, Oxford University Press, Oxford, 1970

Bird R.B. and de Gennes P.G., Discussion about the principle of objectivity, Physica A **118** (1983) 43–47

Bird R.B., Armstrong R.C. and Hassager D., *Dynamics of Polymeric Liquids*, 2nd ed., vol. 1: *Fluid Mechanics*, Wiley, New York, 1987a

Bird R.B., Curtiss C.F., Armstrong R.C. and Hassager D., *Dynamics of Polymeric Liquids*, 2nd ed., vol. 2: *Kinetic Theory*, Wiley, New York, 1987b

Block M.J., Surface tension as the cause of Bénard's cells and surface deformation in a liquid film, Nature **178** (1956) 650

Bodenshatz E., Pech W. and Ahlers G., Recent developments in Rayleigh-Bénard convection, Annu. Rev. Fluid Mech. **32** (2000) 851–919

Bragard J. and Lebon G., Nonlinear Marangoni convection in a layer of finite depth, Europhys. Lett. **21** (1993) 831–836

Braun D. and Libchaber A., Phys. Rev. Lett. **89** (2002) 188103

Brenner H., Rheology of two-phase systems, Annu. Rev. Fluid Mech. **2** (1970) 137–176

Bricmont J., Science of chaos or chaos in science? Physicalia Mag. **17** (1995) 159–221

Bridgman P.W., *The Nature of Thermodynamics*, Harvard University Press, Cambridge, 1941

Busse F.H., Non-linear properties of thermal convection, Rep. Prog. Phys. **41** (1978) 1929–1967

Bustamante C., Liphardt J. and Ritort F., The nonequilibrium thermodynamics of small systems, Phys. Today **58** (2005) 43–48

Callen H.B., *Thermodynamics and an Introduction to Thermostatistics*, 2nd ed., Wiley, New York, 1985

Callen H.B. and Welton T.A., Irreversibility and generalized noise, Phys. Rev. **83** (1951) 34–40

Callen H.B. and Greene R.F., On a theorem of irreversible thermodynamics, Phys. Rev. **86** (1952) 702–706

Camacho J. and Zakari M., Irreversible thermodynamic analysis of a two-layer system, Phys. Rev. E **50** (1994) 4233–4236

Caplan S.R. and Essig A., *Bioenergetics and Linear Nonequilibrium Thermodynamics. The Steady State*, Harvard University Press, Cambridge, 1983

Carrassi M. and Morro A., A modified Navier–Stokes equations and its consequences on sound dispersion, Nuovo Cimento B **9** (1972) 321–343; **13** (1984) 281–289

Casas-Vázquez J. and Jou D., Temperature in non-equilibrium states: a review of open problems and current proposals, Rep. Prog. Phys. **66** (2003) 1937–2023

Casas-Vázquez J., Jou D. and Lebon G. (eds.), *Recent Developments in Non-equilibrium Thermodynamics*, Lecture Notes in Physics, vol. 199, Springer, Berlin Heidelberg New York, 1984

Casimir H.B.G., On Onsager's principle of microscopic irreversibility, Rev. Mod. Phys. **17** (1945) 343–350

Cattaneo C., Sulla conduzione del calore, Atti Seminario Mat. Fis. University Modena **3** (1948) 83–101

Chandrasekhar S., *Hydrodynamic and Hydromagnetic Stability*, Clarendon, Oxford, 1961

Chapman S. and Cowling T.G., *The Mathematical Theory of Non-Uniform Gases*, Cambridge University Press, Cambridge, 1970

Chartier P., Gross M. and Spiegler K.S., *Applications de la thermodynamique du non-équilibre*, Hermann, Paris, 1975 (in French)

Chen G., *Nanoscale Heat Transfer and Conversion*, Oxford University Press, Oxford, 2004

Cimmelli V.A., Boundary conditions in the presence of internal variables, J. Non-Equilib. Thermodyn. **27** (2002) 327–348
Cloot A. and Lebon G., A nonlinear stability analysis of the Bénard–Marangoni problem, J. Fluid Mech. **45** (1984) 447–469
Coleman B.D., Thermodynamics of materials with memory, Arch. Rat. Mech. Anal. **17** (1964) 1–46
Coleman B.D. and Owen D., A mathematical foundation of thermodynamics, Arch. Rat. Mech. Anal. **54** (1974) 1–104
Coleman B.D. and Owen D., On thermodynamics and elastic–plastic materials, Arch. Rat. Mech. Anal. **59** (1975) 25–51
Coleman B.D. and Truesdell C., On the reciprocal relations of Onsager, J. Chem. Phys. **33** (1960) 28–31
Coleman B.D., Owen D. and Serrin J., The second law of thermodynamics for systems with approximate cycles, Arch. Rat. Mech. Anal. **77** (1981) 103–142
Coles D., Transition in circular Couette flow, J. Fluid Mech. **21** (1965) 385–425
Colinet P., Legros J.C. and Velarde M., *Nonlinear Dynamics of Surface-Tension Driven Instabilities*, Wiley, Berlin, 2002
Criado-Sancho M. and Casas-Vázquez J., *Termodinámica química y de los procesos irreversibles*, 2nd ed., Pearson/Addison-Wesley, Madrid, 2004 (in Spanish)
Criado-Sancho M. and Llebot J.E., Behaviour of entropy in hyperbolic heat transport, Phys. Rev. E **47** (1993) 4104–4108
Crisanti A. and Ritort F., Violation of the fluctuation–dissipation theorem in glassy systems: basic notions and the numerical evidence, J. Phys. A: Math. Gen. **36** (2003) R181–R290
Cross M.C. and Hohenberg P.C., Pattern formation outside equilibrium, Rev. Mod. Phys. **65** (1993) 852–1112
Curzon F.L. and Ahlborn B., Efficiency of a Carnot engine at maximum power output, Am. J. Phys. **43** (1975) 22–24
Dauby P.C. and Lebon G., Bénard–Marangoni instability in rigid rectangular containers, J. Fluid Mech. **329** (1996) 25–64
Dauby P.C., Bouhy E. and Lebon G., Linear Bénard–Marangoni instability in rigid circular containers, Phys. Rev. E **56** (1997) 520–530
Davis S.H., Buoyancy-surface tension instability by the method of energy, J. Fluid Mech. **39** (1969) 347–359
Davis S.H., Thermocapillary instabilities, Annu. Rev. Fluid Mech. **19** (1987) 403–435
Day W., *The Thermodynamics of Simple Materials with Fading Memory*, Springer, Berlin Heidelberg New York, 1972
Day W., An objection against using entropy as a primitive concept in continuum thermodynamics, Acta Mechanica **27** (1977) 251–255
Dedeurwaerdere T., Casas-Vázquez J., Jou D. and Lebon G., Foundations and applications of a mesoscopic thermodynamic theory of fast phenomena, Phys. Rev. E **53** (1996) 498–506
Doi M. and Edwards S.F., *The Theory of Polymer Dynamics*, Clarendon, Oxford, 1986
Drazin P.G. and Reid W.H., *Hydrodynamic Stability*, Cambridge University Press, Cambridge, 1981
Dreyer W. and Struchtrup H., Heat pulse experiments revisited, Continuum Mech. Thermodyn. **5** (1993) 3–50
Duhem P., *Traité d'énergétique ou de thermodynamique générale*, Gauthier-Villars, Paris, 1911
Eckart C., The thermodynamics of irreversible processes. I–III, Phys. Rev. **58** (1940) 267–269, 269–275, 919–924; IV, Phys. Rev. **73** (1948) 373–382
Edelen D.G.D. and McLennan J.A., Material indifference: a principle or a convenience, Int. J. Eng. Sci. **11** (1973) 813–817

Elmer F.J., Limit cycle of the ballast resistor caused by intrinsic instabilities, Z. Phys. B **87** (1992) 377–381

Eringen C., *Mechanics of Continua*, Wiley, New York, 1967

Eu B.C., *Kinetic Theory and Irreversible Thermodynamics*, Wiley, New York, 1992

Eu B.C., *Nonequilibrium Statistical Mechanics: Ensemble Method*, Kluwer, Dordrecht, 1998

Evans D.J. and Morriss G.P., *Statistical Mechanics of Non-Equilibrium Liquids*, Academic, London, 1990

Finlayson B., *The Method of Weighted Residuals and Variational Principles*, Academic, New York, 1972

Fowler R.H. and Guggenheim E.A., *Statistical Thermodynamics*, Cambridge University Press, Cambridge, 1939

Frankel N.A. and Acrivos A., The constitutive equations for a dilute emulsion, J. Fluid Mech. **44** (1970) 65–78

García-Colin L.S., Extended irreversible thermodynamics beyond the linear regime: a critical overview, J. Non-Equilib. Thermodyn. **16** (1991) 89–128

Garcia-Colin L.S., Extended irreversible thermodynamics: an unfinished task, Mol. Phys. **86** (1995) 697–706

Gaspard P., *Chaos, Scattering and Statistical Mechanics*, Cambridge University Press, Cambridge, 1998

Gibbs J.W., *The Collected Works of J.W. Gibbs*, vol. 1: *Thermodynamics*, Yale University Press, Yale, 1948

Glansdorff P. and Prigogine I., On a general evolution criterion in macroscopic physics, Physica **30** (1964) 351–374

Glansdorff P. and Prigogine I., *Thermodynamics of Structure, Stability and Fluctuations*, Wiley, New York, 1971

Gollub J.P. and Langer J.S., Pattern formation in nonequilibrium physics, Rev. Mod. Phys. **71** (1999) S396–S403

Gorban A.N., Karlin I.V. and Zinovyev A.Y., Constructive methods of invariant manifolds for kinetic problems, Phys. Rep. **396** (2004) 197–403

Gordon J.M. and Zarmi Y., Wind energy a solar-driven heat engine: a thermodynamic approach, Am. J. Phys. **57** (1989) 995–998

Grad H., Principles of the Kinetic Theory of Gases, in *Hd. der Physik*, vol. XII (Flugge S., ed.), Springer, Berlin Heidelberg New York, 1958

Grandy W.T., *Foundations of Statistical Mechanics*, Reidel, Dordrecht, 1987

Grmela M., Complex fluids subjected to external influences, J. Non-Newtonian Fluid Mech. **96** (2001) 221–254

Grmela M. and Lebon G., Hamiltonian extended thermodynamics, J. Phys. A: Math. Gen. **23** (1990) 3341–3351

Grmela M. and Öttinger H.C., Dynamics and thermodynamics of complex fluids. I. Development of a generic formalism, Phys. Rev. E **56** (1997) 6620–6632

Grmela M., Elafif A. and Lebon G., Isothermal non-standard diffusion in a two-component fluid mixture: a Hamiltonian approach, J. Non-Equilib. Thermodyn. **23** (1998) 312–327

Grmela M., Lebon G. and Lhuillier D., A comparative study of the coupling of flow with non-Fickean thermodiffusion. II. GENERIC, J. Non-Equilib. Thermodyn. **28** (2003) 23–50

Grmela M., Lebon G., Dauby P.C. and Bousmina M., Ballistic-diffusive heat conduction at nanoscale: a GENERIC approach, Phys. Lett. A **339** (2005) 237–245

de Groot S.R. and Mazur P., *Non-Equilibrium Thermodynamics*, North-Holland, Amsterdam, 1962 (republication by Dover in 1984)

Guyer R.A. and Krumhansl J.A., Solution of the linearized Boltzmann phonon equation, Phys. Rev. **148** (1966) 766–788

Gyarmati I., *Non-equilibrium Thermodynamics*, Springer, Berlin Heidelberg New York, 1970

Hafskold B. and Kjelstrup S., Criteria for local equilibrium in a system with transport of heat and mass, J. Stat. Phys. **78** (1995) 463–494

Haken H. (ed.), *Synergetics: A Workshop*, Springer, Berlin Heidelberg New York, 1977

Hänsch W., *The Drift-Diffusion Equation and Its Application in MOSFET Modeling*, Springer, Berlin Heidelberg New York, 1991

Hill T.L., *Free Energy Transduction in Biology, the Steady-State Kinetic and Thermodynamic Formalism*, Academic, New York, 1977

Hoffmann K.H., Burzler J.M. and Schubert S., Endoreversible thermodynamics, J. Non-Equilib. Thermodyn. **22** (1997) 311–355

Hoover W.H., *Time Reversibility, Computer Simulations and Chaos*, World Scientific, Singapore, 1999

Hoover W., Moran B., More R. and Ladd A., Heat conduction in a rotating disk via non-equilibrium molecular dynamics, Phys. Rev. A **24** (1981) 2109–2115

Hutter K. and Jöhnk K., *Continuum Methods of Physical Modelling*, Springer, Berlin Heidelberg New York, 2004

Ichiyanagi M., Variational principles in irreversible processes, Phys. Rep. **243** (1994) 125–182

Ilg P. and Öttinger H.C., Non-equilibrium relativistic thermodynamics in bulk viscous cosmology, Phys. Rev. D **61** (1999) 023510, 1–10

Jaynes E.T., in *Statistical Physics* (Ford W.K., ed.) Benjamin, New York, 1963

Joseph D.D., *Stability of Fluid Motions*, Springer, Berlin Heidelberg New York, 1976

Joseph D.D. and Preziosi L., Heat waves, Rev. Mod. Phys. **61** (1989) 41–74; **62** (1990) 375–392

Jou D. and Llebot J.E., *Introduction to the Thermodynamics of Biological Processes*, Prentice Hall, New York, 1990

Jou D., Casas-Vázquez J. and Criado-Sancho M., *Thermodynamics of Fluids Under Flow*, Springer, Berlin Heidelberg New York, 2000

Jou D., Casas-Vázquez J. and Lebon G., *Extended Irreversible Thermodynamics*, 3rd ed., Springer, Berlin Heidelberg New York, 2001

Jou D., Casas-Vázquez J., Lebon G. and Grmela M., A phenomenological scaling approach for heat transport in nano-systems, Appl. Math. Lett. **18** (2005) 963–967

Jülicher A., Ajdari A. and Prost P., Modeling molecular motors, Rev. Mod. Phys. **69** (1997) 1269–1282

Katchalsky A. and Curran P.F., *Non-Equilibrium Thermodynamics in Biophysics*, Harvard University Press, Cambridge, 1965

Keizer J., Thermodynamics of non-equilibrium steady states, J. Chem. Phys. **69** (1978) 2609–2620

Keizer J, On the relationship between fluctuating irreversible thermodynamics and 'extended' irreversible thermodynamics, J. Stat. Phys. **31** (1983) 485–497

Keizer J., *Statistical Thermodynamics of Nonequilibrium Processes*, Springer, Berlin Heidelberg New York, 1987

Kestin J., *A Course in Thermodynamics*, Blaisdell, Waltham, 1968

Kestin J., A note on the relation between the hypothesis of local equilibrium and the Clausius–Duhem inequality, J. Non-Equilib. Thermodyn. **15** (1990) 193–212

Kestin J., Local-equilibrium formalism applied to mechanics of solids, Int. J. Solids Struct. **29** (1992) 1827–1836

Kestin J. and Bataille J., Thermodynamics of Solids, in *Continuum Models of Discrete Systems*, University of Waterloo Press, Waterloo, 1980

Khalatnikov I.M., *An introduction to the Theory of Superfluidity*, W.A. Benjamin, New York, 1965

Kleidon A. and Lorenz R.D. (eds.), *Non-Equilibrium Thermodynamics and the Production of Entropy. Life, Earths and Beyond*, Springer, Berlin Heidelberg New York, 2005

Kluitenberg G., Thermodynamical theory of elasticity and plasticity, Physica **28** (1962) 217–232

Kondepudi D. and Prigogine I., *Modern Thermodynamics: from Heat Engines to Dissipative Structures*, Wiley, New York, 1998

Koschmieder E.L., *Bénard cells and Taylor Vortices*, Cambridge University Press, Cambridge, 1993

Kratochvil J. and Silhavy M., On thermodynamics of non-equilibrium processes, J. Non-Equilib. Thermodyn. **7** (1982) 339–354

Kreuzer H.J., *Non-Equilibrium Thermodynamics and Its Statistical Foundations*, Clarendon, Oxford, 1981

Kubo R., *Statistical Mechanics*, Wiley, New York, 1965

Landau L.D., *Collected Papers* (Haar T., ed.), Pergamon, Oxford, 1965

Landau L.D. and Lifshitz E.M., *Statistical Physics*, 3rd ed., Pergamon, Oxford, 1980

Landsberg P.T. and Tonge G., Thermodynamic energy conversion efficiencies, J. Appl. Phys. **51** (1980) R1–R18

Lavenda B., *Thermodynamics of Irreversible Processes*, McMillan, London, 1979

Lebon G., *Variational Principles in Thermomechanics*, in CISME Courses and Lecture Notes, vol. 262, Springer, Berlin Heidelberg New York, 1980

Lebon G. and Jou D., Linear irreversible thermodynamics and the phenomenological theory of liquid helium II, J. Phys. C **16** (1983) L199–L204

Lebon G. and Dauby P.C., Heat transport in dielectric crystals at low temperature: a variational formulation based on extended irreversible thermodynamics, Phys. Rev. A **42** (1990) 4710–4715

Lebon G., *Extended Thermodynamics*, in CISME Courses and Lecture Notes, vol. 336, pp 139–204, Springer, Berlin Heidelberg New York, 1992

Lebon G., Casas-Vazquez J. and Jou D., Questions and answers about a thermodynamic theory of the third type, Contemp. Phys. **33** (1992) 41–51

Lebon G., Torrissi M. and Valenti A., A nonlocal thermodynamic analysis of second sound propagation in crystalline dielectrics, J. Phys. C **7** (1995) 1461–1474

Lebon G., Grmela M. and Lhuillier D., A comparative study of the coupling of flow with non-Fickian thermodiffusion. I. Extended irreversible thermodynamics, J. Non-Equilib. Thermodyn. **28** (2003) 1–22

Lebon G., Desaive Th. and Dauby P.C., A unified extended thermodynamic description of diffusion, thermo-diffusion, suspensions, and porous media, J. Appl. Mech. **73** (2006a) 16–20

Lebon G., Desaive Th. and Dauby P.C., Two-fluid diffusion as a model for heat transport at micro and nanoscales, in *Proceedings of 7th International Meeting on Thermodiffusion*, San Sebastian, 2006b

Lebon G., Lhuillier D. and Palumbo A., A thermodynamic description of thermodiffusion in suspensions of rigid particles, Eur. Phys. J.IV **146** (2007) 3–12

Lebowitz J.L., Microscopic origins of irreversible macroscopic behaviour, Physica A **263** (1999) 516–527

Lepri S., Livi R. and Politi A., Thermal conduction in classical low-dimensional lattice, Phys. Rep. **377** (2003) 1–80

Lhuillier D., From molecular mixtures to suspensions of particles, J. Phys. II France **5** (1995) 19–36

Lhuillier D., Internal variables and the non-equilibrium thermodynamics of colloidal suspensions, J. Non-Newtonian Fluid Mech. **96** (2001) 19–30

Lhuillier D. and Ouibrahim A., A thermodynamic model for solutions of deformable molecules, J. Mécanique **19** (1980) 725–741

Lhuillier D., Grmela M. and Lebon G., A comparative study of the coupling of flow with non-Fickian thermodiffusion. Part III. internal variables, J. Non-Equilib. Thermodyn. **28** (2003) 51–68
Liu I.-S., Method of Lagrange multipliers for exploitation of the entropy principle, Arch. Rat. Mech. Anal. **46** (1972) 131–148
Lorenz E.N., Deterministic nonperiodic flow, J. Atmos. Sci. **20** (1963) 130–141
Lumley J., Turbulence modelling, J. Appl. Mech. **50** (1983) 1097–1103
Luzzi R., Vasconcellos A.R. and Ramos J.G., *Statistical Foundations of Irreversible Thermodynamics*, Teubner, Leipzig, 2001
Luzzi R., Vasconcellos A.R. and Ramos J.G., *Predictive Statistical Mechanics: A Non-equilibrium Ensemble Formalism*, Kluwer, Dordrecht, 2002
Lyapounov M., *Stability of Motion*, Academic, New York, 1966 (English translation)
Ma S.K., *Statistical Mechanics*, World Scientific, Singapore, 1985
Madruga S., Pérez-García C. and Lebon G., Convective Instabilities in two laterally superposed horizontal layers heated, Phys. Rev. E **68** (2003) 041607, 1–12
Magnasco M.O., Molecular combustion motors, Phys. Rev. Lett. **72** (1994) 2656–2659
Mahan G., Sales B. and Sharp J., Thermoelectric materials: new approaches to an old problem, Phys. Today **50** (1997) 42–47
Mandel J., *Propriétés Mécaniques des Matériaux*, Eyrolles, Paris, 1978
Maugin G., *The Thermodynamics of Nonlinear Irreversible Behaviors*, World Scientific, Singapore, 1999
Maugin G. and Muschik W., Thermodynamics with internal variables, J. Non-Equilib. Thermodyn. **19** (1994) 217–289
Maurer M.J. and Thomson H.A., Non-Fourier effects at high heat flux, J. Heat Transfer **95** (1973) 284–286
Maxwell J.C., On the dynamical theory of gases, Philos. Trans. R. Soc. Lond. **157** (1867) 49–88
McDougall J.T. and Turner J.S., Influence of cross-diffusion on 'finger' double-diffusive convection, Nature **299** (1982) 812–814
McQuarrie D.A., *Statistical Mechanics*, Harper and Row, New York, 1976
Meixner J., Zur thermodynamik der irreversiblen prozesse in gasen mit chemisch reagierenden, dissozierenden und anregbaren komponenten, Ann. Phys. (Leipzig) **43** (1943) 244–269
Meixner J., The entropy problem in thermodynamics of processes, Rheol. Acta **12** (1973a) 272–283
Meixner J., On the Foundations of Thermodynamics of Processes, in *A Critical Review of Thermodynamics* (Stuart E., Gal-Or B. and Brainard A., eds.), Mono Book Corp., Baltimore, 1973b
Meixner J. and Reik H.G., *Thermodynamik der Irreversible Prozesse*, in Handbuch der Physik, Bd 3/ II, Springer, Berlin Heidelberg New York, 1959
Mongiovì M.S., Extended irreversible thermodynamics of liquid helium II, Phys. Rev. B **48** (1993), 6276–6289
Mongiovì M. S., Proposed measurements of the small entropy carried by the superfluid component of liquid helium II, Phys. Rev. B. **63** (2001) 12501–12510
Morse P.M. and Feshbach H., *Mathematical Methods of Theoretical Physics*, McGraw Hill, New York, 1953
Movchan A., The direct method of Lyapounov in stability problems of elastic systems, J. Appl. Math. Mech. **23** (1959) 686–700
Müller I., On the frame dependence of stress and heat flux, Arch. Rat. Mech. Anal. **45** (1972) 241
Müller I. and Ruggeri T., *Rational Extended Thermodynamics*, 2nd ed., Springer, Berlin Heidelberg New York, 1998
Muschik W., Empirical foundation and axiomatic treatment of non-equilibrium temperature, Arch. Rat. Mech. Anal. **66** (1977) 379–401

Muschik W., *Aspects of Non-Equilibrium Thermodynamics*, World Scientific, Singapore, 1990

Nelson P., *Biological Physics. Energy, Information, Life*, Freeman, New York, 2004

Nicolis G. and Nicolis C., On the entropy balance of the earth-atmosphere system, Q. J. R. Meteorol. Soc. **106** (1980) 691–706

Nicolis G. and Prigogine I., *Self-Organization in Nonequilibrium Systems*, Wiley, New York, 1977

Nicolis G. and Prigogine I., *Exploring Complexity*, W.H. Freeman, New York, 1989

Nield D.A., Surface tension and buoyancy effects in cellular convection, J. Fluid Mech. **19** (1964) 341–353

Noll W., *The Foundations of Mechanics and Thermodynamics*, Springer, Berlin, 1974

Nettleton R.E. and Sobolev S.L., Applications of extended thermodynamics to chemical, rheological and transport processes: a special survey, J. Non-Equilib. Thermodyn. **20** (1995) 200–229, 297–331

Onsager L., Reciprocal relations in irreversible processes, Phys. Rev. **37** (1931) 405–426; **38** (1931) 2265–2279

Onuki A., Phase transitions of fluids in shear flow, J. Phys.: Condens. Mat. **9** (1997) 6119–6157

Onuki A., *Phase Transitions Dynamics*, Cambridge University Press, Cambridge, 2002

Öttinger H.C., A thermodynamically admissible reptation model for fast flows of entangled polymers, J. Rheol. **43** (1999a) 1461–1493

Öttinger H.C., Thermodynamically admissible equations for causal dissipative cosmology, galaxy formation, and transport processes in gravitational collapse, Phys. Rev. D **60** (1999b) 103507, 1–9

Öttinger H.C., *Beyond Equilibrium Thermodynamics*, Wiley, Hoboken, 2005

Öttinger H.C., Non-equilibrium thermodynamics of open systems, Phys. Rev. E **73** (2006) 036126

Öttinger H.C. and Grmela M., Dynamics and thermodynamics of complex fluids. II. Illustrations of a general formalism, Phys. Rev. E **56** (1997) 6633–6655

Ozawa H., Ohmura A., Lorenz R.D. and Pujol T., The second law of thermodynamics and the global climate system: a review of the maximum entropy production system, Rev. Geophys. **41** (2003) 4/1018

Palm E., Nonlinear thermal convection, Annu. Rev. Fluid Mech. **7** (1975) 39–61

Paltridge G.W., Thermodynamic dissipation and the global climate system, Q. J. R. Meteorol. Soc. **107** (1981) 531–547

Parmentier P., Regnier V. and Lebon G., Buoyant-thermocapillary instabilities in medium *Pr* number fluid layers subject to a horizontal temperature gradient, Int. J. Heat Mass Transfer **36** (1993) 2417–2427

Pasmanter R.A., Bedeaux D. and Mazur P., The ballast resistor: an electro-thermal instability in a conducting wire II; fluctuations around homogeneous stationary states, Physica A **90** (1978) 151–163

Pearson J.R.A., On convection cells induced by surface tension, J. Fluid Mech. **4** (1958) 489–500

Peixoto J.P. and Oort A.H., The physics of climate, Rev. Mod. Phys. **56** (1984) 365–429

Ponter A.S., Bataille J. and Kestin J., A thermodynamic model for the time independent plastic deformation of solids, J. Mécanique **18** (1978) 511–539

Prigogine I., *Etude thermodynamique des phénomènes irréversibles*, Desoer, Liège, 1947

Prigogine I., *Introduction to Thermodynamics of Irreversible Processes*, Interscience, New York, 1961

Prigogine I., *La fin des certitudes*, O. Jacob, Paris, 1996

Prigogine I. and Lefever R., Symmetry breaking instabilities in dissipative systems, J. Chem. Phys. **48** (1968) 1695–1700
Prigogine I. and Stengers I., *La nouvelle alliance*, Gallimard, Paris, 1979
Pritchard A.J., A study of the classical problem of hydrodynamic stability, J. Inst. Math. Applics. **4** (1968) 78–93
Pujol T. and Llebot J.E., Extremal principle of entropy production in the climate system, Q. J. R. Meteorol. Soc. **125** (1999) 79–90
Rayleigh Lord, *Scientific Papers*, vol. VI, Cambridge University Press, Cambridge, 1920
Rebhan E., Efficiency of non-ideal Carnot engines with friction and heat losses, Am. J. Phys. **70** (2002) 1143–1149
Regnier V., Dauby P.C. and Lebon G., Linear and nonlinear Rayleigh–Bénard–Marangoni instability with surface deformations, Phys. Fluids **12** (2000) 2787–2799
Reguera D., Mesoscopic nonequilibrium kinetics of nucleation processes, J. Non-Equilib. Thermodyn. **29** (2004) 327–344
Reguera D., Rubí J.M. and Vilar J.M., The mesoscopic dynamics of thermodynamic systems, J. Phys. Chem. B **109** (2005) 21502–21515
Reichl L.E., *A Modern Course in Statistical Physics*, University Texas Press, Austin, 1998
Rivlin R., Forty Years of Non-Linear Continuum Mechanics, in *Proceedings of the 9th International Congress on Rheology*, Mexico, 1984
Rosenblat S., Davis S.H. and Homsy G.M., Nonlinear Marangoni convection in bounded layers, J. Fluid Mech. **120** (1982) 91–138
Rubí J.M., The non-equilibrium thermodynamics approach to the dynamics of mesoscopic systems, J. Non-Equilib. Thermodyn. **29** (2004) 315–325
Rubí J.M. and Casas-Vázquez J., Thermodynamical aspects of micropolar fluids. A non-linear approach, J. Non-Equilib. Thermodyn. **5** (1980) 155–164
Ruelle D., *Chance and Chaos*, Princeton University Press, Princeton, 1991
Salamon P. and Sieniutycz S. (eds.), *Finite-Time Thermodynamics and Optimization*, Taylor and Francis, New York, 1991
Salamon P., Nitzan A., Andresen B. and Berry R.S., Minimum entropy production and the optimization of heat engines, Phys. Rev. A **21** (1980) 2115–2129
Scriven L.E. and Sternling C.V., On cellular convection driven by surface tension gradients: effect of mean surface tension and viscosity, J. Fluid Mech. **19** (1964) 321–340
Segel L., Non-linear hydrodynamic stability theory and its application to thermal convection and curved flows, in *Non-Equilibrium Thermodynamics, Variational Techniques and Stability* (Donelly R.F., Herman R. and Prigogine I., eds.), pp 165–197, University of Chicago Press, Chicago, 1966
Serrin J., Conceptual analysis of the classical second law of thermodynamics, Arch. Rat. Mech. Anal. **70** (1979) 355–371
Shannon C., A mathematical theory of communication, Bell System Tech. J. **27** (1948) 379–423, 623–656
Shlüter A., Lortz D. and Busse F.H., On the stability of steady finite amplitude convection, J. Fluid Mech. **23** (1965) 129–144
Sidoroff F.J., Variables internes en viscoélasticité, J. Mécanique **14** (1975) 545–566, 571–595; **15** (1976) 95–118
Sieniutycz S., *Conservation Laws in Variational Thermohydrodynamics*, Kluwer, Dordrecht, 1994
Sieniutycz S. and Farkas H. (eds.), *Variational and Extremum Principles in Macroscopic Systems*, Elsevier Science, Oxford, 2004
Sieniutycz S. and Salamon P. (eds.), *Extended Thermodynamic Systems*, Taylor and Francis, New York, 1992

Silhavy M., *The Mechanics and Thermodynamics of Continuous Media*, Springer, Berlin Heidelberg New York, 1997
Snider R.F. and Lewchuk K.S., Irreversible thermodynamics of a fluid system with spin, J. Chem. Phys. **46** (1967) 3163–3172
Snyder G.J. and Ursell T.S., Thermoelectric efficiency and compatibility, Phys. Rev. Lett. **91** (2003) 148301
Sparrow C., *The Lorenz Equations: Bifurcations, Chaos and Strange Attractors*, Springer, Berlin Heidelberg New York, 1982
Spencer A.J.M. and Rivlin R.S., Further results in the theory of matrix polynomials, Arch. Rat. Mech. Anal. **4** (1959) 214–230
Stuart J.T., On the non-linear mechanics of hydrodynamic instability, J. Fluid Mech. **4** (1958) 1–21
Swift J. and Hohenberg P.C., Hydrodynamic fluctuation at the convective instability, Phys. Rev. A **15** (1977) 319–328
Thom R., *Stabilité structurelle et morphogénèse*, Benjamín, Reading, MA, 1972
Tisza L. and Manning I., Fluctuations and irreversible thermodynamics, Phys. Rev. **105** (1957) 1695–1704
Truesdell C., Thermodynamics for Beginners, in *Irreversible Aspects of Continuum Mechanics* (Parkus H. and Sedov L., eds.), Springer, Berlin Heidelberg New York, 1968
Truesdell C., *Rational Thermodynamics*, 2nd ed., Springer, Berlin Heidelberg New York, 1984
Truesdell C. and Noll W., *The Non-Linear Field Theories*, in Handbuch der Physik, Bd. III/3, Springer, Berlin Heidelberg New York, 1965
Truesdell C. and Toupin R., *The Classical Field Theories*, in Handbuch der Physik, Bd. III/1, Springer, Berlin Heidelberg New York, 1960
Tsallis C., Some thoughts on theoretical physics, Physica A **344** (2004) 718–736
Turcotte D.L., *Fractals and Chaos in Geology and Geophysics*, Cambridge University Press, Cambridge, 1992
Turing A., The chemical basis of morphogenesis, Philos. Trans. R. Soc. Lond. B **237** (1952) 37–72
Tzou D.Y., *Macro-to-Microscale Heat Transfer. The Lagging Behaviour*, Taylor and Francis, New York, 1997
Valanis K.C., A gradient theory of internal variables, Acta Mechanica **116** (1996) 1–14
Valenti A., Torrisi M. and Lebon G., Shock waves in crystalline dielectrics at low temperature, J. Phys. Condens. Matter **14** (2002) 3553–3564
Van den Broeck C., Taylor dispersion revisited, Physica A **168** (1990) 677–696
Van den Broeck C., Thermodynamic efficiency at maximum power, Phys. Rev. Lett. **95** (2005) 190602
Van Kampen N.G., *Views of a Physicist*, World Scientific, Singapore, 2000
Vernotte P., La veritable equation de la chaleur, Compt. Rend. Acad. Sci. Paris **247** (1958) 2103–2107
Veronis G., Cellular convection with a finite amplitude in a rotating fluid, J. Fluid Mech. **5** (1959) 401–435
Vidal C., Dewel G. and Borckmans P., *Au delà de l'équilibre*, Hermann, Paris, 1994
Vilar J.M.G. and Rubí J.M., Thermodynamics beyond local equilibrium, Proc. Natl Acad. Sci. USA **98** (2001) 11081–11084
de Vos A., *Endoreversible Thermodynamics of Solar Energy Conversion*, Oxford University Press, Oxford, 1992
Westerhoff N.V. and van Dam K., *Thermodynamics and Control of Biological Free-Energy Transduction*, Elsevier, Amsterdam, 1987
Wilmanski K., *Thermodynamics of Continua*, Springer, Berlin Heidelberg New York, 1998

Woods L., *The Thermodynamics of Fluid Systems*, Clarendon, Oxford, 1975
Woods L., The bogus axioms of continuum mechanics, Bull. Inst. Math. Appl. **17** (1981) 98–102; **18** (1982) 64–67
Zemansky M.W., *Heat and Thermodynamics*, 5th ed., McGraw-Hill, New York, 1968

Further Readings

There exists a multiplicity of textbooks and popular books on the subject and it would be unrealistic to mention all of them. The list given below is therefore limited and, of course, subjective.

Atkins P.W., *The Second Law*, Scientific American Library, W.H. Freeman, New York, 1984

Bailyn M., *A Survey of Thermodynamics*, AIP, New York, 1993

Coveney P. and Highfield R., *The Arrow of Time*, Flamingo, London, 1990

Davies P.C.W., *The Physics of Time Asymmetry*, University of California Press, Berkeley, 1974

Denbigh K. and Denbigh J., *Entropy in Relation with Incomplete Knowledge*, Cambridge University Press, Cambridge, 1985

Feynman R.P. and Weinberg S., *Elementary Particles and the Laws of Physics*, Cambridge University Press, Cambridge, 1987

Feynman R.P., Leighton R.B. and Sands M., *The Feynman Lectures on Physics*, Addison-Wesley, 1963 (Chaps. 39–46)

Goodwin B., *How the Leopard Changed Its Spots: The Evolution of Complexity*, Weidenfeld and Nicolson, London, 1994

Harman P.M., *Energy, Force and Matter: The Conceptual Development of Nineteenth-Century Physics*, Cambridge University Press, Cambridge, 1982

Kestin J., *A Course in Thermodynamics*, 2 vols., Hemisphere, Washington, 1979

Longair M.S., *Theoretical Concepts in Physics: An Alternative Views of Theoretical Reasoning in Physics*, 2nd ed., Cambridge University Press, Cambridge, 2003

Maxwell J.C., *Theory of Heat*, Dover, New York, 2001

Pippard A.B., *The Elements of Classical Thermodynamics*, Cambridge University Press, London, 1957

Prigogine I., *From Being to Becoming*, Freeman, San Francisco, 1980

Prigogine I. and Stengers I., *Entre le temps et l'éternité*, Fayard, Paris, 1988

Segrè G., *Einstein's Refrigerator*, Penguin Books, London, 2004

Tisza L. and Manning T., *Fluctuations and irreversible thermodynamics*, Phys. Rev. **105** (1957) 1695–1704

Vidal Ch. and Lemarchand H., *La réaction créatrice*, Herman, Paris, 1988

Walgraef D., *Spatio-Temporal Pattern Formation*, Springer, Berlin Heidelberg New York, 1997

Winfree A., *The Geometry of Biological Time*, Springer, Berlin Heidelberg New York, 1980

Winfree A., *The Timing of Biological Clocks*, Scientific American Library, Freeman, New York, 1987a

Winfree A., *When Time Breaks Down*, Princeton University Press, Princeton, 1987b

Index

Made in the USA
Lexington, KY
22 August 2012